Groundwater Quality

Price: £39.95

AF616588

ASSOCIATION OF GEOSCIENTISTS FOR INTERNATIONAL DEVELOPMENT

AGID Special Publication Series
THE GEOSCIENCES IN INTERNATIONAL DEVELOPMENT

Other Titles in the Series

1. **Geoscientists and the Third World: a Collective Critique of Existing Aid Programs**, edited by A.R. Berger and published as Paper 74-5e by the Geological Survey of Canada, 1975, 32 pages.
2. **Hidden Waters in Arid Lands: Report of a Workshop on Groundwater Research Needs in Arid and Semi-arid Zones, Paris, 1974**, edited by L.A. Heindl and published by International Development Research Centre, Ottawa, Canada (IDRC-057e), 1975, 19 pages.
3. **New Directions in Mineral Development Policies: Report of the AGID Bagauda International Workshop, Nigeria**, edited by D.J.C. Laming and D.E. Ajakaiye, AGID, 1977, 224 pages.
4. **Preliminary Bibliography on Groundwater in Developing Countries, 1970 to 1976**, edited by D.A.V. Stow, J. Skidmore and A.R. Berger, AGID, 1976, 305 pages.
5. **Geoscience Education in Developing Countries**, edited by P.G. Cooray, AGID, 1977, 56 pages.
6. **Mineral Resource Management in Developing Countries: State Participation, Private Enterprise, or Both?** Proceedings of an International Symposium held at Sydney, Australia, 1976, edited by M.E. Woakes, 1978, 28 pages.
7. **Hidden Wealth: Mineral Exploration Techniques in Tropical Forest Areas**. Proceedings of an international symposium held in Caracas, Venezuela, 1977, edited by D.J.C. Laming and A. Gibbs, 1977, 221 pages.
8. **Strategies for Small-Scale Mining and Mineral Industries: Report of a regional workshop held at Mombasa, Kenya, 1980**, edited by J.M. Neilson, 1980, 200 pages.
9. **Geochemistry in Zambia: Report of the AGID Geochemical Workshop held in Zambia, with case histories, 1977**, edited by D.C. Turner, 1977, 94 pages.
10. **AGID Guide to Mineral Resources Development**, edited by M.E. Woakes and J.S. Carman, 1983, 504 pages.
11. **Geosciences for Development – the Australian Role**: Proceedings of the AGID-ILP Workshop held at Sydney, Australia, 1984: edited by M.B. Katz and E.J. Langevad, 1985, 374 pages.
12. **Landplan II: Role of Geology in Planning and Development of Urban Centres in Southeast Asia**. Proceedings of the Symposium held at Kuala Lumpur, Malaysia, 1984: edited by B.K. Tan and J.L. Rau, 1986, 92 pages.
13. **Environmental Geology and Natural Hazards of the Andean Region**, edited by M. Hermelin, 1990.
14. **Geosciences in Development**. Proceedings of an International Conference on the Application of the Geosciences in Developing Countries, 1988, edited by D.A.V. Stow and D.J.C. Laming, AGID/Balkema Publishers, 1990, 322 pages.
15. **Geohazards**, edited by G.J.H. McCall, D.J.C. Laming and S.C.Scott, Chapman & Hall, 1991, 227 pages.
16. **Environmental Geology and Applied Geomorphology in Colombia**, edited by J.E. Lopez-Rendon, 1992 (in Spanish).
17. **Groundwater Quality**, edited by H. Nash and G.J.H. McCall, Chapman & Hall, 1994, 216 pages.
18. **Industrial Minerals in Developing Countries**, edited by S.J. Mathers and A.J.G. Notholt, ODA/BGS/AGID publication, 1994.

Further information on AGID activities, publications and membership from: AGID Headquarters, c/o Instituto de Geosciencias, Universidade de São Paulo, Caixa Postal 20, 899, CEP 01498, São Paulo, Brazil.

Groundwater Quality

17th Special Report

Edited by

H. Nash

Consulting Hydrogeologist, Leeds, UK

and

G.J.H. McCall

Consultant Geologist, Gloucestershire, UK

CHAPMAN & HALL

London · Glasgow · Weinheim · New York · Tokyo · Melbourne · Madras

Published by Chapman & Hall, 2–6 Boundary Row, London SE1 8HN, UK

Chapman & Hall, 2–6 Boundary Row, London SE1 8HN, UK

Blackie Academic & Professional, Wester Cleddens Road, Bishopbriggs, Glasgow G64 2NZ, UK

Chapman & Hall GmbH, Pappelallee 3, 69469 Weinheim, Germany

Chapman & Hall USA, One Penn Plaza, 41st Floor, New York NY 10119, USA

Chapman & Hall Japan, ITP Japan, Kyowa Building, 3F, 2–2–21 Hirakawacho, Chiyoda-ku, Tokyo 102, Japan

Chapman & Hall Australia, Thomas Nelson Australia, 102 Dodds Street, South Melbourne, Victoria 3205, Australia

Chapman & Hall India, R. Seshadri, 32 Second Main Road, CIT East, Madras 600 035, India

---

First edition 1995

© 1995 H. Nash and G.J.H. McCall

Typeset in 10/12 Times by Photoprint, Torquay, Devon
Printed in Great Britain at the Alden Press, Oxford

ISBN 0 412 58620 7

Apart from any fair dealing for the purposes of research or private study, or criticism or review, as permitted under the UK Copyright Designs and Patents Act, 1988, this publication may not be reproduced, stored, or transmitted, in any form or by any means, without the prior permission in writing of the publishers, or in the case of reprographic reproduction only in accordance with the terms of the licences issued by the Copyright Licensing Agency in the UK, or in accordance with the terms of licences issued by the appropriate Reproduction Rights Organization outside the UK. Enquiries concerning reproduction outside the terms stated here should be sent to the publishers at the London address printed on this page.

The publisher makes no representation, express or implied, with regard to the accuracy of the information contained in this book and cannot accept any legal responsibility or liability for any errors or omissions that may be made.

A catalogue record for this book is available from the British Library

Library of Congress Catalog Card Number: 94–71821

AGID Special Publication No. 17

Printed on permanent acid-free text paper, manufactured in accordance with ANSI/NISO Z39.48–1992 and ANSI/NISO Z39.48–1984 (Permanence of Paper).

Contents

List of contributors

Barghash bin Ghalib Al-Said Ministry of Water Resources, Sultanate of Oman.

R.P. Ashley Principal, Ashley Associates, 32 Station Road, Cambridge CB1 2JH, UK.

Fatma Abdel Rahman Attia Research Institute for Groundwater, Water Research Center, El Kanater, El Khairia 13621, Egypt.

J. Bartram Robens Institute, University of Surrey, Guildford, Surrey GU2 5XH, UK.

M.J. Burley Water Supply and Waste Water, Mott MacDonald, Demeter House, Station Road, Cambridge CB1 2RS, UK.

P.J. Chilton British Geological Survey, Maclean Building, Wallingford, Oxon OX10 8BB, UK.

R.J. Connelly Steffen Robertson and Kirsten (UK), Summit House, 9/10 Windsor Place, Cardiff CF1 3BX, UK.

E.P. Daly Geological Survey of Ireland, Beggars Bush, Haddington Road, Dublin 4, Ireland.

J. Davies British Geological Survey, Maclean Building, Wallingford, Oxon OX10 8BB, UK.

W.M. Edmunds British Geological Survey, Maclean Building, Wallingford, Oxon OX10 8BB, UK.

S.S.D. Foster British Geological Survey, Keyworth, Nottingham NG12 5GG, UK.

J.A. Heathcote Hydrotechnica Ltd, 160–162 Abbey Foregate, Shrewsbury SY2 6AL, UK.

R. Herbert Advisor on Hydrogeology to the Overseas Development Organisation (ODA), British Geological Survey, Wallingford, Oxon OX10 8BB, UK.

G. Howard Robens Institute, University of Surrey, Guildford, Surrey GU2 5XH, UK.

K.W.F. Howard Groundwater Research Group, University of Toronto, Scarborough Campus, 1265 Military Trail, Scarborough, Ontario, Canada M1C 1A4.

G.A. Kew Ministry of Water Resources, PO Box 5575, Ruwi, Sultanate of Oman.

A.R. Lawrence British Geological Survey, Maclean Building, Wallingford, Oxon OX10 8BB, UK.

D.N. Lerner Hydrogeology Research Group, School of Earth Sciences, University of Birmingham, Edgbaston, Birmingham B15 2TT, UK.

J.W. Lloyd Hydrogeology Research Group, School of Earth Sciences, University of Birmingham, Edgbaston, Birmingham B15 2TT, UK.

G.J.H. McCall 44 Robert Franklin Way, South Cerney, Gloucestershire GL7 5UD, UK.

P.G. Macumber Ministry of Water Resources, PO Box 5575, Ruwi, Sultanate of Oman.

L.M. Markussen R.H. and H. Consult, Bredevej 2, DK 2830, Denmark.

J. Mather Department of Geology, Royal Holloway University of London, Egham, Surrey TW20 0EX, UK.

H.-M.F. Møller R.H. and H. Consult, Bredevej 2, DK 2830, Denmark.

H. Nash 91, Bayswater Place, Leeds LS8 5LS, UK.

R. Rojas Robens Institute, University of Surrey, Guildford, Surrey GU2 5XH, UK.

M.E. Stuart British Geological Survey, Maclean Building, Wallingford, Oxon OX10 8BB, UK.

Sukrisno Directorate of Environmental Geology, Bandung, Indonesia.

S.E. Sutton S.W.L. Consultants,The Gables, 14 Kennedy Road, Shrewsbury, Shropshire SY37 7AB, UK.

D. Taussig Formerly Steffen Robertson and Kirsten (UK), Summit House, 9/10 Windsor Place, Cardiff CF1 3BX, UK.

C. Tavitian Land Reclamation Service, Ministry of Agriculture, 92 P. Pigadion and Zakynthou Street, 264–41 Patras, Greece.

R.G. Taylor Groundwater Research Group, University of Toronto, Scarborough Campus, 1265 Military Trail, Scarborough, Ontario, Canada M1C 1A4.

T.B. Tennakoon Ministry of Water Resources, PO Box 5575, Ruwi, Sultanate of Oman.

W. Wagner Federal Institute for Geosciences and Natural Resources, Stillweg 2, Postfach 510153, 3000 Hannover 51, Germany.

C.F. Ward Institute of Irrigation Studies, University of Southampton, Southampton SO9 5NH, UK.

M. Woodhouse African Medical Research Foundation, PO Box 30125, Nairobi, Kenya.

L. Woods K.T. Cullen and Co. Ltd, Hydrogeological and Environmental Consultants, Block 2, Beech Hill, Clonskeagh, Dublin 14, Ireland.

P.L. Younger Department of Civil Engineering, University of Newcastle upon Tyne, NE1 7RU, UK.

Foreword

Dorrik A.V. Stow

Editor-in-Chief, Association of Geoscientists for International Development (AGID)

Nowhere on earth is the need for water so acute, nowhere on earth is the very act of existence such constant battle, than in the very heart of our deserts. Some 20% of the world's land area is now classified as desert and a further 10% teeters on the edge of aridity, mainly due to our own poor use of land and resources. Under such conditions the need for groundwater is evident.

However, it is by no means only where surface water is absent or ephemeral that the location and continued supply of good quality groundwater is essential for the health and development of nations. The World Health Organization has repeatedly insisted that the single major factor adversely influencing the general health and life expectancy of a population in many countries of the developing world is ready access to clean drinking water. Highly populated areas, both urban and rural, inevitably place a strain on water supply in terms of excess demand and severe pollution. Coastal regions are always under threat of saline contamination of aquifers by sea-water intrusion. Crystalline basement rocks and volcanic terrains each yield their own special problems of geochemical enrichment of groundwater. And so the list goes on.

We hope that this latest Special Publication in the AGID Geosciences in International Development series will make a significant contribution to the vital challenge of groundwater quality control that faces hydrogeologists, land planners and developers everywhere.

AGID is a network of geoscientists throughout the world concerned with the application of the geosciences to development. As such, we have had a long history of involvement in groundwater issues, as evidenced by two of our early publications in this series: *Hidden Waters in Arid Lands* (Heindl, 1975) and *Preliminary Bibliography of Groundwater in Developing Countries* (Stow et al. 1976). The topic has remained very much alive in several more recent publications (see list p. ii) and through a number of grassroots projects around the world.

The collection of papers in this volume grew out of those presented at a meeting jointly sponsored by AGID(UK) and the Hydrogeological Specialist Group of the Geological Society. These papers have been supplemented by cogent reviews and topical syntheses specially prepared for this volume. It is undoubtedly one of the best and most comprehensive of its kind, and is a credit to the energy and enthusiasm of the meeting conveners, the publication editors and the many contributors involved.

Dorrik A.V. Stow

Preface

This volume is based on a meeting on 'Groundwater quality in developing countries' held at the Geological Society, Burlington House, London on 25 March 1992, and sponsored by the Joint Association of Geoscientists for International Development and the Hydrogeological Group of the Society. To broaden the scope of the book the two editors called for further contributions from both the UK and overseas. Thus only seven of the twenty papers in this volume are based on the original one-day symposium, and the urban context has been added to the dominantly rural theme of the original symposium. The volume has been structured in four sections: Groundwater chemistry and development, Salinity, the Impact of human activity, and Urban environments. This subdivision is not rigorous since some of the articles cover more than one of these topics. Brief introductions have been written for each section, drawing the text together, and introducing broader issues not covered in the individual articles. The original introduction by Stephen Foster to the symposium has been modified to embrace the broader coverage of this volume.

We are not intending to present a textbook on 'Groundwater quality' but rather, by presenting a number of case histories and studies from different countries, to provide a set of articles of value to the numerous and growing body of workers in this field throughout the world. The coverage is not exhaustive, in the sense that it embraces all topics, but it is reasonably comprehensive. Only a single article promised failed to materialize – the editors would have liked to include one on salinity problems related to irrigation in Pakistan.

The editors are indebted to the authors who have borne with their demands and queries with good humour and patience, and the referees who have ensured that the articles published are of a high standard.

Harriet Nash, Chair, Hydrogeological Group, Geological Society

Joe McCall, Vice-Chair, Joint Association of Geoscientists for International Development, Geological Society, and Commissioning Editor for AGID Special Publications generated in the UK.

Referees

The following have acted as referees:

David Allen	Charles Jones
Paul Ashley	John Heathcote
Sarah Beeson	Joe McCall
Trevor Bestow	Rodney Mitchell
Simon Bottrell	Michael Morrey
Jeffery Davies	Harriet Nash
Mark Fermor	Steve Parsons
Steve Fletcher	Catherine Ward

Introduction: Groundwater for development – an overview of quality constraints

S.S.D. Foster

General situation

Groundwater is immensely important for human water supply in both the urban and rural areas of developing nations. Estimates for Asia and Latin America alone suggest that more than 1 000 million people are directly dependent upon this source. Innumerable large towns and many new megacities of these continents derive a major component of their domestic and industrial water supply from aquifers, both by municipal well-fields and by very large numbers of private boreholes.

In those parts of the tropics with a pronounced dry season or more generally arid climate, groundwater is also widely used as a source of primary or supplementary irrigation in agricultural development.

Groundwater is naturally of excellent microbiological quality and generally of adequate chemical quality for all these uses, although the latter can be problematic in certain hydrogeological conditions. The three most common causes of unacceptable groundwater quality are:

- anthropogenic pollution of vulnerable, and inadequately protected, aquifers;
- saline intrusion in inadequately managed aquifers; and
- naturally occurring problems related to hydrogeochemical evolution in certain types of strata.

A fourth, and sadly all too frequent, cause of water quality problems is inadequate design, construction, operation and maintenance of wells themselves. In complex multi-layered alluvial formations, where the shallowest phreatic aquifer is often the most vulnerable to anthropogenic pollution and most susceptible to saline intrusion, polluted groundwater will be drawn deeper if wells are not completed so as to line-out the shallow aquifer. Poor sanitary sealing is more generally a very common cause of microbiological quality deterioration, and inappropriate construction materials sometimes lead to unnecessarily high iron concentration.

Naturally occurring quality problems

Naturally occurring groundwater quality problems are relatively widespread in arid regions and in volcanic areas. They include:

- elevated concentrations of iron and manganese whose solubilities are related to pH–Eh variations;
- occurrence of high salinity, at locations remote from the coastline, due to inadequate flushing of marine sediments and solution of evaporitic minerals;
- elevated concentrations of fluoride, and occasionally arsenic, especially in some volcanic regions;
- possible aluminium mobility in unbuffered soil profiles, subject to acidic infiltration; and
- high nitrate concentrations related to former palaeoclimatic conditions.

While fluoride, nitrate and aluminium present significant health problems, elevated concentrations of iron, manganese and certain salts (such as magnesium sulphate) can cause taste problems and lead to the abandonment of wells of otherwise excellent quality.

Saline intrusion in inadequately managed aquifers

The problem of saline intrusion/upconing in coastal aquifers and some intermontane valleys, together with

Groundwater Quality Edited by H. Nash and G.J.H. McCall. Published in 1994 by Chapman & Hall. ISBN 0 412 58620 7

land subsidence under certain situations, are serious, and effectively irreversible, consequences of inadequate management of groundwater resources.

Aquifer management in developing nations presents special problems as a result of:

- the reluctance of governments to allow a realistic price to be levied for urban water supply and to grant excessive subsidies for agriculture irrigation;
- the implementation of controls on groundwater exploitation make organizational, legal and technical demands which are often beyond the institutional capacity of regulatory departments or agencies, especially where private well drilling is concerned in situations of water-supply deficiency; and
- the control of abstraction presupposes that groundwater resources have been reliably evaluated; in developing nations many uncertainties remain regarding the size of the groundwater resource in terms of both replenishment and storage and the potential effects of over-exploitation.

Anthropogenic pollution

Groundwater quality problems consequent upon man's activities at the land surface are being reported with increasing frequency in the more rapidly developing and industrializing nations, due to the absence of positive aquifer protection policies. Many more are occurring as yet unobserved, as a result of the generally inadequate level of quality monitoring that is normally achieved.

Impact of urbanization

The impact of urbanization will be particularly critical in developing nations in coming decades. By the year 2000 it is predicted that more than 50% (3 200 million) of the world's population will have become urban dwellers, compared with less than 15% at the turn of century.

Most growth of urban areas will be concentrated in the less-developed nations. Urbanization, coupled with almost inseparable industrial development, has profound impacts on the hydrological cycle, including major, but not always readily identified, changes in the frequency, volume and quality of groundwater recharge; modifying existing recharge mechanisms and introducing new ones. These are a consequence of the importation of large volumes of water from the surrounding areas to meet urban demands, together with surface impermeabilization due to the construction of roofs and paved or compacted areas and stormwater drainage arrangements.

If drainage and sanitation are achieved through on-site arrangements such as soakaways, septic tanks, cesspools and latrines, urbanization will lead to major increases in the overall groundwater recharge rate but will cause significant deterioration in groundwater quality as a result of increasing concentrations of nitrate, organic carbon (including the possibility of toxic synthetic compounds) and, in certain hydrogeological conditions and/or with inadequate sanitation design, to contamination with pathogenic bacteria and viruses.

In those areas where a significant proportion of the waste water disposal is via mains sewerage, then large volumes of normally untreated waste water are generated. In the more arid climates this waste water is used very widely, either directly or indirectly, for agricultural irrigation. In many situations this presents both direct public health risks and can lead to serious chemical contamination of local groundwater, since irrigation efficiencies with waste water are invariably low, leading to high rates of infiltration.

While all waste water infiltration normally results in major improvements in microbiological quality, it can lead to permanent deterioration in chemical quality. This presents a fundamental dilemma. From the public health point of view it is desirable to dispose of waste water as rapidly and effectively as possible to the ground, since this obviates the risk of major epidemics of water-borne disease and at the same time presents an opportunity of water resource recovery. However, recharge to aquifers will invariably lead to relatively serious deterioration in chemical quality, which while not sufficient to prevent the use of groundwater for agricultural irrigation, will prejudice its use for public water supply.

Impact of agriculture

Radical changes in farming practice, aimed at increasing productivity and reducing reliance on imported food, have led to the widespread introduction of agricultural monocultures, sustained by major increases in the use of agrochemicals and irrigation, especially in Asia and Latin America. Trends (not surprisingly) are mirroring those which occurred in the industrialized world during 1950–70.

This intensification of agricultural production can lead (and has led) to serious deterioration in groundwater quality in some pedological, geological and climatic conditions. The principal problems are the leaching of nitrate and pesticide compounds, together with increasing salinity in the more arid environments. This has impacts in terms of the continued use of aquifers for human water supply in neighbouring towns and in the rural areas themselves. Under some conditions it can affect the sustainability of agricultural irrigation itself.

General conclusions

With the pressing need for rapid development of new water supplies, adequate attention is rarely given to quality issues and protection requirements, despite the fact that in the longer term these can be a serious constraint on sustainable development. Too many groundwater exploitation schemes in developing nations are designed without due attention to quality issues – as if quality were a stable unchanging parameter.

The urban situation is becoming especially critical. Administrators in the land and water sector need to take much fuller account of the impacts on groundwater when making decisions on the choice of sanitation and drainage arrangements, and the location of certain industries.

Groundwater is one of the most valuable natural resources that many developing nations possess and without improved management and closer protection it will suffer irreversible deterioration on a widespread basis.

There is an urgent need for rapid surveys of groundwater utilization, aquifer pollution vulnerability and subsurface contaminant load, to be undertaken. Groundwater pollution risk and susceptibility to over-exploitation effects can then be assessed and protection measures prioritized and initiated.

Part One

Groundwater Chemistry and Development

W.M. Edmunds

There are many reasons why it is necessary to investigate the chemistry of groundwater. At the outset it is important to know the change in mineral character which will vary from place to place dependent upon the geology of the terrain. Hard or soft water, mineralized or non-mineralized water are characteristics that depend on the extent of reactions with the rocks through which the water has passed. A geochemical understanding of the processes controlling these baseline characteristics forms the starting point from which to conduct hydrogeological investigations and other applied studies.

The natural chemistry can often have an important bearing on human health or on livestock. A detailed analysis of the major, minor and trace constituents (including organics) of groundwater is a prerequisite for commissioning public supplies in developing countries. Guidelines are given by international bodies (e.g. the World Health Organization and European Community) as to the standards for drinking water supplies. An excess of certain species (e.g. of Fe, F, Mn) or a deficiency (e.g. I, Se) of certain elements in the natural system can have an adverse impact on health locally, and/or the acceptability of the groundwater for drinking and domestic use. In developing countries is also desirable that the same standards are adhered to, although comprehensive sampling for many constituents will not be practicable.

For the hydrogeologist, an understanding of the geochemical (including the isotopic) characteristics of groundwater systems can be an important aid to determining physical properties of flow systems. Hydrochemical data can be used to help estimate such properties as the amount of recharge, the extent of mixing, the circulation pathways, maximum circulation depths, temperatures at depth, as well as residence time.

Once the baseline hydrogeochemistry is well understood, then the superimposed impact of humans can be quite easily recognized and the chemical information on the unpolluted waters is thus needed to assist in groundwater protection. The activities of humans may magnify this natural background (the increase in salinity by excessive pumping of saline water or the accelerated oxidation of sulphide minerals by water-level drawdown, for example). The impact of diffuse agricultural or urban pollutants such as nitrate and pesticides has become a serious risk in many parts of the developing world. It may also be the case that the introduction of some pollutants can also be used to advantage as a 'tracer' of flowpaths and to determine transit times.

In order to optimize sampling and analysis campaigns it is important to decide the purpose of the study. This is especially important in developing countries, where resources and access to facilities may be limited. Much relevant information may be obtained by simple chemical analysis. For example, from the analysis of chloride alone (or even specific electrical conductance as a proxy for chloride) it may be possible to derive information on recharge, mixing or transit times. At the other end

Groundwater Quality Edited by H. Nash and G.J.H. McCall. Published in 1994 by Chapman & Hall. ISBN 0 412 58620 7

of the scale, precise and accurate analysis on a number of ionic constituents plus *in situ* measurements of pH, and other parameters is required to carry out proper geochemical interpretation and modelling. Isotopic analyses (O, H, C) have proved of immense value, especially in arid zone studies, in characterizing groundwaters of different origin and in estimating residence times.

Many developing countries now possess facilities to conduct comprehensive studies of groundwater chemistry. Regional facilities also exist for specialist studies (e.g. for isotopic analysis), and through bilateral projects with collaborating partners more detailed work may be possible. Nevertheless, it is always going to be an advantage for the individual geochemist or hydrogeologist to have control over the 'basics' of field chemistry, sampling procedures and the day-to-day analysis of a key set of parameters, notably major ion analysis. It is of special importance that the field scientist (geochemist/hydrogeologist) is fully aware of all the pitfalls of sampling groundwaters, including filtration and storage of samples. It is important to recognize that nearly every water sample will be a mixture of water from different depths.

The five chapters presented in the following section address a number of the areas outlined above, from low technology investigations to more sophisticated studies. Each of them provides not only an idea of how to proceed in chemical investigations of groundwaters, but often some original ideas that can be adapted to other situations.

One approach to basic hydrochemical analysis, independent of external laboratories, is provided in the chapter by **Younger**. In this study, hydrochemical investigations of groundwaters from sandstones and lacustrine deposits were carried out 'for reasons of cost, speed of analysis' and independence from external support. Intercomparison was, however, made with commercial laboratory results and comparable results (ionic balance $\pm 10\%$) in terms of precision and accuracy were obtained. This approach allows the possibility of conducting on-the-spot water quality surveys, with the benefit of rapid feedback for resources investigation – e.g. such results could influence the course of borehole drilling programmes.

During the 'International Drinking Water Supply and Sanitation Decade' a major goal was to provide clean water supplies within 1.6 km of every citizen. Borehole construction has formed a major activity in this campaign in an attempt to provide safer groundwater from hand pumps as opposed to surface water supplies. In many cases the new supplies were perceived by the population to have an inferior quality to the abandoned surface water supplies. In the chapter by **Taylor and Howard**, this problem is addressed in relation to Uganda where over 150 wells in weathered basement rocks were examined. Whilst some of the poorer water quality could be traced locally to bad well siting (in relation to pit latrines and well corrosion for example) several more widespread problems are described in relation to metals occurring in excess of WHO limits.

In a careful study of the alluvial aquifer system of central Bangladesh, **Davies** has illustrated the value of good geological data in providing the basis for hydrogeochemical interpretation. Data collected from 250 tube wells has provided a 3-dimensional understanding of a very young aquifer system that has been forming throughout the late Quaternary. The aquifer itself is unusual, being formed essentially of fine sands and detrital micas but containing no clays. A sequence of redox changes and chemical gradients are observed within the aquifer, which can also be related to the diagenetic processes in the aquifer minerals, such as the weathering of biotite. The water quality problems relate mainly to the widespread reducing conditions in the groundwaters, notably the high occurrence of iron.

The chapter by **Ashley and Burley** is a case study of the occurrence of high fluoride in groundwaters in Ethiopia and can be regarded as representative of the widespread problem of high fluoride in natural waters in the East African rift valley area. The problem is made acute by two processes – the calcium deficient nature of the volcanic and other igneous rocks and the extent of tropical weathering whereby base cations have been removed. The chapter emphasizes the essential control on F^- concentrations by fluorite and that the highest F^- concentrations are to be found where the Ca^{2+} is lowest. The authors consider the options for the optimal siting of wells and boreholes and the possibility of defluoridation.

The chapter by **Edmunds** highlights the way in which minor elements may be used to provide information on various processes and reactions taking place along flow paths in groundwaters. Examples are given from the arid and semi-arid zones of northern Africa. Information can be obtained on the redox conditions in aquifers by studying the relative concentrations of Fe and NO_3, for example. Bromine and Br/Cl ratio can be used to determine the likely sources of salinity. In the absence of radiocarbon dating or other age indicators, the concentrations of certain elements may be used as indicators of residence time (e.g. Li, F, Sr). Some elements (e.g. B, Sr) may be used to determine origins of groundwater and consequently may be useful for establishing mixing.

With the increased use of groundwater for potable supplies in tropical countries has come the growing evidence of pollution. Once thought of as a problem of developed countries, pollution from agricultural sources has rapidly become a serious problem in many developing countries. In the chapter by **Chilton, Lawrence and Stuart** (Chapter 10) the question of agricultural pollution in tropical groundwaters is reviewed comprehensively, citing the factors controlling the degree of deterioration of water quality. The case studies used show that groundwater nitrate pollution is often widespread, but more sinister is the risk of pesticides entering the water supply. It is almost impossible to monitor this problem in developing countries for a number of reasons, not least the sophistication and costs of analysis. The problems are exacerbated by the intensification of agriculture, especially by means of irrigation where the triple problems of salinization, nitrate and pesticide increases can be encountered.

1 The hydrogeochemistry of alluvial aquifers in central Bangladesh

J. Davies

Abstract Hydrogeological and hydrochemical characteristics of Pleistocene and Late Quaternary alluvial aquifers underlying the area between Dhaka and the confluence of the Ganges and Brahmaputra rivers in central Bangladesh have been studied. The older Madhupur aquifer occurs within Dupi Tila Formation sediments that underlie the Madhupur Tract Pleistocene terrace. The younger Jamuna aquifer system occurs within grey non-indurated alluvial sediments of the Dhamrai Formation that infill the Jamuna, Brahmaputra and Ganges river valleys. Groundwater is abstracted by means of hundreds of nominal 56 l s^{-1} capacity deep tubewells, from both the Madhupur and coarse grained Lower Jamuna aquifers. Groundwater is also obtained from the finer grained Upper Jamuna aquifer by means of many 15 l s^{-1} capacity shallow tubewells.

Hydrochemical analyses of 77 water samples obtained during two surveys from a series of deep tubewells, shallow tubewells and observation wells at 23 sites were determined. These baseline data are used to describe the hydrochemical characteristics of the Madhupur and Jamuna aquifers and genesis of Late Quaternary alluvial aquifer groundwaters in the context of palaeoclimatic and sea-level change. The groundwaters are all of $Ca(HCO_3)_2$ type, suitable for crop irrigation and domestic usage. Major ion concentrations increase from northeast to southwest, indicative of the direction of groundwater flow.

Introduction

Bangladesh, with an area of approximately 144 000 km^2, encompasses the combined Ganges, Brahmaputra and Meghna delta system. The low-lying central parts of that system are regularly inundated by flood waters during the annual monsoon period. A large proportion of the 110 million population of Bangladesh is employed within the agricultural sector, many at subsistence level. Groundwater has been extensively developed for irrigation purposes, during the dry season, and as a source of 'clean' water for domestic use, especially by the rural population. Many shallow and deep tubewells have been constructed for the supply of irrigation water, especially for farmer co-operative groups.

Four thousand deep tubewells were installed for irrigation purposes during the Deep Tubewell II Project (DTWII Project) within central and northeastern Bangladesh. This multidonor project, funded partly by the UK Overseas Development Administration (ODA), was undertaken by Mott MacDonald International (MMI) and the Bangladesh Agricultural Development Corporation (BADC) during 1983–1992 (Mott MacDonald International, 1992).

Two hydrochemical studies were undertaken as part of the DTWII Project. The first was part of a pilot study into optimum deep tubewell design undertaken by the British Geological Survey (BGS) during 1984–88 (Herbert, Barker and Davies, 1989). The second, undertaken by BGS during 1992, was a reconnaissance hydrochemical survey of the DTWII Project area (Davies and Exley, 1992). Data from these two studies are used to describe the hydrogeochemical characteristics of the Jamuna and Madhupur alluvial aquifer systems within a study area located between Aricha, at the junction of the Brahmaputra and Ganges rivers, and Dhaka (Figure 1.1).

Groundwater Quality Edited by H. Nash and G.J.H. McCall. Published in 1994 by Chapman & Hall. ISBN 0 412 58620 7

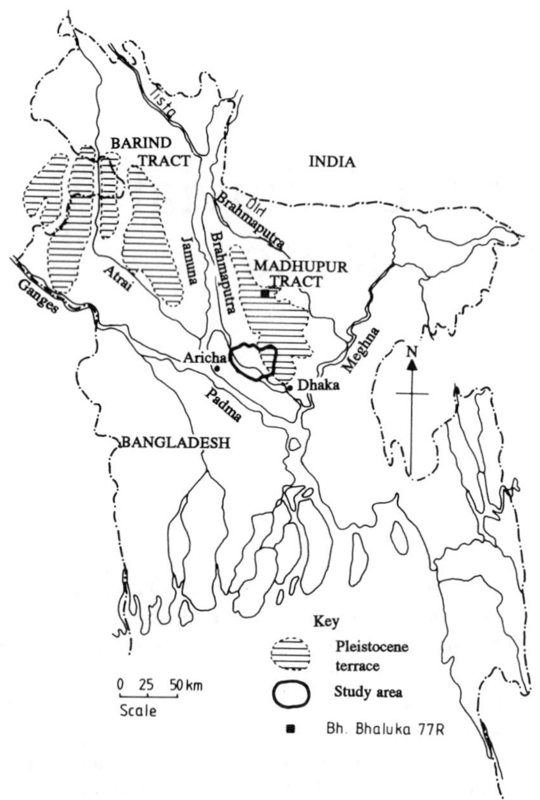

Figure 1.1 Study area location map.

Geomorphological setting

Coleman (1969) described the geomorphology and hydrology of the Brahmaputra/Ganges delta system, using aerial photograph interpretation and field surveys. Morgan and MacIntire (1959) compared the geomorphology of the Ganges–Brahmaputra Delta with that of the Mississippi Delta, describing the surface expression of Late Quaternary sediments within the major valleys of the former and the Pleistocene terraces of the Madhupur and Barind Tracts (Figure 1.1). During the Late Pleistocene, sea-levels changed in response to successive glacial events and periodically increased river gradients, resulting in valley incision and the formation of river terraces. The sediments that form the terraces and valley infill were derived from the Himalayas via the Indo-Gangetic plain (Davies, 1989).

Hydrogeology

Data from 250 deep tubewell records were used to outline the geological and hydrogeological characteristics of the two alluvial aquifer systems present within the study area (Figure 1.2).

Madhupur aquifer

MMI (1992) described the Madhupur aquifer, which occurs beneath the Madhupur Tract within red-brown alluvial fine to coarse grained sands and gravels of the Lower Pleistocene Dupi Tila Formation. Monsur (1990) dated the Dupi Tila Formation as being more than 900 000 y old. These alluvial sands and gravels are highly weathered; the mafic minerals originally present were oxidized during post-depositional periods of weathering to form red-brown ferric oxide cements and secondary clays, thus reducing intergranular porosity and permeability. The Madhupur aquifer has transmissivities of 500–2000 $m^2 d^{-1}$ and permeabilities of 15–30 $m d^{-1}$. This aquifer is confined by the Madhupur clay, a residual soil horizon composed of red-brown silty clays, that has been dated on the basis of reversed magnetic polarity as between 730 000 and 900 000 y old (Monsur, 1990). The alluvial sediment facies distribution and lithological variation of the Dupi Tila Formation is illustrated using the geological log of borehole 'Bhaluka 77R', drilled within the Madhupur Tract to the north of the study area (Figures 1.1, 1.3).

Jamuna aquifer

The Jamuna aquifer system (Figure 1.2) occurs within non-indurated grey alluvial sands and gravels of the Late Quaternary age Dhamrai Formation, which is divided into two parts about a discontinuous palaeosol layer. The Dhamrai Formation underlies the low-lying flood plain of the Ganges/Brahmaputra river system and its distributaries, which is inundated during the annual monsoon when water-levels rise by at least 6 m. The hydrogeological characteristics of the Jamuna aquifer system were assessed during the BGS pilot study into optimum deep tubewell design. At each of 16 test sites a 2 cusec capacity test deep tubewell and nine observation boreholes were installed. The results of comprehensive test pumping and flow logging are described by Davies *et al.* (1988). The test sites were located within the Manikganj, Saturia and Dhamrai areas (Figure 1.2).

Methodologies developed by Walker (1979) and Reading (1986) were used to recognize six environments or facies of sediment deposition using data from grain-size distribution analyses and geological borehole logs (Davies, 1989). The Dhamrai Formation can be divided using the facies units recognized as shown in a geological cross section of the study area (Figure 1.4). The Lower Dhamrai Formation is a fining upwards succession of coarse sands and gravels deposited by strongly flowing braided rivers. The coarse to medium sands of the Upper Dhamrai Formation were also deposited as a series of upwards fining units, from smaller braided and, latterly, meandering rivers.

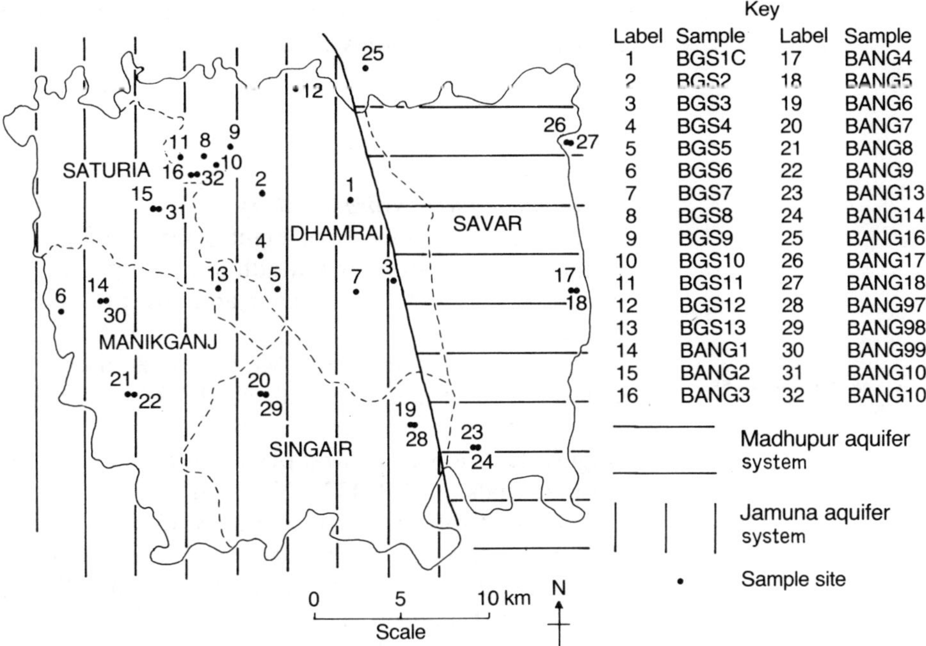

Figure 1.2 Sample site location and aquifer distribution map.

Together, these coarse-grained, thick bedded alluvial deposits form a very productive aquifer system. The fine sediments at the top of both the Upper Dhamrai Formation and the Lower Dhamrai Formation contain much biotite mica and disseminated red wood fragments. Within the eastern part of the area they form a well-defined palaeosol layer capping the Lower Dhamrai Formation. Here the layer is composed of highly weathered orange, red, brown and purple silty clays interbedded with thin bands of ferricrete and clayey silts, indicating that the sediments have been intensely oxidized. These oxidizing conditions promote organic carbon decomposition through bacteriological action producing acidic conditions where iron is readily leached from the biotite mica. The biotite micas are converted to vermiculite and kaolinite type clays while the mobilized iron is redeposited as orange brown ferric oxide cement and concretions of ferricrete (Acker and Bricker, 1992; Lee, 1992; Velde, 1992).

Transmissivities of 1000–3000 $m^2\ d^{-1}$ from the lower and 2000 $m^2\ d^{-1}$ from the upper parts of the Jamuna aquifer system were obtained from vertical permeability determinations (Davies and Herbert, 1990) and test pumping data. The lower coarser part of the Jamuna aquifer is confined by a palaeosol layer in the east, but is leaky confined in the west where the palaeosol is absent. The upper part of the Jamuna aquifer system is leaky confined. Many of the deep tubewells installed only partially penetrate the aquifer system.

Radiocarbon ages obtained from the analysis of wood fragments provide approximate dates of deposition of Dhamrai Formation sediments (Table 1.1). The basal coarse sediments of the Lower Dhamrai Formation were deposited between 20 000 and <48 000 yr BP, while those of the Upper Dhamrai Formation were deposited between 7 000 yr BP and the present day (Table 1.1). A sample obtained from the Dupi Tila Formation was dated at <39 400 yr BP supporting the deposition dates of Monsur (1990). The history of sediment deposition and weathering within the Dhamrai Formation, adjacent to the Dupi Tila Formation, is illustrated in Figure 1.5.

Hydrochemistry

Data and sample collection

Two hydrochemical surveys were undertaken by BGS within the study area at sites shown on Figure 1.2. Fifty-eight groundwater samples were obtained for hydrochemical analysis at 13 test sites during the 1986 and 1987 field seasons, during long-term test pumping and from hand-pumped observation boreholes. A representative group of 24 analyses is presented in Table 1.2. During 1992 an additional 19 water samples were obtained from shallow and deep tubewells during a reconnaissance hydrochemical survey of central and northeastern Bangladesh (Davies and Exley, 1992). Field determinations of pH, redox potential (Eh), specific electrical conductance (SEC), temperature and bicarbonate content were undertaken according to methods and criteria described by Cook, Edmunds and Miles (1979).

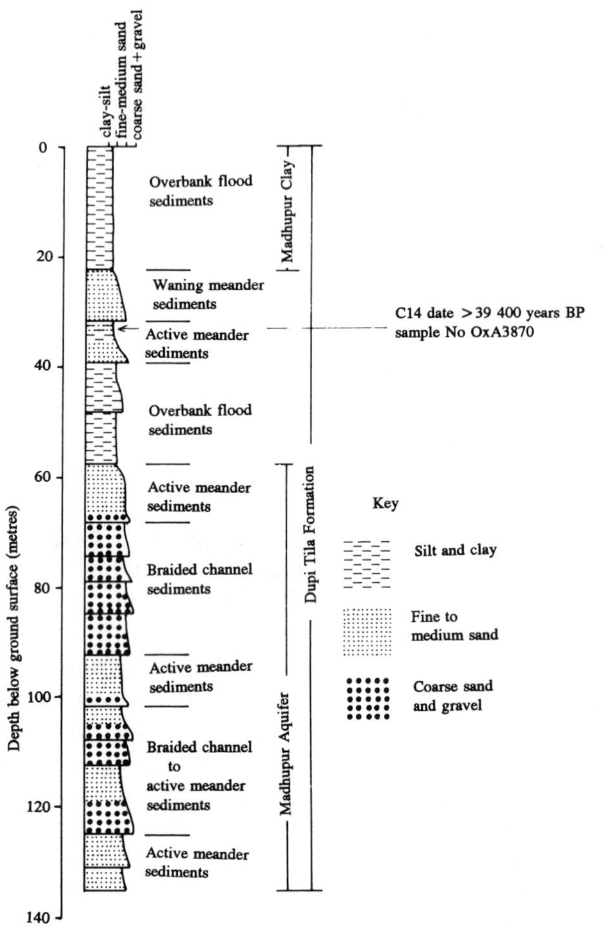

Figure 1.3 Borehole No. Bhaluka 77R – geological log and facies distribution.

Laboratory analyses

Samples were submitted to the analytical laboratory at BGS Wallingford, acidified samples being analysed for 31 major, minor and trace elements. Determinations of cations, SO_4 and selected trace elements were undertaken by inductively coupled plasma optical emission spectrometry (ICP-OES). Chlorine concentrations were determined by automated colorimetry. Of the elements determined, Sc, Be, Y, Co, V, Cd, La, Cu, Zr, Cr, Ni, Mo and Pb were found to be absent or below detectable levels. The concentrations of Ca, Mg, Na, K, HCO_3, SO_4, Cl, total Fe, Mn, Si, NO_3–N and P in samples analysed as well as field determinations are presented in Table 1.2. Ranges of concentrations of Sr, Ba, Li, B, Zn, Al, Br and I determined within the samples analysed are listed in Table 1.3.

Data collected during the BGS surveys have been used to produce summary descriptions of the hydrochemical characteristics of the two aquifer systems present within the study area.

Results

Madhupur aquifer hydrochemistry

Hydrochemical analyses of eight samples of groundwater collected from the Madhupur aquifer at sites within or adjacent to the Savar area are presented in Table 1.2.

The groundwaters of the Madhupur aquifer are of $Ca(HCO_3)_2$ trending towards $NaHCO_3$ type (Figure 1.6). They have low conductivities (83–370 $\mu S\ cm^{-1}$) (Figure 1.7), low to neutral pH levels (6–6.8) and high Eh potentials, of 233–437 mV (Figure 1.8).

Ranges of major, minor and trace element concentrations determined for the Madhupur aquifer are listed in Table 1.3. Ionic concentrations are generally low, with all but total iron and manganese being below the recommended WHO levels for human consumption. Under conditions of neutral pH and high Eh potentials, such as occur within the Madhupur aquifer, both iron and manganese tend to precipitate out as orange ferric hydroxide and black manganese oxide respectively. Silica concentrations are high at 17–30 mg l^{-1} (Figure 1.9), indicative of fairly acid groundwater conditions. Fish mortalities due to the application of tubewell water to fish ponds have been reported from the Madhupur Tract area to the north of Savar. Being acidic, the groundwaters of the Madhupur aquifer can corrode metal borehole and pump fittings as indicated by moderately high levels of zinc recorded at several sites. Of the trace elements, fluoride is high at over 0.5 mg l^{-1} while iodine is very low. The low levels of iodine indicate that the development of goitre amongst the human population could become a problem should the standard of nutrition decline.

Jamuna aquifer hydrochemistry

Hydrochemical analyses of 35 samples of groundwater collected from the Jamuna aquifer at sites within the study area are also presented in Table 1.2.

The groundwaters of the Jamuna aquifer are of $Ca(HCO_3)_2$ type (Figure 1.6). They are chemically reducing in nature, with high conductivities (258–675 $\mu S\ cm^{-1}$ (Figure 1.7), neutral to high pH levels (5.7–8.2) and low Eh potentials (48–222 mV) (Figure 1.8).

Levels of conductivity and ionic concentration in the study area generally increase from east to west and with depth within the Jamuna aquifer system. The ranges of ionic concentrations determined for the Jamuna aquifer are presented in Table 1.3. Only total iron and manganese levels exceed WHO limits for human consumption. The hydrochemistry of iron and manganese within

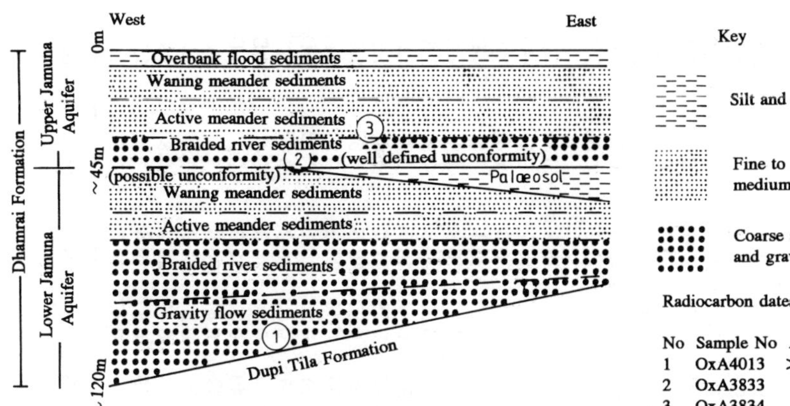

Figure 1.4 A schematic geological cross-section showing sediment facies distribution across the study area, with radiocarbon dates.

Table 1.1 Radiocarbon age determinations

Sample No. (see Figure 1.4)	Sample No.	Bh No.	Depth (m)	Age (yr)
1	OxA-4013	15/TW	62.0	>48 500
2	OxA-3833	4/100	34.4	6 400 + 85
3	OxA-3834	13/100	26.0	6 290 + 75
4	OxA-3870	Bhal77R	30.0	>39 400

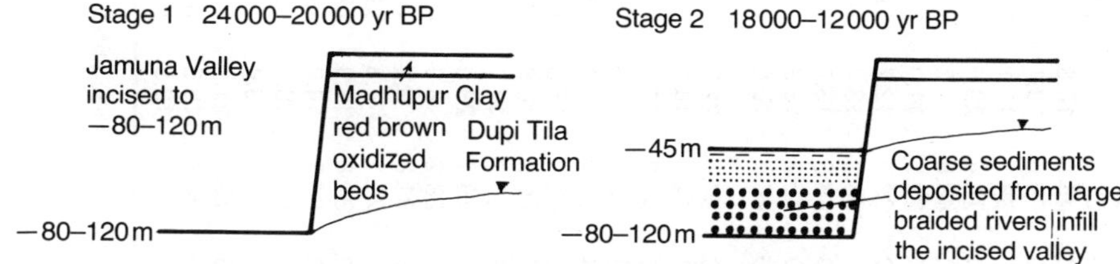

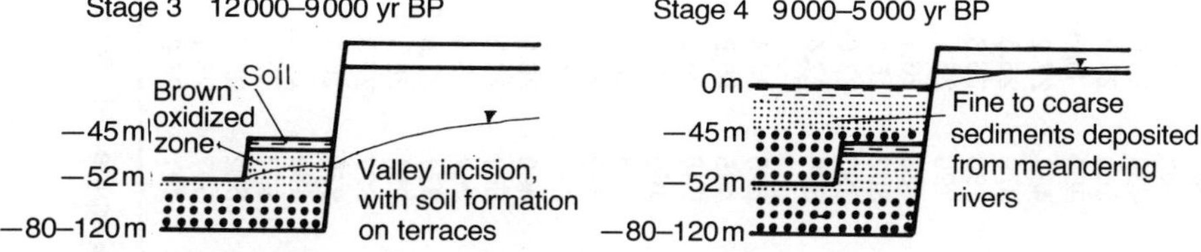

Figure 1.5 Late Quaternary sequence of Dhamrai Formation deposition.

anoxic aqueous conditions is controlled by the neutral pH and low Eh potentials recorded. Under these chemically reducing conditions, iron leached from abundant biotite micas remains in solution in the Fe^{2+} state. Manganese occurs in a similar form to iron. Both iron and manganese compounds are oxidized on contact with the atmosphere at the pump head, precipitating out to form, in the case of iron, a characteristic orange sludge seen coating many of the irrigation channels of central Bangladesh. Dissolved total iron concentrations are high (< 10 mg l^{-1}) (Figure 1.10), as are those of phosphorus at > 1 mg l^{-1}, but iodine concentrations are low.

Discussion

The results of the detailed geological and hydrogeological investigations undertaken have been used to construct a conceptual model to explain the deposition of alluvial sediments of the Dhamrai and Dupi Tila Formations. The impact of climatic and sea-level change controlling the distribution of sediment facies of deposition has also been examined. The radiocarbon date of wood from the basal part of the Dhamrai Formation implies that these alluvial sediments were deposited during the interglacial periods preceding the last Ice Age. The Dupi Tila Formation appears to have

Table 1.2 Hydrochemical analyses of groundwater samples from central Bangladesh

Bh. No.	Type	Aquifer Unit	Date Collected	Depth (m)	SEC (μS cm^{-1})	Temp. (°C)	pH	Eh (mV)	Ca	Mg	Na	K	HCO_3	SO_4	Cl mg l^{-1}	Total Fe	Mn	Si	NO_3–N	P
BGS1C	DTW	LJ	8/6/86	52–77	275	26.9	?	N/D	33	12.2	17	2.2	226	0.6	1.8	0.78	N/D	25.1	0.27	N/D
BGS2	DTW	LJ	22/1/86	56–80	565	26.9	6.8	N/D	65.5	23.3	20	3.1	366	1.6	5.8	1.23	N/D	26.2	0.05	N/D
2/25/3	HPW	UJ	29/12/85	37–40	562	26.1	N/D	N/D	57.8	19.8	21	3.4	311	5.4	8.8	0.22	N/D	11.6	0.05	N/D
BGS3	DTW	LJ/M	15/3/86	52–70	321	26.15	N/D	N/D	31	13.8	13.4	1.0	205	1.2	2.4	0.13	N/D	28.0	0.15	0.2
3/50/3	HPW	UJ	18/2/86	15–19	258	24.6	7.1	N/D	24.9	9.1	18.6	1.9	176	5.3	3.2	0.12	N/D	14.2	0.5	0.2
BGS4	DTW	LJ	9/4/86	46–77	465	26.4	N/D	N/D	62	16.3	14.9	3.7	317	2.7	14	8.46	N/D	21.9	0.5	1.5
4/25/3	HPW	UJ	2/4/86	12–15	600	25.95	N/D	N/D	82	36	26	3.2	422	60	6.7	1.77	N/D	12.8	0.5	0.2
BGS5	DTW	LJ	30/4/86	44–87	575	28.4	N/D	N/D	63.8	25	43	3.5	436	0.5	14.5	5.32	N/D	26.6	0.38	0.8
5/50/3	HPW	UJ	28/12/86	18–21	565	25.9	7.3	N/D	54.3	23	37.1	4.5	370	2.5	7.3	0.89	N/D	6.30	0.05	0.2
BGS6	DTW	LJ	22/12/86	28–50	670	26.6	7.0	N/D	82	35	15.3	3.5	427	4.3	11.4	8.8	N/D	21.8	0.08	N/D
6/50/3	HPW	UJ	22/12/86	31–34	580	26.3	7.38	N/D	82	31	15	4.4	422	5.4	12.7	0.57	N/D	12.5	0.08	N/D
BGS7	DTW	L/UJ	3/5/86	31–80	362	26.8	N/D	N/D	51.7	15.5	23	3	258	2.1	7.2	3.04	N/D	27.8	0.79	0.4
BGS8	DTW	L/UJ	16/3/86	39–75	458	24.6	6.7	N/D	52.5	17.8	27.2	3.9	280	5.0	15.0	6.9	N/D	21.9	2.1	1.2
8/25/3	HPW	UJ	16/3/86	18–21	430	26.4	7.3	N/D	45.6	18.8	18.7	4.1	248	9.5	14.9	1.0	N/D	14.5	0.05	0.3
BGS9	DTW	L/UJ	27/2/87	31–71	360	25.9	6.3	N/D	39.5	13.1	14.1	3	205	7.9	9.6	15.8	N/D	23.3	0.06	0.6
9/50/3	HPW	UJ	27/2/87	24–27	350	N/D	6.7	N/D	39.2	13.1	14.2	2.5	158	59	2.8	5.7	N/D	17.4	0.06	0.2
BGS10	DTW	L/UJ	29/3/87	29–71	484	27.2	6.4	N/D	74.4	23.5	24	4.1	411	4.2	5.2	5.9	N/D	22.1	0.05	0.8
10/25/3	HPW	UJ	29/3/87	24–27	370	26.6	7.0	N/D	48.8	17.3	12.3	3.1	244	17.1	2.6	5.9	N/D	12.3	0.05	0.2
BGS11	DTW	L/UJ	22/1/87	29–71	290	26.6	6.94	N/D	27.5	11.0	11.9	2.5	176	0.5	2.5	7.0	N/D	21.9	0.06	1.1
11/25/3	HPW	UJ	22/1/87	18–21	290	26.1	7.29	N/D	29.9	12.7	11.1	2.8	187	2.2	1.9	0.12	N/D	17.1	0.06	0.2
BGS12	DTW	LJ	12/2/87	52–77	338	26.1	6.75	N/D	33.2	13.4	24.4	2.7	236	0.5	2.1	0.84	N/D	30.2	0.06	0.2
12/25/3	HPW	UJ	12/2/87	12–15	302	25.1	7.3	N/D	43.6	15.2	17.4	3.4	219	19.3	4.4	0.32	N/D	14.2	0.41	0.2
BGS13	DTW	UJ	24/4/87	28–56	392	27.0	6.7	N/D	49.2	18.2	9.7	3.5	223	23.1	6.4	5.8	N/D	19.6	0.1	1.0
13/25/3	HPW	UJ	24/4/87	24–27	378	27.4	6.7	N/D	46.7	15.6	11.7	3.5	219	24.9	9.4	0.11	N/D	12.3	0.05	0.2
BANG1	HPW	UJ	18/1/92	N/D	675	24.4	6.7	133	60.4	22.5	14.2	2.7	381	0.38	28.7	24.7	0.74	19.3	<0.3	1.3
BANG2	HPW	UJ	18/1/92	N/D	465	25.8	6.9	222	69	14.9	8.1	3.5	307	3.66	11.5	0.48	1.42	18.2	0.6	N/D
BANG3	HPW	LJ	18/1/92	N/D	546	25.2	6.66	111	49.9	27.3	14.2	2.3	268	N/D	25.8	15.0	0.63	20.0	N/D	1.1
BANG4	DTW	M	18/1/92	45–97	146	26.3	6.37	403	12.9	3.5	13.3	1.3	102	1.52	4.0	0.09	0.01	30.1	1.6	N/D
BANG5	HPW	M	18/1/92	18–24	83	26.0	6.03	304	4.9	1.6	7.5	1.6	31	0.31	7.0	0.67	0.01	16.8	1.9	N/D
BANG6	HPW	LJ	19/1/92	55–61	392	25.3	7.10	138	42.8	15.3	19.7	1.5	302	0.38	4.5	1.07	0.32	20.8	0.9	0.8
BANG7	HPW	UJ	19/1/92	43–46	465	24.3	6.73	84	61.9	16.8	12.1	3.1	344	6.22	5.0	12.90	0.44	18.7	0.7	1.0
BANG8	DTW	LJ	19/1/92	50–74	490	26.5	6.80	48	56.7	17.7	17.7	7.4	319	7.11	11.8	8.10	0.26	21.4	1.4	1.3
BANG9	HPW	UJ	19/1/92	25–28	553	25.3	6.80	90	59.7	29.0	11.6	3.8	371	28.38	2.8	9.85	1.37	18.5	0.5	1.0
BANG13	DTW	M	19/1/92	30–55	267	25.4	6.48	437	28.2	8.9	26.5	1.7	212	0.63	2.8	0.04	0.10	25.9	<0.3	N/D
BANG14	HPW	M	21/1/92	N/D	331	21.5	6.76	323	31.8	9.3	41.5	1.6	295	N/D	2.9	0.91	0.11	20.1	N/D	N/D
BANG16	HPW	M	21/1/92	40–42	291	26.1	6.85	349	26.3	7.3	31.1	1.3	210	0.39	2.1	0.07	0.03	20.3	<0.3	0.3
BANG17	DTW	M	21/1/92	52–95	327	26.3	6.85	400	31.8	11.9	27.5	1.1	244	0.48	1.4	0.04	0.16	24.6	0.4	N/D
BANG18	HPW	M	21/1/92	49–55	370	26.1	6.81	233	34.3	14.8	29.8	1.0	268	0.38	6.4	0.49	0.05	22.5	<0.3	N/D
BANG97	DTW	LJ	21/2/92	52–88	489	26.0	7.09	149	52.8	18.5	18.4	3.1	366	0.1	4.0	7.11	0.47	20.5	1.6	0.6
BANG98	DTW	L/UJ	21/2/92	41–65	577	26.6	6.92	132	65.0	23.4	20.9	3.1	390	6.46	7.0	11.9	0.31	21.7	1.1	1.2
BANG99	DTW	L/UJ	21/2/92	35–59	572	25.7	6.77	129	73.1	21.7	13.7	2.4	390	0.64	11.3	14.2	0.49	20.2	0.8	1.2
BANG100	DTW	L/UJ	21/2/92	31–67	507	26.1	6.84	N/D	68.4	19.5	10.9	3.4	361	3.53	2.8	6.69	0.81	19.4	1.4	1.2
BANG101	DTW	LJ	21/2/92	51–76	427	26.0	6.73	130	39.5	19.5	13.1	3.5	297	14.21	12.5	10.3	0.47	20.7	0.6	1.1

UJ – Upper Jamuna aquifer, LJ – Lower Jamuna aquifer, M – Madhuper aquifer; DTW – Deep Tubewell, DPW – Hand-pumped well, 2/25/3 – Shallow observation borehole 7.62 m from test bh. BGS 2; N/D – not determined.

Table 1.3 Summary of ionic concentrations and field parameter determination ranges per aquifer

Aquifer unit	Jamuna aquifer (mg l^{-1})	Madhupur aquifer (mg l^{-1})
Lab determinations		
Calcium (Ca)	25–82	5–34
Magnesium (Mg)	9–36	2–15
Sodium (Na)	8–43	8–42
Potassium (K)	1.0–7.4	1.0–1.7
Manganese (Mn)	0.3–1.4	0.002–0.16
Total iron (Fe)	0.05–24.7	0.04–0.91
Zinc (Zn)	0.03–27	0.03–0.19
Aluminium (Al)	0.1–0.4	0
Strontium (Sr)	0.18–0.28	0.01–0.21
Lithium (Li)	0.008–0.02	0
Boron (B)	0.03	0.03–0.04
Bicarbonate (HCO_3)	165–436	31–295
Sulphate (SO_4)	0.4–60	0.3–1.5
Chloride (Cl)	1.8–28.7	1.4–7
Bromide (Br)	0.02–0.04	0.01–0.07
Fluorine (F)	0.1–0.44	0.05–0.57
Iodine (I)	0.005–0.015	0.004–0.092
Nitrate (NO_3–N)	0.05–3.4	0.15–1.9
Silicon (Si)	5.6–30.2	16.8–30.1
Phosphorus (P)	0.2–1.5	0
Field determinations		
pH	5.7–8.2	6.0–6.8
Eh (mV)	48–222	233–437
SEC ($\mu S\ cm^{-1}$)	258–675	83–370
Temperature (°C)	24.3–28.4	21.5–26.3

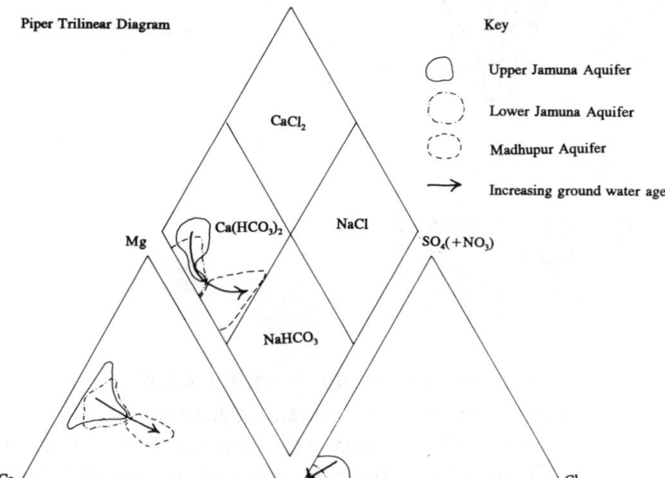

Figure 1.6 Aquifer hydrochemistry.

undergone several periods of intense oxidation and weathering.

Within the Lower Jamuna aquifer, the anoxic nature of the groundwater promotes bacteriological attack of organic carbon that leads to the formation of various humic and fulvic compounds. Under these chemically reducing conditions, iron is readily leached from biotite mica passing into solution in the ferrous state, a process that takes place at a slow, possibly steady, rate. With increased depth and age the quantity of organic carbon appears to diminish while the colour of the biotite gets steadily lighter, changing from black through brown and orange to white. Ultimately, biotite decomposes to form vermiculite and kaolin type clays.

Within the oldest Dupi Tila Formation there are substantial amounts of grey clay and orange ferric hydroxide cements present. Mica is present in a very weathered orange to clear form but wood fragments are generally absent. In such a well-weathered formation there is very little material left to buffer the acidity of recently recharged rainwater. Consequently the groundwaters are of good quality, of $Ca(HCO_3)_2$ type, but of low pH and high Eh character typical of oxidizing conditions. There is increased ion exchange and a trend towards an $NaHCO_3$ type water.

The Madhupur and Jamuna aquifers have distinct groundwater chemistries as shown in Figure 1.6, permitting their mapping on a regional scale through the plotting of hydrochemical parameters, e.g. Eh, SEC, Si and total Fe (Figures 1.7–1.10). The distribution of ionic concentrations, as shown in Figures 1.9 and 1.10, would seem to indicate a general groundwater flow pattern from northeast to southwest. However, this pattern contradicts that implied using the distribution of groundwaters characterized by bulk hydrochemistry as shown in Figure 1.6. The groundwater chemistry of each aquifer block is controlled by sediment mineralogy, reactable constituents such as wood fragments, and the periodic intrusion of brackish waters into aquifer systems adjacent to the major rivers. These factors hinder identification of groundwater flow patterns on the basis of groundwater chemistry alone.

Conclusions

The hydrochemistry of the sedimentary aquifers of central Bangladesh is dependent upon a number of factors each of which is imperfectly understood at the present time.

(1) The alluvial sedimentary aquifers are composed of rock fragment material derived from micaceous metamorphic and acid igneous rock bodies within the Himalayas. Present day and recent medium to fine sands, deposited by the Brahmaputra river and its tributaries, contain much detrital biotite mica and wood fragment material, but contain no detrital clay deposits.

(2) The initial post-depositional interaction of organic carbon and sediments under waterlogged, anoxic

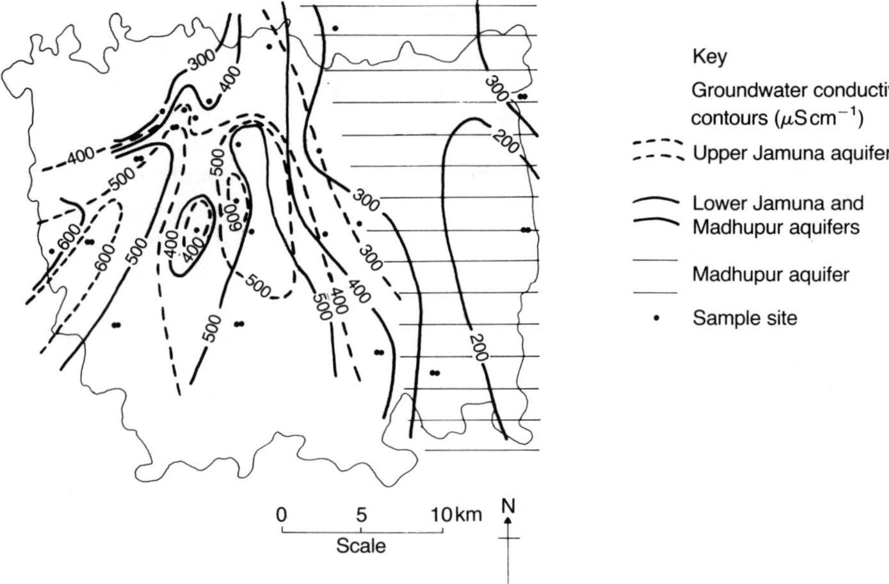

Figure 1.7 Groundwater conductivity contour map ($\mu S\ cm^{-1}$).

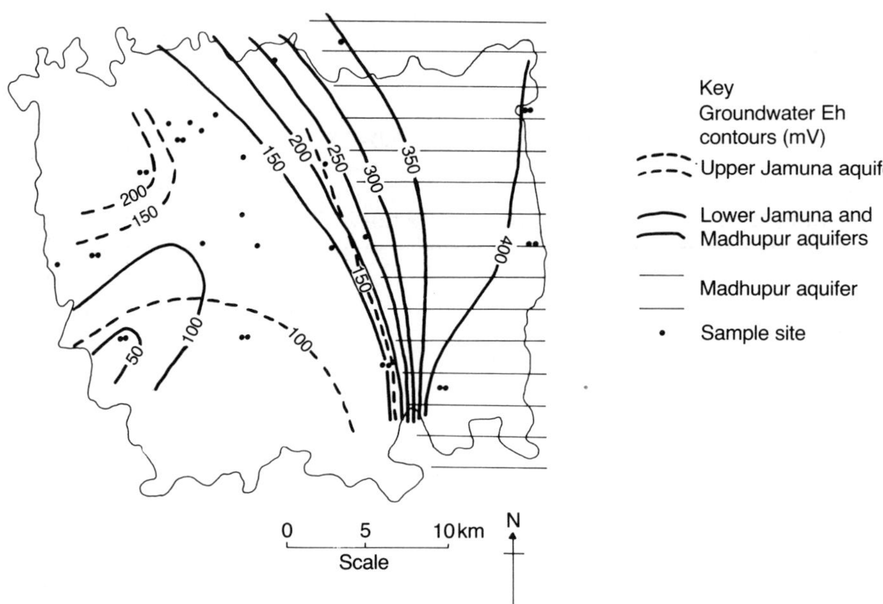

Figure 1.8 Groundwater Eh contour map (mV).

and reducing conditions promotes the mobilization of iron in the ferrous form, leached from biotite mica, and the degeneration of feldspars. This initial form of diagenesis is accelerated within the near surface zone where the water table rises and falls with the annual monsoon season, the wetting and drying of sediments promoting rapid weathering under oxidizing conditions.

(3) Late Pleistocene sea-level changes, related river incision and river terrace deposition have been noted and dated approximately. The available radiocarbon dates from the Dhamrai Formation suggest that there may have been several periods of river incision and infilling within what is now the valley of the Brahmaputra. Additional age datings are required to produce a deposition model. Within the Dupi Tila Formation changes in water tables within the interfluves resulted in the oxidation of iron from the soluble ferrous form to the ferric state to be precipitated as orange brown ferric hydroxide cement.

(4) The nature, intensity and depth of weathering within the interfluves during prolonged cycles of climatic and sea-level changes during the Upper Pleistocene prompted the removal of organic carbon from the Dupi Tila Formation, precipitation of

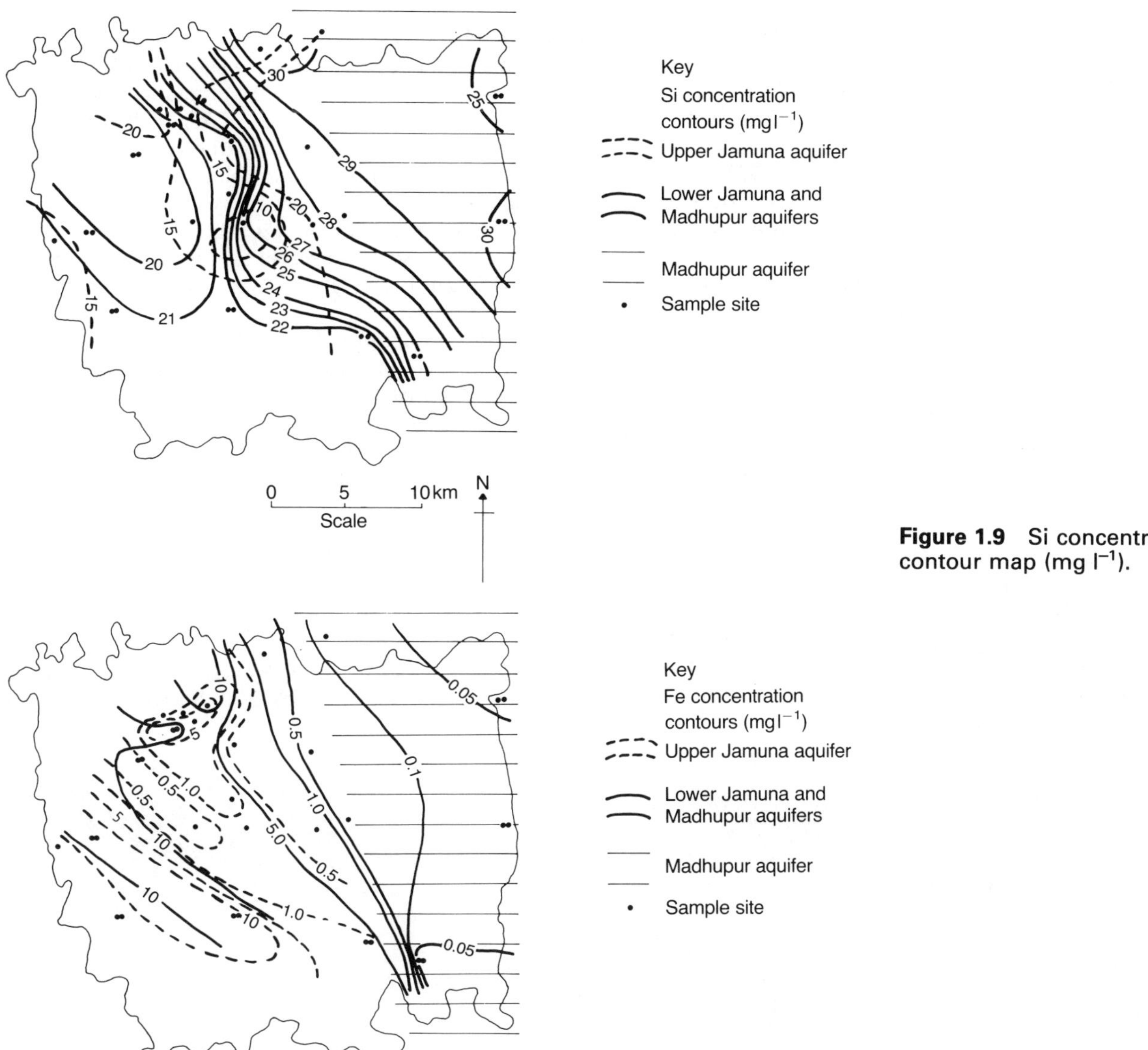

Figure 1.9 Si concentration contour map (mg l^{-1}).

Figure 1.10 Total dissolved iron concentration contour map (mg l^{-1}).

iron as ferric oxide cements and concretions and the degradation of feldspars and micas to kaolinitic clays. It is possible that the proportion of mica to clay in these and the inter Dhamrai Formation palaeosol deposits could be used to date these deposits approximately.

Therefore, each block of alluvial aquifer located above the 20 000 yr BP level, i.e. 120 m below that of the present day, should have a distinctive sediment mineralogy, hydrogeological parameters and groundwater chemistry that reflect the age of deposition and weathering history experienced.

The geological, hydrogeological and hydrochemical factors discussed above can control the distribution of minor elements, some of which can be toxic or helpful to aspects of agricultural, health and aquacultural systems.

Acknowledgments

The author is grateful to Peter Ravenscroft for his helpful discussions in Bangladesh and Carole Sharratt for her help in preparing the text. This chapter is published by permission of the Director, British Geological Survey (NERC).

References

Acker, J.G. and Bricker, O.P. 1992. The influence of pH on biotite dissolution and alteration kinetics at low temperature. *Geochimica et Cosmochimica Acta*, vol. 56, pp. 3073–3092.

Coleman, J.M., 1969. Brahmaputra River: Channel processes and sedimentation. *Sedimentary Geology*, vol. 3, pp. 129–239.

Cook, J.M., Edmunds, W.M. and Miles, D.L., 1979. Water sampling and field chemical measurements for overseas hydrogeological surveys. *British Geological Survey Technical Report, WD/ST/79/11.*

Davies, J., 1989. The pilot study into optimum well design: IDA 4000 Deep Tubewell II Project. Volume 2 The geology of the alluvial aquifers of Central Bangladesh. *British Geological Survey Technical Report WD/89/9.*

Davies, J. and Exley, C., 1992. Short-term BGS pilot project to assess the 'hydrochemical character of the main aquifer units of Central and North-eastern Bangladesh and possible toxicity of groundwater to fish and humans', Final Report. *British Geological Survey Technical Report WD/92/43R.*

Davies, J. and Herbert, R., 1990. Field determination of vertical permeability. *British Geological Survey Technical Report, WD/90/47R.*

Davies, J. Herbert, R., Nuruzaman, M.D., *et al.*, 1988. The pilot study into optimum well design: IDA 4000 Deep Tubewell II Project. Volume 1 Fieldwork – Results. *British Geological Survey Technical Report WD/88/21.*

Herbert, R., Barker J.A. and Davies, J., 1989. The pilot study into optimum well design: IDA 4 000 Deep Tubewell II Project. Volume 6 Summary of the programme and results. *British Geological Survey Technical Report WD/89/14.*

Lee, C., 1992. Controls on organic carbon preservation: The use of stratified water bodies to compare intrinsic rates of decomposition in oxic and anoxic systems. *Geochimica et Cosmochimica Acta*, vol. 56, pp. 3323–3335.

Monsur, H., 1990. Some aspects of the Quaternary sediments of the Bengal Basin. PhD Thesis, Free University of Brussels, Belgium.

Morgan, J.P. and McIntire, W.G., 1959. Quaternary geology of the Bengal Basin, East Pakistan and India. *Geological Society of America Bulletin*, vol. 70, pp. 319–342.

Mott MacDonald International Limited, 1992. Deep Tubewell II Project, Main Report. Dhaka, Bangladesh.

Reading, H.G. (ed.) 1986. *Sedimentary Environments and Facies*. Blackwell Scientific Publications, Oxford, 615 pp.

UNDP, 1982. Groundwater survey: the hydrogeological conditions of Bangladesh. UNDP Technical Report DP/UN/BGD-74-009/1, 113pp.

Velde, B., 1992. *Introduction to Clay Minerals*. Chapman & Hall, London, 198 pp.

Walker, R.G. (ed.) 1979. *Facies Models. Geoscience Canada Reprint Series*, No. 1, Geological Society of Canada, Waterloo, 317 pp.

2 Characterization of groundwaters in semi-arid and arid zones using minor elements

W.M. Edmunds

Abstract The information on groundwater chronology and recharge history from semi-arid and arid regions provided by environmental isotopes often gives a framework within which geochemical processes can be established. The major ion chemistry provides the basic information on the geochemical reactions taking place, but minor elements can be used to give important additional information on the recharge history and on processes occurring along flow paths in the aquifers of semi-arid regions. Results from studies of minor elements (NO_3, Fe, Mn, Br, Sr, F, Ba, Li, F, B) are given from groundwaters in Sudan and Libya. Redox characteristics can be inferred using the concentrations of NO_3 and Fe_T (total iron) in the absence of measurements of dissolved oxygen or Eh. Oxygen has persisted in much of the groundwater in the eastern Sahara for periods of many thousands of years and within these waters variable nitrate concentrations are considered to reflect different palaeo-environments. Bromide and Br/Cl ratios may be used to provide information on present and past recharge sources as well as sources of salinity. The age of groundwater, determined using isotopic methods, may be used to determine the length of time taken for attainment of equilibrium with various minerals. Relative timescales may be constructed for groundwater residence using concentrations of Sr, F and Li, where these are below limits of mineral saturation. Strontium and fluoride are important indicators of facies changes (marine transgression) as, for example, in the Miocene aquifer of central Libya. In the continental Tertiary and Cretaceous aquifer system in Libya, hydrogeochemical changes are related to recharge sources, residence time, or mixing and reaction along flow paths. High boron concentrations (and B/Cl) were used to characterize recharge from shallow wadi systems in Sudan and could be used to distinguish this water from recharge from the River Nile or from palaeowaters.

Introduction

Environmental isotopes occupy a unique role in arid zone hydrogeological studies in providing information on the recharge history and residence time (Fontes and Edmunds, 1989). In the major aquifers of the Sahara and in other desert regions it has been possible to demonstrate conclusively the existence of palaeowaters recharged during Pleistocene and early Holocene pluvial episodes and to determine meteorological and palaeoclimatic features of these water bodies using distinctive isotopic fingerprints. At the present day, isotopic data have helped to demonstrate that the amount of water that is being safely recharged from highly erratic rainfall regimes is often small; this provides a warning of over-development in groundwater projects. Large scale development schemes are often at the expense of smaller sustainable traditional schemes.

For a fuller understanding of the evolution of the groundwater system, and particularly in view of the many quality constraints on development, it is equally important to have parallel hydrogeochemical data. The isotopic data can then be considered as the framework for the aquifer hydrology on which the solute chemistry provides much important information on processes, either reinforcing the isotopic results or giving quite

Groundwater Quality Edited by H. Nash and G.J.H. McCall. Published in 1994 by Chapman & Hall. ISBN 0 412 58620 7

separate lines of diagnostic evidence. It should also be remembered that isotopic studies are of value in geochemical process studies to provide time frames of reaction. In this chapter some examples of the use of minor and trace elements are given, which provide additional understanding of the groundwater evolution.

Limitations of major ions in hydrogeological studies

The major ions Ca, Na, Mg, K, HCO_3, SO_4 and Cl constitute ~ 98% of the total mineralization of most groundwaters. The major ion chemistry is a statement of the principal reactions or processes that have taken place during recharge and subsurface flow. It will indicate whether carbonate minerals or sulphate minerals have been involved in the flow path and, using suitable computer programs, it is possible to determine to what extent equilibrium with these minerals has been reached. It will also suggest whether significant evapotranspiration or mixing with saline sources has occurred. However, it is frequently impossible to derive unambiguous information on groundwater evolution using major elements alone.

The total mineralization (or specific electrical conductivity), representing the sum of all major ions, can serve as a powerful index of evolution in conjunction with stable isotopes to indicate the extent of evapotranspiration or the extent of mixing with saline marine or continental waters. Chloride is of special value in recharge estimation, and in hydrogeochemical studies in general, since it usually can be treated as an inert reference element (Allison and Hughes, 1978; Edmunds and Walton, 1980) of value in recharge studies and to monitor additions from the aquifer matrix. Sulphate is of importance in tracing flow paths and mixing, especially where evaporite sequences are present, but is generally not inert. It may sometimes be used to estimate inputs to hydrological systems but additions may take place as a result of sulphide oxidation as well as mineral dissolution; elsewhere *in situ* reduction of sulphate to sulphide may occur. The other major ions taken alone, are often of little diagnostic value. In the case of HCO_3, Ca and Mg, maximum concentrations may be reached rapidly which are the result of carbonate mineral equilibria. Increases in major element ratios, such as Mg/Ca, may nevertheless be indicative of incongruent dissolution of carbonates or the precipitation of calcite, and increases in Na/Cl ratio may indicate clay mineral transformations or dissolution of rock forming minerals. These trends are often time-dependent and of special importance when used together with isotopic studies (Edmunds *et al.*, 1988).

Minor elements and their significance

Minor elements and some trace element concentrations may sometimes be of more value individually as diagnostic tools than major ions. This is because they are often released slowly as a result of incongruent reactions or have undergone fractionation as a result of hydrological as well as geochemical processes. Minor elements and minor/major element ratios may therefore assist interpretation in relation to residence times, trends towards groundwater maturity, identification of major facies changes, interpretation of redox conditions, identifying recharge regimes including palaeoclimatic indices and, of special importance, defining the presence (or absence) of hydraulic interconnection and flow paths. Several of these are important in relation to groundwater development.

In the present chapter, data are presented for several elements – NO_3, Fe, Mn, Br, B, Li, Sr, Ba and F – from hydrogeological studies in Libya and Sudan, usually where there is a good isotopic framework against which interpretations can be made.

Hydrogeological and isotopic framework – Libya and Sudan

The present chapter is based mainly upon studies of two regions of the eastern Sahara for which a good isotopic framework exists, and for which there is also a reasonable hydrogeological and geochemical understanding. The region is shown in Figure 2.1 where the common link – the Nubian sandstone system – can be seen.

Libya

The hydrogeology of eastern Libya has been summarized by Wright *et al.* (1982). Two major sedimentary basins – the post-Eocene in the upper Sirte Basin and the Cretaceous Nubian in the Kufra Basin – constitute regional hydrogeological systems. In the former, mainly fluviatile sands and clays grade into marine facies forming an important boundary in the north of the basin. The Nubian sandstone includes mainly cross-bedded sandstones with some fluvial or lacustrine shales and conglomerates. Groundwater flow is mainly north-eastwards towards topographic depressions in Egypt. The post-Eocene aquifer may be subdivided into a lower (Lower and Middle Miocene – LMM) and an upper (post-Middle Miocene – PMM) unit which may or may not be hydraulically connected either by upward leakage or lateral flow. The overall hydrogeochemistry is illustrated in Figure 2.2 where the total mineralization in the two aquifers – LMM and PMM – is shown. The 3-dimensional water quality is more complicated than shown and the map has been built up

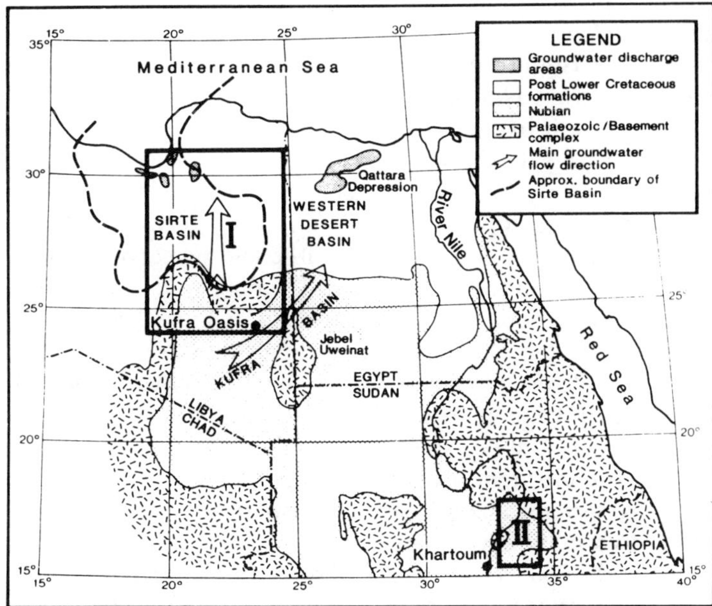

Figure 2.1 Location map of northeast Africa showing the two areas investigated – eastern Libya (I) and Butana, Sudan (II).

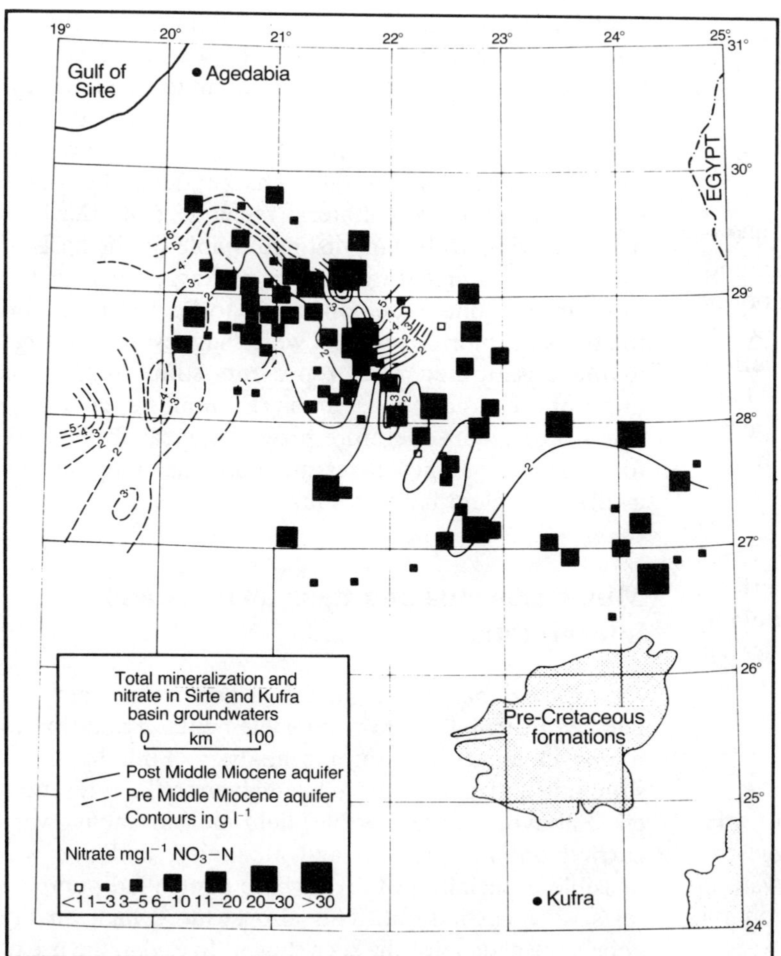

Figure 2.2 Total mineralization of groundwaters in eastern Libya with contours in mg l^{-1}. Nitrate concentrations are also shown by solid squares.

from individual analyses of groundwaters from a range of depths below the water table. There is also some cross-formational flow from the Nubian to the post-Eocene aquifers. The region is characterized by water of relatively low salinity occurring south of 28°N and in a lobe extending northwards to 29°N.

The present day water quality has been shaped by four important factors:

(1) Lithology – especially the transition northwards from continental to marine sediments, which is important in restricting the large scale occurrence of fresh water to the south of 28°N;
(2) Topography – which has allowed the 'fresh' water lobe to develop and which has also been the prime cause of water quality deterioration in the low land to the southeast and north of Jalu (21°30′E 29°02′N);
(3) Mixing and leakage – which have produced modification at the boundaries between the aquifer units in both basins; and
(4) Palaeohydrology – which has maintained the hydraulic gradient from south to north and also from the west. It has also given rise to late, superimposed features such as the main freshwater lens southeast of Jalu together with a thin, fresh water layer detected over much of the aquifer system.

The isotope results (Edmunds and Wright, 1979) in particular, have enabled the recharge history and palaeohydrology of the region to be reconstructed. Radiocarbon 'ages' have been corrected on the basis of their stable carbon isotope ratios, hydrogeochemistry and on environmental samples from the area. Two main groups can be recognized:

(1) Low ^{14}C activity groundwaters (13 000–34 000 yr BP) with $\delta^{13}C$ of −5.6 to −11.7 %0. These groundwaters correspond to the main Würm glacial period of the northern hemisphere. The period of 35 000–25 000 yr BP is known from this period. Only one sample with an age between 25 000 and 15 000 yr BP is known from the area. The period 15 000–8000 yr BP corresponded to one of hyper-arid activity with no apparent recorded recharge.
(2) Higher ^{14}C activity groundwaters (5000–7800 yr BP) enriched in carbon-13 ($\delta^{13}C = -3.2\%$). These represent recharge during warm humid periods. Residual major wadi features, in places overlain by dune sands of the current arid phase, can be seen in the centre, south and west of the Sirte Basin. Fresh-water lenses defining the courses of these wadis have been identified and one has been dated to this period.

In the Kufra area, the stable isotope ($\delta^{18}O$, $\delta^{2}H$) relationship (Edmunds and Wright, 1979) is suggestive of a major palaeorecharge source with compositions much lighter than the present day but lying on the meteoric water line. Further north, all groundwaters, including those from the Holocene, lie on a line with slope of 4.5 which may indicate a single evaporation trend or a composite of several different evaporation sequences. No evidence of current recharge was found on the basis of tritium.

Sudan

The Butana region lies at the southern margin of the extensive Nubian sandstone system (Figure 2.1, area II) where it overlies crystalline basement rocks of the African Shield. The sandstone aquifer lies to the southeast of the River Nile, from which it is isolated as an embayment. Isotope studies ($\delta^{13}C$, $\delta^{14}C$, $\delta^{18}O$, $\delta^{2}H$) have enabled the sources of recharge and recharge history of this region to be established (Darling *et al.* 1987). These may be summarized in the delta diagram (Figure 2.3a) and in the idealized cross-section south to north through the River Nile and featuring the local situation near the control site at Abu Delaig (Figure 2.3b).

The Nubian sandstone aquifer contains a group of groundwaters isotopically lighter and distinct from contemporaneous shallow groundwaters. There are two main age subgroups within this main group, one, 5500–8000 yr BP (uncorrected age) and one with an uncorrected age range from 10 000 to 21 000 yr which may also be hydrogeologically distinct from the younger group. Shallow groundwaters derived from infiltration via wadis are the only source of current recharge, except for the Nile (Edmunds *et al.* 1992). Both these recharge sources are isotopically distinct enabling them to be distinguished from palaeowaters. Tritium shows that lateral recharge from the wadis extends at least 1 km. Lateral recharge from the Nile may extend up to 10 km, distinguishable on the basis of the $\delta^{18}O$, $\delta^{2}H$ signature. Direct recharge through the unsaturated zone is demonstrated to be 1 mm yr^{-1} on the basis of chloride profiles, which suggest that storage in the unsaturated zone represents some 2000 yr or more of recharge in most areas (Edmunds *et al.* 1992). The isotopic data therefore provide a clear framework for hydrogeochemical assessment and also pinpoint the available replenishable resources.

Minor elements and their analysis and interpretation

During regional studies in Libya and Sudan, analysis has been carried out for minor elements in addition to major element and isotopic analysis. Full details of sampling and analysis have been given in the references cited above. Where feasible, field measurements were carried out for pH, Eh and alkalinity, and this was possible especially in Libya where controlled pumping tests were carried out. This also ensured high quality geochemical data for the sites shown. In Sudan the use of submersible pumps was infrequent and samples collected by reciprocating or hand pumps were unsuitable for certain measurements such as Eh or dissolved oxygen.

Analysis for Libyan groundwaters was carried out by atomic absorption for Fe, Mn, Li, Sr and Ba, by auto-

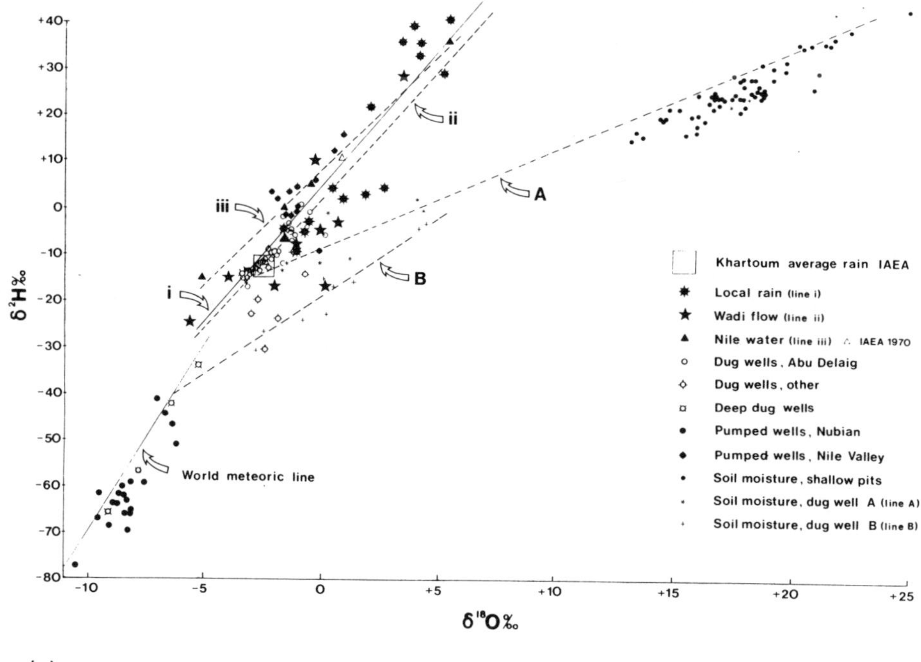

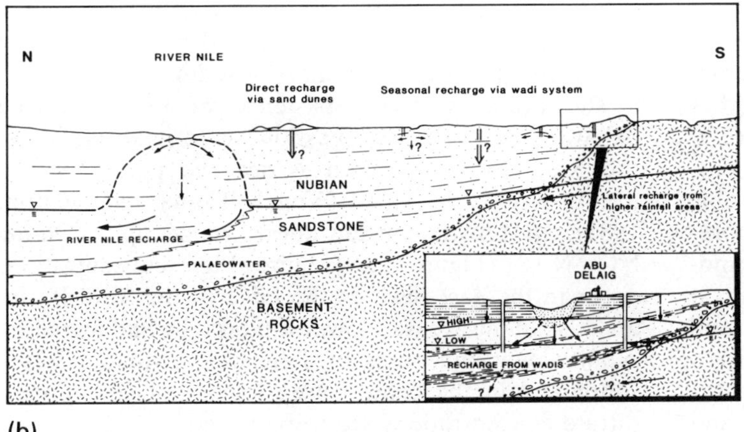

Figure 2.3 Summary diagram (a) of stable isotope ratios of rains, surface waters and groundwaters in the Butana area of Sudan. The principal groups of waters can be identified using the schematic cross-section (b).

mated colorimetry for nitrate and by direct colorimetry for Br and B. Fluoride was analysed by ion selective electrode in the case of groundwaters from both countries. In Sudan, Fe, Mn, Li, Sr, Ba and B were analysed by ICP optical emission spectrometry and bromide and nitrate by automated colorimetry.

Nitrate and other redox indicators (Fe, Mn, Eh)

Since it often proves difficult to measure Eh or dissolved oxygen in groundwaters abstracted from wells in semi-arid or arid regions, a different approach may be adopted to determine the redox status of the groundwater. As groundwater moves along flow paths, oxygen is gradually consumed by microbiological processes in the soil and by reducing agents, notably Fe^{2+} and organic carbon in the aquifer. Once the oxygen is consumed, nitrate then acts as the next oxygen source and generally reacts quite rapidly either as a result of bacterial denitrification (Heaton, 1986) or more widely by further inorganic reaction with Fe^{2+} (Edmunds *et al.* 1984). A sequence of reactions is predicted theoretically (Stumm and Morgan, 1981) and may be accompanied by changes in the redox potential (Eh) which can also be used in the field, under favourable circumstances, to monitor certain changes in the system especially the disappearance of oxygen.

Nitrate is seldom found alone without dissolved oxygen, but once oxygen is consumed, nitrate is quite rapidly reduced, usually via NO_2 to N_2 gas. The presence or absence of nitrate can therefore be used as a coarse indicator of oxidizing (O_2, NO_3 present) or reducing (O_2, NO_3 absent) conditions. Once these oxygen sources have reacted, Eh falls to near or below zero in groundwater, which allows Fe^{2+} solubility to

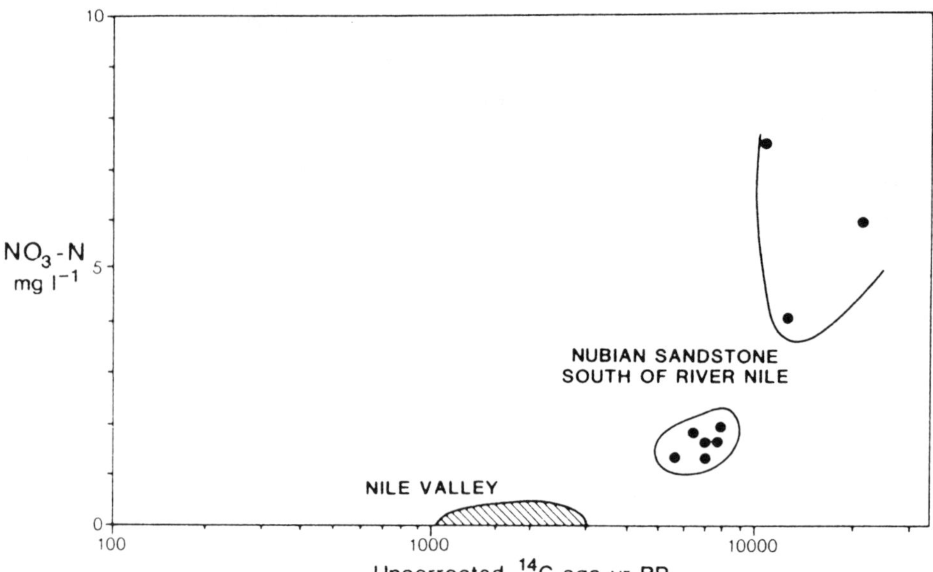

Figure 2.4 Nitrate concentrations vs age for those groundwaters from Sudan where both were measured.

increase. Therefore, NO_3–N and Fe^{2+} can be used as complementary redox indicators.

Data from Libya and Sudan are given in Table 2.1. As a general rule it can be seen that low nitrate (NO_3 – N <1 mg l^{-1}) occurs in the groundwater recharged from the Nile, as identified isotopically. Complementary high $Fe^{2+} > 100$ μg l^{-1} and low NO_3 together indicate reducing conditions. In a few cases (e.g. Kaboushir) both nitrate and reduced iron are present. This is interpreted as indicating that mixing of aerobic and anaerobic groundwater is occurring in the boreholes. The dissolved Fe^{2+}, even in the aerobic groundwaters, is still much higher than should occur at equilibrium with fresh $Fe(OH)_3$ (Hem, 1985), and indicates the likely presence of some colloidal as well as dissolved iron. From Table 2.1 it can be seen that a significant Eh drop occurs between the aerobic and anaerobic (nitrate and oxygen free) groundwaters.

Nitrate concentrations have been measured extensively in both Libya and Sudan. Widespread aerobic conditions may be inferred for the phreatic groundwater aquifer in eastern Libya. Thus, oxygen and nitrate-oxygen has persisted with only a handful of exceptions for up to and greater than 30 000 yr in the phreatic Nubian and post-Eocene aquifers, indicating the virtual absence of reducing material. The same is true for the Nubian sandstone aquifer in Sudan and compares with other semi-arid regions (Winograd and Robertson, 1982). Nitrate results are shown (Figure 2.4) plotted against uncorrected radiocarbon ages for those sites where both sets of data are available. Nitrate concentrations of between 4 and 7 mg NO_3–N l^{-1} are found in the oldest set of groundwaters. A much closer grouping of data (1–2 mg NO_3–N l^{-1}) is found for the six waters thought to represent the most recent Holocene recharge episode and indicates the relative homogeneity of this group of waters. The younger groundwaters representing recharge from the Nile nearly all have negligible or zero NO_3–N indicating that reducing conditions have developed probably during flow through organic rich alluvial sediments.

The presence and persistence of nitrate in the Saharan aquifers poses questions as to its origin(s). In the Kalahari of southern Africa the persistence of nitrate is also described (Vogel, Talma and Heaton, 1981). In Libya, concentrations range up to 30 mg NO_3–N l^{-1} (Figure 2.2); quite high concentrations occur in the east towards the edge of the Sirte Basin, and also in the areas of lower topography and increased salinity near the oases of Jalu and Awjilah. To the south and in the central area lower but still modest nitrate concentrations are to be found.

Four possible sources for these high nitrate concentrations were originally considered (Wright and Edmunds, 1969):

(1) *Nitrate-rich sediments*
Nitrate shales containing up to 6% $NaNO_3$ occur in Palaeocene rocks in the Nile valley (Hume, 1915; Duncan, 1945).

(2) *Evaporites*
The lower Eocene succession contains thick evaporites which may give rise locally to high nitrate in association with more saline groundwaters. One of the highest nitrate concentrations in southeastern Libya is recorded from a well in which the casing is perforated at between 210 and 225 m in a succession containing Eocene evaporites. This is, however, an isolated occurrence and cannot be considered a general explanation.

Table 2.1 Redox status of palaeowaters in Libya and Sudan. Sample localities refer to those used in earlier papers. Groundwaters deduced to be oxygenated are indicated (+) and those oxygen-free (−). Mixed groundwaters are also indicated

	NO_3–N (mg l^{-1})	Fe_T (μg l^{-1})	Eh (mV)	Redox Status
Libya (Sirte Basin)				
J(FF1-65)	2.4	3	+249	+
JA-P	5.2	21	+186	+
JB-P	5.9	<5	+195	+
JC-P	<0.5	7	−15	−
JD-P	9.3	5	+194	+
JE-P	14.1	5	+179	+
JF-P	9.1	24	+144	+
JF1-97	16.1	44	−	+
JKK1-12	9.8	400	−	+
JC1-95A	5.4	16	−	+
J(A1-LP3C)	16.1	490	−	Mix
Sudan (Nile Valley)				
Shendi	0.6	96	−	+
Kaboushir 1	1.4	1740	−	Mix
Kaboushir 2	14.0	<7	−	+
Mtemma 1	12.4	40	−	+
Mtemma 2	9.9	10	−	+
Mtemma 3	<1	540	−	−
Kumeir	<1	1300	−	−
Diem El Gray	<1	410	−	−
Taragma	<1	10	−	−
El Guweir	<1	510	−	−
Zakiap	<1	70	−	−
Tondb	<1	<7	−	−
Mesaktab	<1	280	−	−
Sudan (Nubian Sandstone)				
El Geheid	1.3	61	−	+
Wad El Hamad	1.3	9	−	+
Wad Hassuna	13.3	174	−	+
Umm Durwa	1.5	<6	−	+
El Shein	1.0	71	−	+

(3) *Thermal waters*
Evaporites, containing 8.8% nitrate, have been found in Libya in association with Bzema hot spring (Mahrholz, 1968). This deep-seated meteoric water may be recycling nitrate locally, but again can only be a minor source.

(4) *Palaeosols and past environments*
The occurrence of high nitrate in the vicinity of the oases south of Jalu is linked to a regional topographic low. In this area the water table would have been several tens of metres higher during the Holocene probably supporting extensive vegetation cover. Thick palaeosols from this now deflated area have been dated at no more than 2000 yr BP (Edmunds and Wright, 1979). Nitrate concentrations could have built up in these soil horizons via leguminous plants, especially during drier interludes and leaching to the water table would have occurred especially during the more humid episodes.

Some support for the latter interpretation has been provided from Sudan. High concentrations occur in several soil profiles (Figure 2.5) from the Butana area, where acacia trees and shrubs are found currently. Recharge rates in these areas are ≤1 mm yr^{-1} and there is up to 2 000 yr of storage of water and solutes in some 30 m of profile. In some profiles however, very little nitrate is found and this may be related either to local reducing conditions or to the local absence of leguminous vegetation. It is envisaged that pulses of nitrate and other solutes are leached periodically to the saturated zone during more humid episodes.

The palaeosol explanation is the most likely for the high nitrate occurrences in groundwaters in Libya but the possibility of nitrate shales acting as a secondary source in the southeast of the country needs further investigation. Further detail on the palaeobotany of the Sahara is also needed to determine the former extent of leguminous plants. This is also a topic for further study using nitrogen isotopes, building on the work in the Kalahari (Vogel, Talma and Heaton, 1981).

Bromide

In saline groundwaters, the bromide/chloride ratio is well established as an index of the source of chloride; depleted Br/Cl ratios denote that an evaporite source has been involved, whilst enrichment in Br/Cl may indicate that organic rich sediments have been involved (Rittenhouse, 1967). Trends may be related to the marine dilution line. In dilute groundwaters also, the Br/Cl ratio may be indicative of geochemical processes and more particularly, of recharge conditions.

The available data from Libya and Sudan have been combined in Figure 2.6. All of the groundwaters from Sudan as well as the River Nile are considerably enriched in Br/Cl in relation to sea-water. The trend for rainfall over continental areas having higher Br/Cl ratios than sea-water possibly due to the higher volatility of Br or to freezing effects in clouds has been observed (Winchester and Duce, 1967). Although no rainfall data are available for either region, it is reasonable to assume that the present day groundwaters have similar Br/Cl ratios to local rainfall, that is about 0.013 compared with 0.0034 in sea-water. The Sudan palaeowaters are rather less enriched in bromide than present day groundwater with Br/Cl ratios of about 0.006, although still indicating a continental influence for rainfall but with a recharge regime different from that at present. Some of the Nile valley groundwaters appear to be mixtures of contemporaneous water and palaeowater as would be expected from the hydrogeological model.

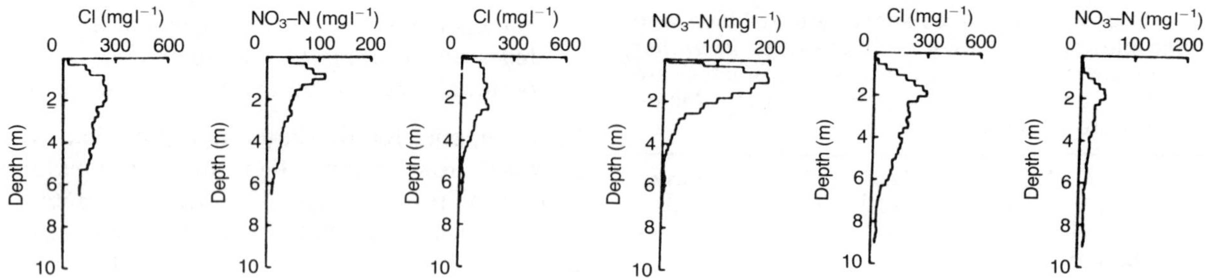

Figure 2.5 Nitrate and chloride concentrations in pore water solutions in three profiles of the unsaturated zone from near Abu Delaig, Sudan. Concentrations have been determined on moisture elutriated from core material.

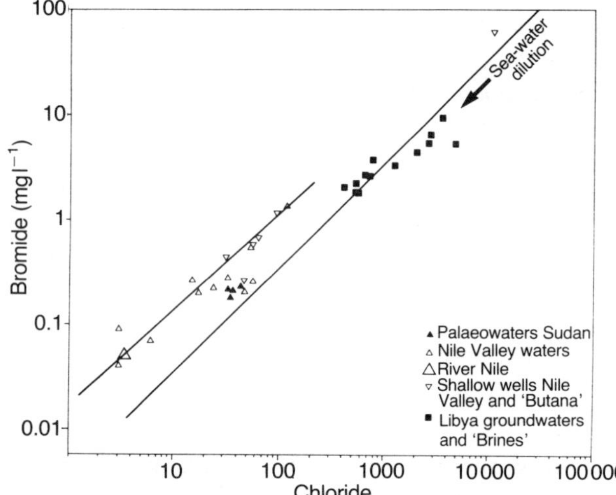

Figure 2.6 Bromide/chloride ratios in various groundwaters from Sudan and Libya.

The Sirte Basin palaeogroundwaters in Libya are rather more saline than their counterparts in Sudan. The fresher groundwaters still have Br/Cl enriched relative to sea-water but the most saline groundwaters are distinctly depleted in bromide. This is interpreted as showing that evaporite chloride has been dissolved by saline groundwaters along flow paths, or that mixing by upward leakage may have occurred. The enrichment in Br/Cl in the most dilute groundwaters must reflect the recharge source. This continental signature is therefore taken as additional to the isotopic evidence (Edmunds and Wright, 1979) that the rainfall source during the Holocene and late Pleistocene was tropical rain derived from the south or west rather than Mediterranean rains.

Strontium

The use of strontium to characterize groundwaters from the Sirte Basin has been described elsewhere (Edmunds, 1981). An outline of the results is given here in view of the importance of strontium in tracing groundwater evolution; the results of palaeowaters from Sudan are added to give a wider perspective.

In sedimentary aquifers, strontium concentrations are primarily governed by the reaction of carbonate or sulphate minerals. Freshwater diagenesis of limestones, especially marine limestones, with higher initial strontium, will result in the progressive release of trace element impurities including strontium from the lattice by incongruent dissolution and the recrystallization of a purer and more stable carbonate (Kinsman, 1969; Bath and Edmunds, 1981).

The distribution of strontium in the central Sirte Basin is shown in Figure 2.7 and it can be seen that there is a dramatic increase in Sr^{2+} along the flow line at around 28°40′N where Sr^{2+} increases from < 1 mg l^{-1} to above 10 mg l^{-1}; this effect is more marked in the LMM than the PMM aquifer and reflects the change from continental to marine facies within the Miocene succession. Groundwater acts as a very sensitive marker of this change and picks up the earliest appearance of aragonite or high Mg-calcite in the formation. The relationship is shown more clearly in the Sr/Cl diagram (Figure 2.8) where characteristic Sr/Cl ratios are found for each of the PMM and LMM over a range of salinities, but much lower Sr/Cl ratios are found in the continental (southern) group of LMM groundwaters. The Mg/Ca in contrast to strontium shows a decrease across the continental to marine line. This is due to an increase in dissolution of gypsum and gradual dilution of the ratio by Ca and SO_4. Both Sr and the Mg/Ca have been used in the hydrogeological interpretation to indicate that there is little or no interconnection between the two Miocene aquifer units (PMM and LMM).

The Sr/Cl ratios in the Nubian continental palaeowaters from Sudan, also shown in Figure 2.8, have a characteristic ratio which is lower than the PPM continental facies. Groundwaters from the Nubian aquifer at Kufra also have a low Sr/Cl ratio. Thus, steady-state ratios are developed in each lithology within the time-scale of the groundwater evolution which can be used to identify water from each formation. These values reflect the partitioning between carbonate or silicate minerals and groundwater. The strontium concentration remains below saturation with respect to

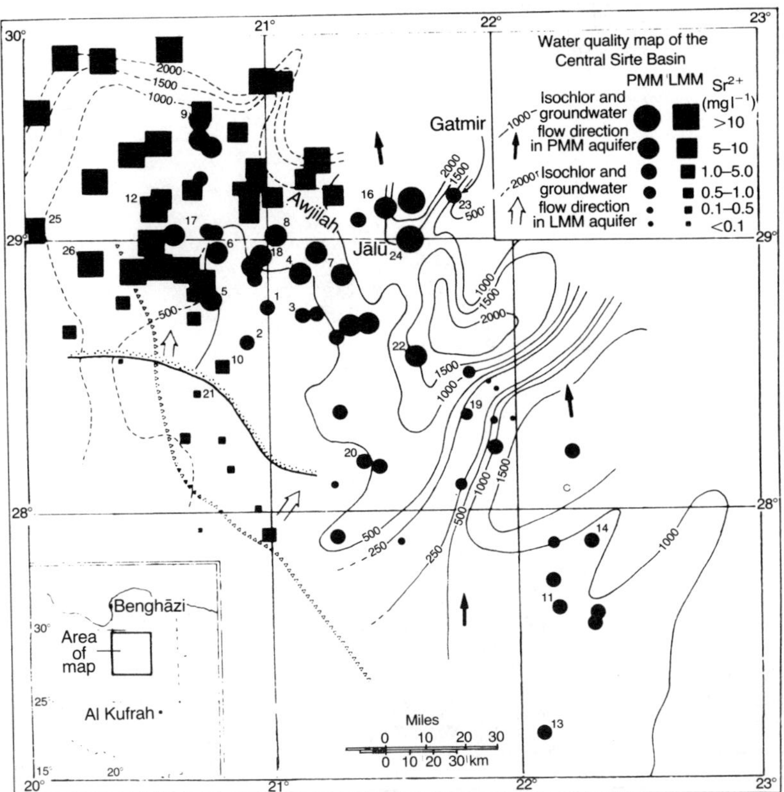

Figure 2.7 Hydrogeochemical map of the central Sirte Basin showing isochlors in the post-Middle Miocene (PMM) and Lower and Middle Miocene (LMM) aquifers. Strontium values for both groups of groundwater are also distinguished. The approximate outcrop of the base of the PMM (Calanscio Formation) is indicated by ^^^ and the southern limit of significant marine limestone development in the LMM (Marada Formation) is shown stippled.

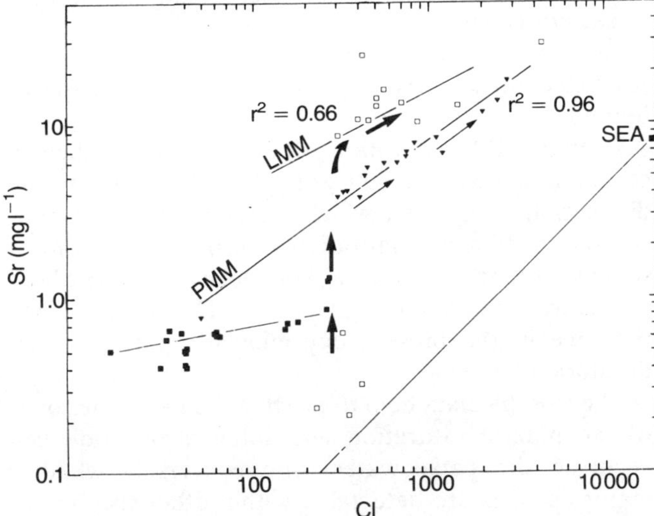

Figure 2.8 Strontium/chloride ratios for groundwater from Libya (PMM and LMM) and Sudan palaeowaters (solid squares). Arrows indicate evolution of water quality along flow lines in the two Libyan aquifers.

celestite, the most probable ultimate mineral control in all the groundwaters studied; in theory, therefore, still higher concentrations could occur.

Barium

Barium is a relatively common minor element. In contrast to strontium, barium concentration limits in groundwater are controlled by mineral solubility. Barium concentrations in groundwaters from Sudan (Table 2.2) are usually limited to below 1 mg l^{-1} Ba by the solubility of barite: in one sample, a higher value 1.4 mg l^{-1} was found in the River Nile with a sulphate concentration of 4 mg l^{-1} but the water is close to barite saturation. Barium therefore cannot often be used in the study of groundwater evolution since the solubility limits appear to be rapidly reached in most lithologies.

Fluoride

In contrast to barium, fluoride is a relatively rare element in sedimentary rocks, except where concentrated in specific areas of mineralization. Fluoride abundance is, however, generally higher in marine compared with continental sediments, and fluoride slowly dissolves during freshwater diagenesis until saturation with CaF_2 is reached; fluoride tends to remain in solution as long as undersaturation is maintained. Maximum concentrations will depend mainly on the amount of CaF_2 in solution.

Some fluoride concentrations for palaeowaters from Libya and Sudan are given in Table 2.3, together with saturation indices for fluorite calculated using WATEQF (Plummer and Jones, 1976). Nearly all groundwaters are at or rather below fluorite saturation.

The distribution of fluoride in groundwaters from eastern Libya is shown in Figure 2.9. Concentrations are uniformly low (0.8 mg l^{-1}) in the continental Nubian and PMM aquifers but there is a sharp increase

Table 2.2 Barium and sulphate concentrations with SI barite for various waters from Sudan

	Ba ($\mu g\ l^{-1}$)	SO_4 ($mg\ l^{-1}$)	SI barite
River Nile			
Shendi (7.4.82)	1400	4	0.36
Palaeowaters			
Tomeid Haj El Tair	69	115	−0.32
Essawad	395	52	−0.28
Umm Durwa	139	29	−0.40
Shein	122	25	−0.56
Wadi Hassuna	75	105	−0.18
Nile Valley waters			
Zakiab	90	228	0.01
Mtemma Artoli	130	51	−0.26
Tondb	380	26	−0.06

SI = saturation index

Table 2.3 Saturation indices for fluorite and fluorapatite calculated from groundwater chemistry (data from Libya and Sudan) in mg l^{-1}

	Aquifer	F^-	SI fluorite	SI fluorapatite
Libya				
JA-P	PMM	3.6	0.35	11.31
JB-P	PMM	1.51	−0.53	4.99
JC-P	PMM	1.40	−0.48	7.28
JD-P	PMM	1.36	−0.37	6.07
JE-P	PMM	1.50	−0.44	6.60
JF-P	PMM	1.50	−0.47	6.44
J(F1-97)	PMM	1.20	−0.37	6.81
J(KK1-12)	PMM	1.35	−0.37	6.73
T(U2-65)	PMM	0.73	−1.18	–
T(T2-65)	PMM	0.68	−1.01	–
T(FF1-65)	PMM	0.71	−1.07	–
Sudan				
El Shein	Nubian	0.29	−1.90	
Bir El Geheid	Nubian	0.32	−1.71	
Umm Durwa	Nubian	0.29	−1.84	
Tomeid Haj El	Nubian	0.85	−0.81	
Abu Delaig 1	Nubian	1.50	−0.63	
Abu Delaig 3	Nubian	6.20	0.27	
Abu Delaig 5	Nubian	3.10	−0.02	
Abu Delaig 20	Nubian	0.62	−1.12	
Abu Delaig 23	Nubian	2.90	−0.12	

in fluoride at the same latitude (about 28°40′) as for strontium. Fluoride distribution therefore also reflects the lithofacies change to marine or at least to shoreline conditions in the Miocene. Despite the significant increase in fluoride, only in one sample from the PMM (JB-P) is fluorite saturation reached. In view of the long

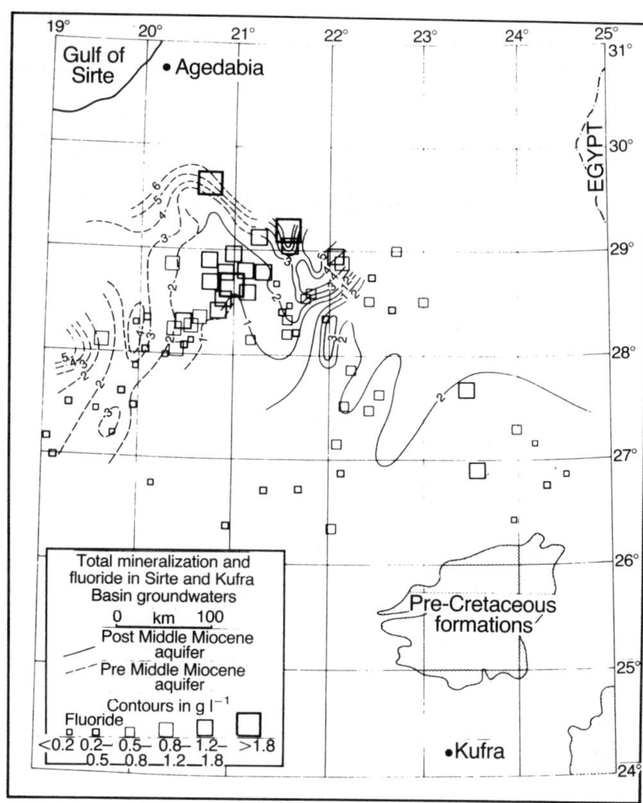

Figure 2.9 Distribution of fluoride in the groundwaters of eastern Libya.

residence times, this implies that fluoride is source limited.

In the LMM, saturation with fluorite is reached in one oasis sample at a value of 1.1 mg l^{-1} F, due to the dissolution of gypsum which increases the Ca^{2+} concentration. Fluorite saturation is also reached in most shallow groundwaters from Abu Delaig (Sudan) where concentrations up to 6 mg l^{-1} occur. This must indicate a source in the present day alluvial aquifer, soil or unsaturated zone.

The isotopic data help to provide the timescales over which mineral saturation and solute acquisition can occur in the palaeowaters and conversely whether major changes are detectable within otherwise homogeneous aquifers. In the present study, several anomalies are found in southeastern Libya where deeper boreholes have penetrated different lithofacies beneath the Nubian or PMM aquifers. Fluoride remains well below saturation in the continental sandstone aquifers over periods of 30 000 yr, but approaches saturation within 20 000 yr in the marine facies. Fluoride may therefore be used together with strontium to characterize LMM and PMM aquifers and to identify any mixing between them.

Lithium

Lithium typically increases rapidly relative to Cl and Na during the early stages of infiltration, in response to the

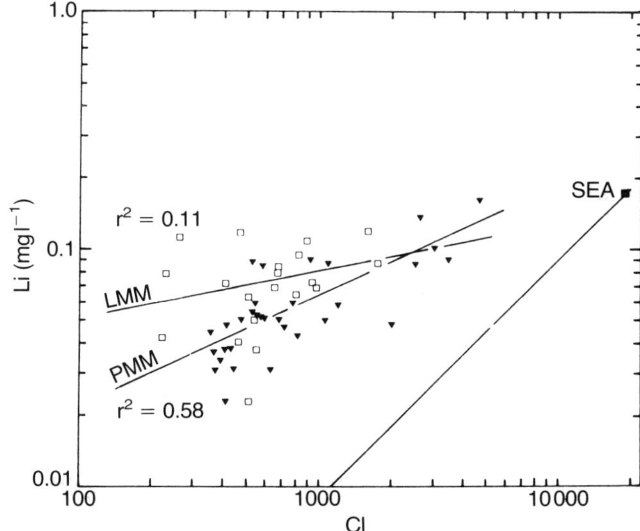

Figure 2.10 Lithium/chloride relationships in groundwaters from eastern Libya.

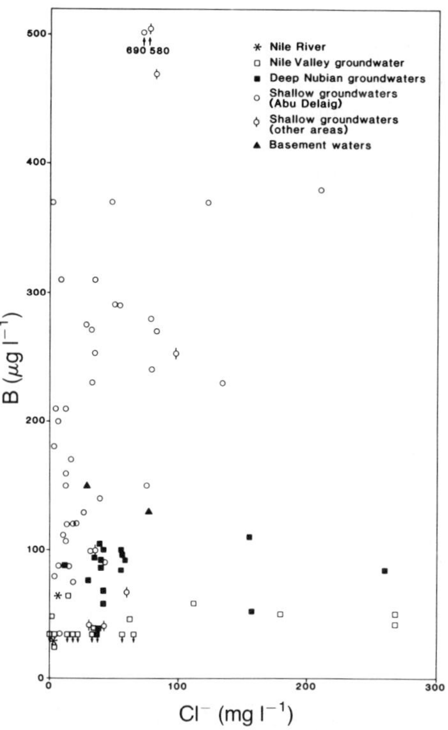

Figure 2.11 Boron/chloride relationships in groundwaters from the Butana area, Sudan.

release from clays and rock forming minerals. The increase in Li/Cl is often proportional to residence time as demonstrated in flow through confined aquifers (Edmunds, Cook and Miles, 1986). After long residence times (e.g. 10^4–10^5 yr) a typical Li/Cl ratio is reached for a particular formation, which signifies a new equilibrium between water and rock. The Li/Cl ratios for the Sirte Basin groundwaters (Figure 2.10) however do not show a clear relationship; strong correlations do not exist for either the PMM or LMM aquifer, neither is there any sharp delineation in Li/Cl between the marine LMM and non-marine PMM aquifers. The lack of good correlation in the example given is taken as evidence of diverse depositional and recharge environments although in other major sedimentary basins the Li/Cl (or Li/Na) ratio may be used as an indicator of residence time.

Boron

Boron exists mainly as uncharged boric acid in neutral, natural waters and is generally rather insensitive to processes involving water–rock interaction. Data from Libya and Sudan are used to illustrate trends in the boron to chloride ratio which appear to have some diagnostic value in semi-arid zone groundwater studies.

In the regional Nubian sandstone aquifer (Figure 2.11) the concentration of boron remains at or below 100 $\mu g\ l^{-1}$ B and there is no evidence of an increase with chloride. The River Nile contains a low boron concentration (up to 64 $\mu g\ l^{-1}$) and groundwaters recharged from the Nile also contain low boron with comparable concentrations to those in the river over a range of chloride. Very large increases in boron and in B/Cl are however found in the recent shallow groundwaters associated with wadi systems. High boron (above 120 $\mu g\ l^{-1}$) can therefore be used to characterize the recent recharge from wadis, but it is not possible as yet to identify the mechanism for the increase in mobility in boron in the near surface environment which contrasts strongly with its apparent removal in deeper, older groundwater. The palaeowaters from the Sirte Basin, Libya, all demonstrate a trend towards lower B/Cl ratios with salinity increase, which may be due to mixing (see bromide) involving a fresher (younger) groundwater with a relatively high B/Cl ratio.

Conclusions

Information derived from environmental isotopes in Libya and Sudan has provided important information on residence time and recharge history. The minor elements, taken in conjunction with major elements, provide additional or confirmatory information on the recharge, groundwater evolution and geochemical processes generally.

Bromide and boron provide evidence of recharge history and in particular boron enables recent recharge via wadi systems to be distinguished from palaeowater and also from Nile-derived groundwater.

Within palaeowaters, which may retain a rather uniform isotopic signature, it is possible to identify different groundwater types which relate to lithofacies changes, mixing and possibly to varying residence times using minor element concentrations or minor/major element ratios, for example Sr, Br, F.

New insight into past recharge conditions and palaeoclimates is also provided by the minor elements and the related geochemistry of the groundwater. The variations in nitrate, preserved in a predominantly oxidizing environment, may give evidence of vegetation and land use history and the use of nitrogen isotopes may further assist with such studies.

All of these results point to the importance of multiparameter geochemical studies of recent and palaeowaters from semi-arid and arid environments, where the present day hydrological conditions often differ dramatically from those of previous centuries or millennia.

Acknowledgments

This study was carried out in part with the financial support of the British Overseas Development Administration. I am grateful to George Darling and John Andrews for their helpful comments on the manuscript. This chapter is published with the permission of the Director, British Geological Survey (Natural Environment Research Council).

References

Allison, G.B. and Hughes M.W., 1978. The use of environmental chloride and tritium to estimate total recharge to an unconfined aquifer. *Australian Journal Soil Research*, vol. 16. pp. 139–157.

Bath, A.H. and Edmunds, W.M., 1981. Identification of connate water in interstitial solution of chalk sediment. *Geochimica et Cosmochimica Acta*, vol 45. pp. 1449–1461.

Darling, W.G., Edmunds, W.M., Kinniburgh, D.G. and Kotoub, S., 1987. Sources of recharge to the basal Nubian sandstone aquifer, Butana region, Sudan. In: *Isotope Techniques in Water Resources Development*. IAEA, Vienna. pp. 205–224.

Duncan, R., 1945. Report on the exploitation of the Upper Egyptian nitrate shales, *Middle East Supply Centre Report No 1178*.

Edmunds, W.M., 1981. The hydrogeochemical characterisation of groundwaters in the Sirte basin, using strontium and other elements. In: *Geology of Libya* (M.J. Salem, and M.T. Busrewil, eds). Academic Press, London. pp. 703–714.

Edmunds, W.M., Cook, J.M. and Miles, D.L., 1986. Lithium mobility and cycling in dilute continental waters. (Extended Abstracts of Fifth Symposium on Water–Rock Interaction, Reykjavik, Iceland), Orkustofnun, Reykjavik. pp. 183–187.

Edmunds, W.M., Darling, W.G. and Kinniburgh, D.G. 1988. Solute profile techniques for recharge estimation in semi-arid and arid terrain. In: *Estimation of Natural Groundwater Recharge* (I. Simmers, ed.), NATO ASI Series C222, Reidel, Dordrecht. pp. 139–157.

Edmunds, W.M., Darling, W.G., Kinniburgh, D.G. *et al.*, 1992. Sources of recharge at Abu Delaig, Sudan. *Journal of Hydrology*, vol. 131, pp. 1–24.

Edmunds, W.M., Miles, D.L. and Cook, J.M., 1984. A comparative study of sequential redox processes in three British aquifers. In: *Hydrochemical Balances of Freshwater Systems* (E. Eriksson, ed.). IAHS Publication No 150, pp. 55–70.

Edmunds, W.M. and Walton, N.R.G., 1980. A geochemical approach to recharge evaluation in semi-arid zones. In: *Arid Zone Hydrology, Investigations with Isotope Techniques*. IAEA, Vienna, pp. 47–68.

Edmunds, W.M. and Wright, E.P., 1979. Groundwater recharge and palaeoclimate in the Sirte and Kufra basins, Libya. *Journal of Hydrology*, vol. 40. pp. 215–241.

Fontes, J.-Ch. and Edmunds, W.M., 1989. The use of environmental isotope techniques in arid zone hydrology. A critical review. *Technical Documents in Hydrology. IHP-III Project 5.2*. UNESCO, Paris.

Heaton, T.H.E., 1986. Isotopic studies of nitrogen pollution in the hydrosphere and atmosphere: a review. *Chemical Geology*, vol. 59. pp. 87–102.

Hem, J.D., 1985. Study and interpretation of the chemical characteristics of natural water. *USGS Water Supply Paper 2254*.

Hume, W.F., 1915. The nitrate shales of Egypt. *Memoires Institute Egypt*, vol. 8. pp. 146–169.

Kinsman, D.J.J., 1969. Interpretation of Sr^{2+} concentrations in carbonate minerals and rocks. *Journal of Sedimentary Petrology*, vol. 39, pp. 486–508.

Mahrholz, W.W., 1968. Geological exploration of the Kufra region. *Geological Section Bulletin Ministry Industry, Libya*.

Plummer, L.N. and Jones, B.F., 1976. WATEQF - A Fortran-IV version of WATEQ, a computer program for calculating chemical equilibrium of natural waters. *USGS Water-Resources Investigations, 76–13*. US Geological Survey, Washington.

Rittenhouse, G., 1967. Bromide in oilfield waters and its use in determining possibilities of origin of these waters. *American Association Petroleum Geologists Bulletin*, vol. 51. pp. 2430–2440.

Stumm, W. and Morgan, J.J., 1981. *Aquatic Chemistry*. Wiley, New York.

Vogel, J.C., Talma, A.S. and Heaton, T.H.E., 1981. Gaseous nitrogen as evidence for denitrification in groundwater. *Journal of Hydrology*, vol. 50, pp. 191–200.

Winchester, J.W. and Duce, R.A., 1967. The global distribution of iodide, bromide and chlorine in marine aerosols. *Naturwissenschaften*, vol. 54, pp. 110–113.

Winograd, I.J. and Robertson, F.N., 1982. Deep oxygenated groundwater: anomaly or common occurrence? *Science*, vol. 216, pp. 1227–1230.

Wright, E.P. and Edmunds, W.M., 1969. Hydrogeological studies in central Cyrenaica, Kingdom of Libya: distribution and origin of nitrate in the groundwaters. Unpublished Report, Institute of Geological Sciences, London.

Wright, E.P., Benfield, A.C., Edmunds, W.M. and Kitching, R. 1982. Hydrogeology of the Kufra and Sirte basins, eastern Libya. *Quarterly Journal Engineering Geology*, vol. 15. pp. 83–103.

3 Groundwater quality in rural Uganda: hydrochemical considerations for the development of aquifers within the basement complex of Africa

R.G. Taylor and K.W.F. Howard

Abstract Groundwater held within fractures of the crystalline basement complex of Africa is widely available and generally free from the sediment and biological impurities that frequently plague surface waters. Consequently, groundwater has become the principal supply of potable water to rural communities living over basement rocks across the continent. Development of the basement aquifer, through the construction of boreholes, has intensified during the last decade as nations have attempted to ensure a clean supply of water within a mile (1.6 km) of every citizen for the United Nations International Drinking Water Supply and Sanitation Decade. Efforts to achieve this goal in Uganda have been hindered however, due to concerns of the user population over the perceived quality of water obtained via borehole handpumps.

As part of continuing research into the basement aquifer system of Uganda, water samples were collected from over 150 sites comprising primarily deep boreholes but also springs and shallow piezometers, tapping into the weathered overburden. The samples were taken from two catchments, in northern and southwestern Uganda, of differing topography, climate and bedrock geology. Each sample was analysed for major ions and a range of trace metals. The majority of groundwater is Ca–HCO_3 in character and, in terms of risks to human health, of acceptable inorganic quality. However, levels of aluminium, chloride, iron, manganese, zinc and hardness were detected in excess of aesthetic limits set by the World Health Organization (WHO) at numerous sites, confirming suspicions held by rural consumers. Nitrate and chromium concentrations surpassing WHO health guidelines were determined at 11 sites.

Trace metals (Al, Cr, Fe, Mn, Zn) in groundwater are derived from two sources: weathering of the bedrock matrix and corrosion of borehole rising mains. The high nitrate values may arise from the proximity to boreholes of sewage waste facilities such as pit latrines. Closer attention to borehole location would avoid nitrate contamination while regular maintenance of pipe fixtures would serve to reduce contamination from corrosion. However, deterioration of groundwater quality through natural weathering of the basement matrix is unavoidable. The potential therefore for undesired metals to exist in concentrations unpalatable to the consumer must be considered in the development of basement aquifers. Without guidance, rural communities may revert to biologically polluted surface sources to meet their water needs.

Introduction

Groundwater held within fractures of the crystalline basement complex of Africa is widely available and generally free from the sediment and biological impurities that frequently plague surface waters. As a result, groundwater has become the principal source of potable water to rural communities living over basement rocks across the continent. During the last decade, development of basement aquifers has become

Groundwater Quality Edited by H. Nash and G.J.H. McCall. Published in 1994 by Chapman & Hall. ISBN 0 412 58620 7

especially intensive as nations attempted to provide a clean supply of water within a mile (1.6 km) of every citizen for the United Nations' International Drinking Water Supply and Sanitation Decade. Although utilization of groundwater resources reduces or eliminates the costs of reticulation, as the supply may be sited at the point of demand, the generation of a large number and dispersed sources makes the quality of the water supply extremely difficult and costly to monitor. Unfortunately, provisions for the monitoring of water quality are rarely included in groundwater development schemes.

Owing to a scarcity of chemical and biological analyses of water sources, due in part to a lack of equipment and trained personnel, daily users at borehole level are resigned to monitoring water quality through its odour, appearance and taste. Hence, the aesthetic quality of the groundwater becomes the key determinant in deciding its adoption over another source. Frequently it is the relatively harmless, inorganic constituents which are likely to offend the senses rather than the more virulent biological contaminants. Fortunately, the overburden material, or regolith, which results from the prolonged, intense weathering of basement rocks, commonly features a clay-rich unsaturated zone. Microbial pathogens that are leached into the sub-surface from pit latrines and other sources, are filtered and adsorbed by the clay before the recharging waters reach the water table. Biological contamination of groundwaters within the weathered mantle has, however, been observed in Malawi and was attributed to routes of recharge that bypass the clay matrix through macro-fissures and tubular channels caused by plant and animal activity in the unsaturated zone (Lewis and Chilton, 1984).

The quality of a groundwater source is not the only factor contributing to disease prevention. The magnitude and reliability of a groundwater supply must also be considered since failure of a safe groundwater source to meet the daily water requirements of users throughout the year will force them to seek other sources. Health benefits gained from the periodic use of a safe supply would consequently be either reduced or eliminated. In Uganda, fractures within the basement complex exhibit both a poor transmissivity and low storativity (Howard *et al.*, 1992; Callist, 1992). The regolith which overlies the basement complex of Africa has been recognized as an alternative to the weak, fractured-basement aquifer in Malawi (Chilton and Smith-Carington, 1984), Nigeria (Omorinbola, 1984), Uganda (Howard and Karundu, 1992) as well as Tanzania and Zambia (Jones, 1985).

Current research activities in Uganda are focused on determining the potential of the regolith to provide rural communities with a reliable and adequate supply of potable water. While the utilization of this shallower source would constitute a considerable saving in development and operating costs, it is acknowledged that both the real and perceived quality of water discharging from pumps will ultimately determine the acceptability of regolith groundwaters to the user community. Because the composition and therefore the quality of groundwater is dependent upon the source, route and timing of recharge as well as the lithology of the aquifer unit, hydrochemical evaluations of groundwaters from the regolith and basement rock aquifers have been a major focus of the study in Uganda.

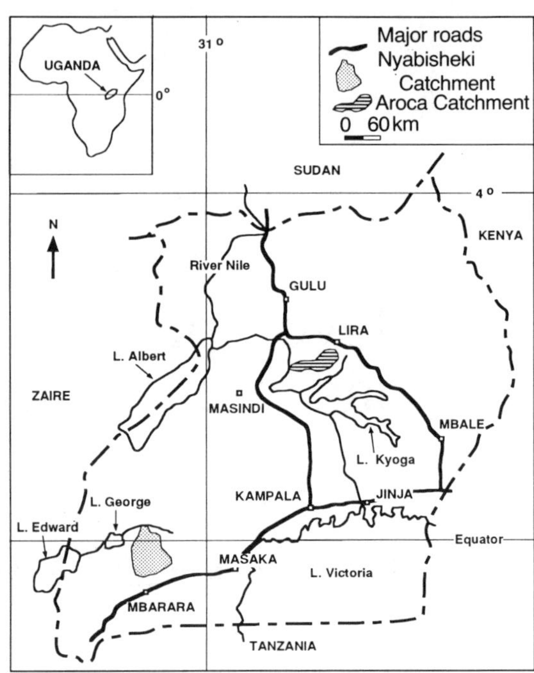

Figure 3.1 A map of the study areas in Uganda.

This chapter describes the results of hydrochemical investigations undertaken in two catchments of Uganda (Figure 3.1). The results from preliminary research into the hydrogeological nature of the regolith aquifer in the Aroca catchment, are also included. In this work, groundwater samples analysed for major ion, minor ion and trace element content enable the quality of groundwater currently being pumped from the fractured-bedrock aquifer to be compared to the quality of water that may be expected from the development of the regolith aquifer. Water quality is defined in terms of aesthetic and health-related guidelines furnished by the World Health Organization (WHO, 1984).

Study areas

The two catchments that were selected for investigation feature different topography, climate and bedrock geology. The Aroca catchment lies to the north of

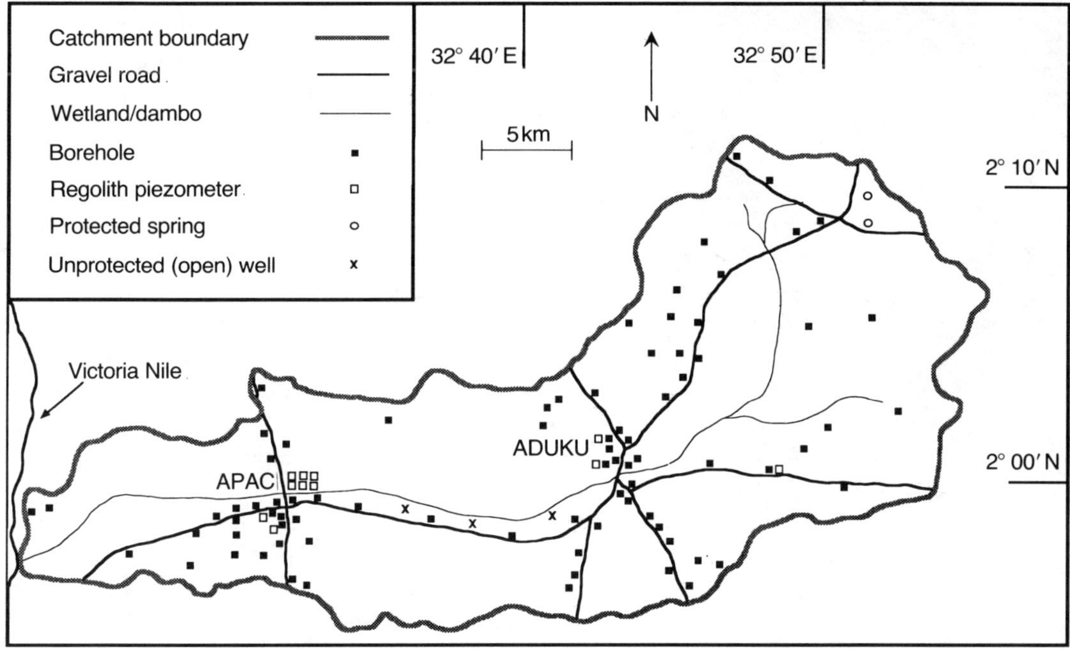

Figure 3.2 A map of the Aroca catchment including groundwater sources.

Kampala and is bounded to the south by Lake Kyoga and to the east by the Victoria Nile. The Nyabisheki catchment is located southwest of Kampala and is bounded by Lake George to the east and the town of Mbarara to the south.

Aroca catchment

The Aroca catchment, shown in Figure 3.2, covers an area of 840 km^2 and slopes gently from 1100 m above sea level (masl) in the northeast to 1030 masl in the southwest where the Aroca wetland meets the Nile. The underlying bedrock is composed of Precambrian undifferentiated gneisses characteristic of the basement complex that spans much of northern Uganda. Joints and fractures within the bedrock have developed from pressure release associated with the erosion and weathering of the gneissic formation. The zone of fracturing extends approximately 30 m below the bedrock surface and provides a weakly transmissive aquifer (Callist, 1992). The bedrock is concealed entirely by a regolith with a roughly uniform depth of 30 m. A cross-sectional representation of the regolith–bedrock system is presented in Figure 3.3. Acworth (1987) described the evolution of the regolith unit as the progressive chemical degradation, involving primarily the leaching of silica, of the basement complex to a lateritic soil cover. Generally, a lateritic crust rests over a clay unit formed from the hydrolysis of parent rock fragments. The water table is then commonly encountered at the base of the clay unit. Below this, hydrolytic processes

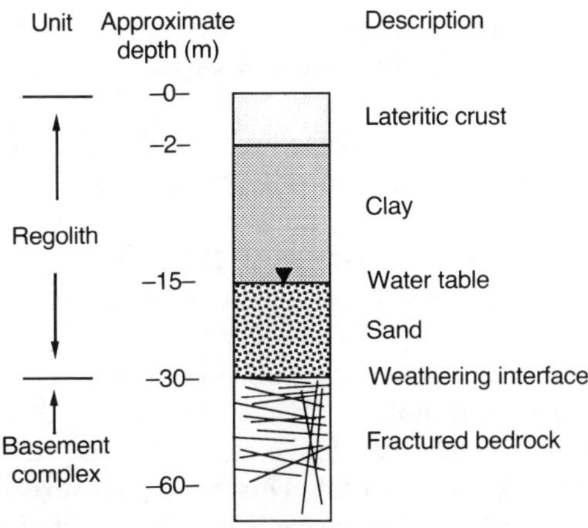

Figure 3.3 A cross-sectional representation of the regolith–bedrock system.

are retarded and sand-sized grains from the basement complex predominate to the bedrock surface. It is within this coarse unit of the regolith that the properties of permeability and porosity yield an aquifer.

The Aroca catchment receives an average annual rainfall of around 1400 mm a^{-1} which falls primarily between the months of April and October. An estimate of recharge to the sub-surface from soil moisture balance techniques yields an average rate of infiltration of 160 mm a^{-1} which is highly variable but dependent on the number of heavy (> 10 mm d^{-1}) rain events (Taylor *et al.* 1993). Examination of stable isotope signatures for rainfall, groundwaters and surface waters

Figure 3.4 A protected spring in the Aroca catchment.

supports this finding by suggesting that replenishment to the sub-surface aquifers occurs solely from direct infiltration of rainfall at the soil surface during the wet season.

Groundwater flow is directed toward the marshy area (or dambo) that runs centrally through the catchment (Figure 3.2). Marshes in this area have been estimated to evapotranspire at a rate of 1800 mm a^{-1} (Olivier, 1961) allowing ponding to occur only at the height of the rainy season.

With a seasonal scarcity of surface water, groundwater is the dominant source of water for 82% of the estimated 276 000 residents that live in the Aroca catchment (Angura, 1992). This demand is served by 80 boreholes, at least 3 open wells and 2 protected springs (encased in concrete as pictured in Figure 3.4). The remaining population relies on a combination of groundwater and surface water that includes marsh ponds and roof run-off to meet its water needs. Over one quarter of the population consider their water source to be of unacceptable quality (Angura, 1992). The overriding complaints observed across the region have been the salty taste of the water and the reddish-brown (iron oxide) stains that appear on clothes and pots from utilizing groundwater sources.

Nyabisheki catchment

The Nyabisheki catchment (Figure 3.5) has an area of 2750 km^2 and encompasses the basin of the River Nyabisheki which runs centrally along the length of the catchment, and its tributary, the River Oruyubu. Numerous outcrops of Precambrian rocks including granites, schists, gneisses, phyllites and quartzites occur throughout the region. Fracturing within the bedrock formations is due both to regional tectonic activity, associated with the western arm of the Rift Valley, and to pressure release resulting from erosion and weathering of the bedrock. The fracture zone extends to a depth of 40 m below the bedrock surface and displays poor aquifer properties similar to those observed in the Aroca catchment, with transmissivities typically in the range of 1–5 m^2 d^{-1} (Howard *et al.*, 1992). The regolith is similar in form to that described for the Aroca basin (see Figure 3.3), yet due to a more rolling and varied topography, weathered materials in some areas have been reworked and deposited as unconsolidated sediments.

Rainfall across the basin varies with altitude. Approximately 1400 mm a^{-1} has been observed in the higher altitude settings near Ibanda (1500 masl) in the west while roughly 750 mm a^{-1} has been recorded at an altitude of 1200 masl in the east around Kiruhura. A catchment water balance estimated the annual rate of groundwater recharge to be 17 mm a^{-1} (Howard and Karundu, 1992). Although patterns of recharge have yet to be defined, the volume of recharge is sufficient to maintain baseflow to both the River Nyabisheki and the River Oruyubu throughout the year.

Groundwater from boreholes and springs is the primary source of water for the rural community. About 100 boreholes, seven protected springs and a large number of unprotected springs are utilized by 99% of the approximately 285 000 people who reside in the Nyabisheki basin (Angura, 1992). Untreated river water is the remaining source employed in the region. Similar to Aroca, a survey of the inhabitants found that around one quarter considered their water supply to be of unacceptable quality (Angura, 1992).

Analytical method

Sampling

A total of 161 water samples were collected in both catchments for major ion and trace metal analysis. Boreholes, springs and in the case of the Aroca basin, open wells and piezometers were all sampled to evaluate regional water quality. Sample bottles were washed with nitric acid, rinsed thoroughly with deionized water and air-dried in advance of sampling events. All water samples were filtered through a 0.2 μm membrane in the field. In an effort to minimize the contribution from pipe fixtures to sample chemistry, boreholes were

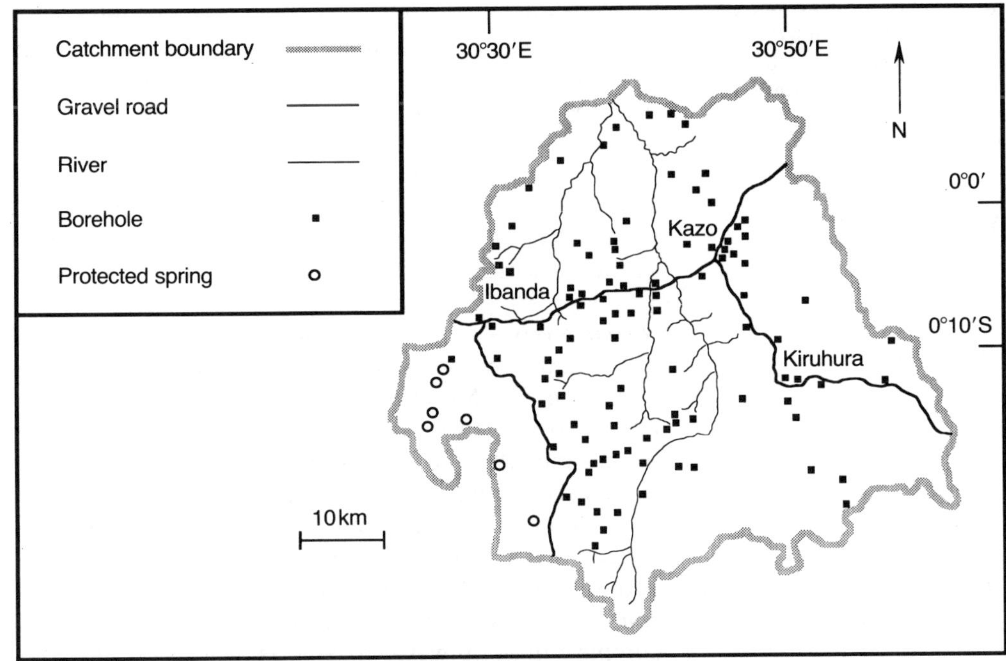

Figure 3.5 A map of the Nyabisheki catchment including groundwater sources.

Figure 3.6 A borehole being pumped in the Nyabisheki catchment.

pumped until the volume resting in the pipes had been discharged, before obtaining the sample. Samples collected for the analysis of cations and trace metals were acidified with nitric acid before being sealed to prevent metals precipitating out of the water sample prior to testing. Field blanks and sample duplicates were obtained to check the accuracy and precision of the analyses. An important drawback from a water quality perspective in the described sampling protocol is that daily users such as those in Figure 3.6 do not enjoy the benefits of having their borehole pumped for an extended period and their water filtered through a fine membrane before consumption. The collected water is, therefore, representative of the medium from which it was extracted, but may not be entirely indicative of the quality of water being consumed.

Analysis

At the point of sample collection, temperature, pH and conductance were measured using electrodes while alkalinity was determined by titration. Collected samples were transported to the University of Toronto for major ion analysis by ion chromatography. Trace metal analysis was conducted in Toronto using inductively coupled plasma-atomic emission spectroscopy and -mass spectrometry (ICP-AES and ICP-MS). Mercury content was determined by atomic absorption spectrometry. Determination limits attained for each trace

Table 3.1 WHO drinking water quality guidelines for trace metals along with the determination limits for trace metal species analysed by inductively coupled plasma-atomic emission spectroscopy (ICP-AES) and inductively coupled plasma-mass spectrometry (ICP-MS)

Element	WHO Limit ($\mu g\ l^{-1}$)	Limit type	ICP-AES ($\mu g\ l^{-1}$)	ICP-MS ($\mu g\ l^{-1}$)
aluminium	200	A	100	5
arsenic	50	H	30	2
barium	none	–	10	2
boron	none	–	N.A.	5
cadmium	5	H	10	0.5
chromium	50	H	10	2
cobalt	none	–	10	1
copper	1 000	A	5	2
iron	300	A	100	10
lead	50	H	20	0.1
lithium	none	–	10	N.A.
manganese	100	A	10	2
mercury	1	H	*	*
molybdenum	none	–	10	2
nickel	none	–	10	2
selenium	10	H	N.A.	2
silver	none	–	5	0.5
strontium	none	–	5	5
tin	none	–	100	2
uranium	none	–	N.A.	0.1
zinc	5 000	A	5	2

Type: A: WHO aesthetic limit; H: WHO health limit
N.A.: Not analysed
* mercury analysed by atomic absorption spectrometry determination limit = 0.1 $\mu g\ l^{-1}$
Note: The improved ability of a mass detector (MS) to resolve the plasma discharge (ICP) over an emission detector (AES) results in a lowering of the instrument's detection limit.

metal through the various analytical techniques are listed in Table 3.1.

Results

Major ions

Results from the major ion analysis of groundwaters in both catchments and their relationship to WHO guidelines for drinking-water quality are presented in Tables 3.2, 3.3 and 3.4. Groundwaters were found to be fresh (total dissolved solids < 1000 mg l^{-1}); however, in the Aroca basin, undesirable levels of chloride and chemical hardness were recorded. The major ion chemistry of groundwaters over both catchments is represented on a Durov diagram in Figure 3.7. Analyses represented on tables and the Durov plot exhibit an anion–cation balance of better than $\pm 10\%$.

Figure 3.7 reveals an anionic spectrum for basement groundwaters that is dominated by the bicarbonate ion. The cationic spectrum is enriched in roughly even proportions of calcium and sodium but displays a slight tendency towards calcium. Similar major ion profiles have been observed in groundwaters from basement rocks in Nigeria (Malomo *et al.* 1990; Loehnert, 1981) and the Ivory Coast (Faillat, 1981). Groundwaters sampled from the regolith in the Aroca basin demonstrate a minor shift away from calcium-bicarbonate to sodium-chloride dominance.

Nitrate levels – Aroca catchment

Field and laboratory analyses of groundwaters for nitrate in the Nyabisheki catchment have yet to be undertaken. Levels of nitrate recorded in the Aroca basin and their relationship to the WHO health standard of 45 mg l^{-1} are listed in Tables 3.2 and 3.3. A broad range in concentrations was found in groundwaters from both fractured-bedrock and regolith aquifers. Such a variation in nitrate content has been similarly observed in groundwaters from the basement complex in southwestern Nigeria where nitrate concentrations ranged from 0.2 mg l^{-1} to 550 mg l^{-1} (Malomo *et al.* 1990). The distribution of nitrate levels over the Aroca basin is given in Figure 3.8. As found in Nigeria, elevations in the nitrate concentration of groundwater correspond with relatively dense settlements and the presence of markets. In the Aroca basin, concentrations of nitrate exceeding the WHO limit were detected at three groundwater sources in the two urban areas, Apac and Aduku.

Trace metals

The results from the trace metal analysis of groundwater samples from the Aroca and Nyabisheki catchments are presented in Tables 3.5, 3.6 and 3.7 and Tables 3.8 and 3.9 respectively. The distribution of sites having metal concentrations in violation of aesthetic and health objectives for drinking-water quality set by the WHO are displayed in Figures 3.9 and 3.10.

It is evident from Tables 3.5–3.9 that groundwater from the fractured bedrock of both catchments is predominantly a safe source of water. No sample from either basin exceeded WHO health guidelines concerning an acceptable concentration of arsenic, lead, mercury or selenium. In the vast majority of samples, these metals either went undetected or were an order of magnitude below WHO health guidelines. Levels of cadmium marginally beyond the WHO limit were recorded in two samples collected from the regolith in the Aroca basin. In the Nyabisheki catchment, 8 of the 87 boreholes sampled registered chromium concentrations higher than the WHO guideline of 0.050 mg l^{-1}.

The levels of trace metals, principally those of iron and manganese, were found to diminish significantly the aesthetic quality of basement groundwaters. In the

Table 3.2 Major ion levels in groundwaters from the fractured bedrock aquifer at 52 sites in the Aroca basin. Their relationship to WHO guidelines for drinking water quality is also given

Parameter	$C_{AVERAGE}$ (mg l^{-1})	S.D. (mg l^{-1})	C_{RANGE} (mg l^{-1})	WHO limit (mg l^{-1})	Limit type	$N_{EXCEEDED}$ (of 52)
Cl^-	66	92	0.7–420	250	A	3
SO_4^{2-}	38	50	0.4–281	400	A	0
NO_3^-	9	18	0.08–115	45	H	2
I^-	0.005	0.012	0.0003–0.068	–	–	–
Br^-	0.20	0.29	0.01–1.2	–	–	–
Na^+	44	36	13–199	200	A	0
Ca^{2+}	55	40	8–184	–	–	–
Mg^{2+}	17	16	2–87	–	–	–
K^+	6	2	2–11	–	–	–
Alkalinity	154	77	55–362	–	–	–
Hardness	200	160	29–776	500	A	3
TDS	320	210	90–987	1 000	A	0

$C_{AVERAGE}$: average ionic concentration in groundwater samples
S.D.: standard deviation in the average ionic concentration
C_{RANGE}: range of ionic concentrations in groundwater samples
Type: A: WHO aesthetic limit; H: WHO health limit
$N_{EXCEEDED}$: Number of sites with an ionic concentration beyond the WHO limit
Alkalinity: carbonate and bicarbonate alkalinity (mg $CaCO_3$ l^{-1})
Hardness: chemical hardness (sum of divalent cations in mg $CaCO_3$ l^{-1})
TDS: total dissolved solids

Table 3.3 Major ion levels in groundwaters from the regolith aquifer at 7 sites in the Aroca basin. Their relationship to WHO guidelines for drinking water quality is also given

Parameter	$C_{AVERAGE}$ (mg l^{-1})	S.D. (mg l^{-1})	C_{RANGE} (mg l^{-1})	WHO limit (mg l^{-1})	Limit type	$N_{EXCEEDED}$ (of 7)
Cl^-	43	35	2–110	250	A	0
SO_4^{2-}	41	61	2–174	400	A	0
NO_3^-	10	20	0.02–61	45	H	1
Br^-	0.18	0.12	0.09–0.36	–	–	–
Na^+	43	27	10–103	200	A	0
Ca^{2+}	28	36	6–114	–	–	–
Mg^{2+}	11	15	2–48	–	–	–
K^+	7	4	1–14	–	–	–
Alkalinity	103	111	21–345	–	–	–
Hardness	142	187	28–590	500	A	1
TDS	190	160	22–562	1 000	A	0

$C_{AVERAGE}$: average ionic concentration in groundwater samples
S.D.: standard deviation in the average ionic concentration
C_{RANGE}: range of ionic concentrations in groundwater samples
Type: A: WHO aesthetic limit; H: WHO health limit
$N_{EXCEEDED}$: Number of sites with an ionic concentration beyond the WHO limit
Alkalinity: carbonate and bicarbonate alkalinity (mg $CaCO_3$ l^{-1})
Hardness: chemical hardness (sum of divalent cations in mg $CaCO_3$ l^{-1})
TDS: total dissolved solids

Nyabisheki basin, the average concentration of iron in borehole water was 4 mg l^{-1} with 66 of 87 sites exceeding the WHO aesthetic limit of 0.3 mg l^{-1}. Manganese and zinc were found beyond aesthetic quality objectives at 53 boreholes. Levels of aluminium were higher than aesthetic guidelines at 20 of the 87 locations. In contrast, only one of seven spring-water sites were found to exceed a WHO aesthetic standard for a single parameter, manganese. Violations of aesthetic guidelines in the Aroca catchment were observed in groundwater from both the regolith and fractured-bedrock aquifers. Iron and manganese concentrations from both units were in the vicinity of 1 and 0.1 mg l^{-1} respectively. However, aluminium levels in regolith

Table 3.4 Major ion levels in groundwaters from the fractured bedrock aquifer at 17 sites in the Nyabisheki basin. Their relationship to WHO guidelines for drinking water quality is also given

Parameter	$c_{AVERAGE}$ (mg l^{-1})	S.D. (mg l^{-1})	c_{RANGE} (mg l^{-1})	WHO limit (mg l^{-1})	Limit type	$N_{EXCEEDED}$ (of 17)
Cl^-	36	32	7–114	250	A	0
SO_4^{2-}	42	28	6–123	400	A	0
Na^+	42	30	14–110	200	A	0
Ca^{2+}	42	22	6–88	–	–	–
Mg^{2+}	17	10	3–42	–	–	–
K^+	4	2	1–9	–	–	–
Alkalinity	222	76	66–352	–	–	–
Hardness	176	76	56–288	500	A	0
TDS	436	153	167–686	1 000	A	0

$c_{AVERAGE}$: average ionic concentration in groundwater samples
S.D.: standard deviation in the average ionic concentration
c_{RANGE}: range of ionic concentrations in groundwater samples
Type: A: WHO aesthetic limit; H: WHO health limit
$N_{EXCEEDED}$: Number of sites with an ionic concentration beyond the WHO limit
Alkalinity: carbonate and bicarbonate alkalinity (mg $CaCO_3$ l^{-1})
Hardness: chemical hardness (sum of divalent cations in mg $CaCO_3$ l^{-1})
TDS: total dissolved solids

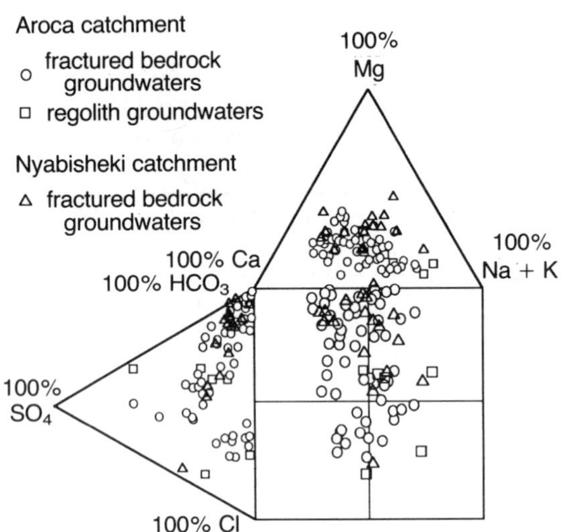

Figure 3.7 A Durov plot relating the major ion chemistry of groundwaters from both the Aroca and Nyabisheki catchments.

groundwaters were notably higher than those recorded in groundwater from the underlying bedrock and may be indicative of the secondary weathering of clay minerals in the regolith (McFarlane, 1992). Deterioration of groundwater quality in basement terrain from the presence of iron and manganese has also been observed through a geochemical study in Nigeria (Malomo *et al.*, 1990). Finally, barium and strontium were detected in many samples across both basins. Guidelines for these parameters have yet to be defined so their impact on water quality, other than to the extent to which they contribute to chemical hardness, is not considered.

Discussion

Concerns expressed over groundwater quality by one quarter of the rural consumers in both catchments were confirmed by chemical analysis. Values exceeding WHO aesthetic drinking-water quality guidelines at sites throughout the Aroca and Nyabisheki basins were common. Groundwater sources in violation of WHO health limits were, on the other hand, considerably less widespread. Although the priority of remedial measures to sites that exceed health guidelines should be observed, aesthetically displeasing sources should not be neglected. The salty taste of groundwater and excessive scaling and staining caused during use, deter communities from utilizing groundwater facilities. This results in a reduction in groundwater consumption and the adoption of biologically polluted surface sources, both of which can affect the health of a community through water-washed and water-borne disease. In the absence of a comprehensive and on-going system of monitoring the quality of rural water supplies, consideration of the link between user perception and aesthetic water quality may yield a viable substitute. To address concerns over water quality, an understanding is required of the source, effect and if possible, remediation of constituents violating drinking-water guidelines.

The chemical hardness of groundwaters from the fractured-bedrock and regolith aquifers is beyond the WHO aesthetic limit at 4 of the 59 sites tested in the Aroca catchment. Chemical hardness is defined as the sum of divalent cations in solution but depends largely upon the concentrations of aqueous calcium and magnesium. The relative enrichment of calcium over

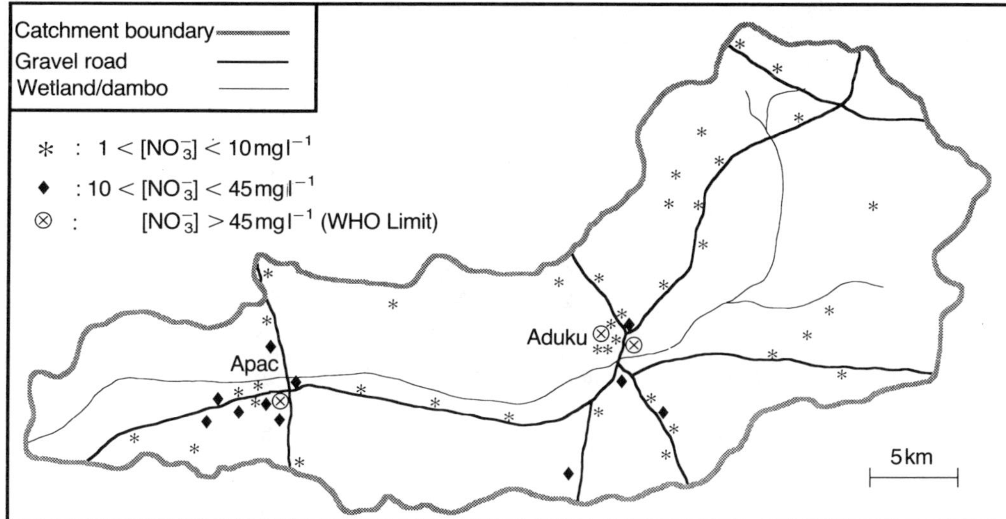

Figure 3.8 The distribution of nitrate concentrations determined at sites over the Aroca catchment.

Table 3.5 Trace metal levels determined by ICP-AES in groundwaters from the fractured bedrock aquifer at 42 sites in the Aroca basin

Element	$N_{DETECTED}$ (of 42)	$C_{AVERAGE}$ ($\mu g\ l^{-1}$)	S.D. ($\mu g\ l^{-1}$)	C_{RANGE} ($\mu g\ l^{-1}$)	$N_{EXCEEDED}$ (of 42)
barium	39	140	170	20–1 000	–
iron	4	280	100	200–400	1
lithium	12	19	12	10–60	–
manganese	15	130	120	10–490	8
mercury	2	0.2	0.1	0.1–0.2	0
nickel	4	15	10	10–30	–
silver	1	6	–	–	–
strontium	42	340	290	66–1 200	–
zinc	28	250	550	6–3 400	0

$N_{DETECTED}$: Number of sites with an elemental concentration beyond the detection limit (Table 3.1)
$C_{AVERAGE}$: average elemental concentration in samples exceeding the detection limit
S.D.: standard deviation in the average elemental concentration
C_{RANGE}: range of elemental concentrations in samples exceeding the detection limit
$N_{EXCEEDED}$: Number of sites with an elemental concentration beyond the WHO limit (Table 3.1)

other metals in the groundwaters from basement rocks has been attributed to the higher susceptibility of calcium-bearing minerals to hydrolytic weathering processes (Acworth, 1987; Jones, 1985). With the bicarbonate ion also being generated from hydrolytic reactions, the dominance of this anion in groundwaters (Figure 3.7) is similarly explained. Scaling of cooking pots through $CaCO_3$ deposition and a salty taste are two commonly observed by-products of utilizing chemically hard waters. Although water softening can be achieved through ion-exchange filtration where calcium ions, Ca^{2+}, are substituted by sodium ions, Na^+, such techniques are unlikely to be available to rural consumers. Remediation at the source is problematic since the calcium rich water is derived from the aquifer matrix.

Analysis for nitrate in groundwaters of the Aroca catchment reveals that three sites exceed the WHO health guideline of 45 mg l^{-1}. Nitrate originates in the sub-surface through the microbial degradation of organic nitrogenous material including human and animal waste as well as agricultural fertilizers and plant refuse. The ammonium ion, NH_4^+, generated from microbial decomposition, is oxidized to yield nitrite, NO_2^-, and nitrate, NO_3^-. Of the three sources found to surpass the WHO nitrate limit in the Aroca basin, one is located 20 m from an exposed sewage pit while the other is situated at the centre of the region's weekly market. Similar links to probable contaminant sources have been identified in Nigeria where the presence of cattle markets were implicated in elevating the nitrate concentrations in adjacent wells (Owoade *et al.*, 1989; Malomo *et al.*, 1990).

The health risks associated with nitrate and other

Table 3.6 Trace metal levels determined by ICP-MS in groundwaters from the fractured bedrock aquifer at 15 sites in the Aroca basin

Element	$N_{DETECTED}$ (of 15)	$c_{AVERAGE}$ (μg l^{-1})	S.D. (μg l^{-1})	c_{RANGE} (μg l^{-1})	$N_{EXCEEDED}$ (of 15)
aluminium	15	62	58	20–115	1
arsenic	1	5	–	–	0
barium	15	150	120	19–570	–
boron	11	9	2	5–12	–
cadmium	4	1.1	0.1	1.0–1.3	0
chromium	15	4	2	3–6	0
cobalt	2	1	0	1.0–1.5	–
copper	2	6	3	3–8	0
iron	15	1 100	800	222–3 580	14
lead	7	0.7	0.6	0.1–1.9	0
manganese	15	89	91	7–290	5
mercury	3	0.1	0	0.1–0.2	0
nickel	15	8	6	3–40	–
selenium	1	2	–	–	0
strontium	15	230	230	53–1 030	–
uranium	8	0.6	0.6	0.1–1.8	–
zinc	15	1 400	3 100	41–12 600	1

$N_{DETECTED}$: Number of sites with an elemental concentration beyond the detection limit (Table 3.1)
$c_{AVERAGE}$: average elemental concentration in samples exceeding the detection limit
S.D.: standard deviation in the average elemental concentration
c_{RANGE}: range of elemental concentrations in samples exceeding the detection limit
$N_{EXCEEDED}$: Number of sites with an elemental concentration beyond the WHO limit (Table 3.1)

Table 3.7 Trace metal levels determined by ICP-MS in groundwaters from the regolith aquifer at 10 sites in the Aroca basin

Element	$N_{DETECTED}$ (of 10)	$c_{AVERAGE}$ (μg l^{-1})	S.D. (μg l^{-1})	c_{RANGE} (μg l^{-1})	$N_{EXCEEDED}$ (of 10)
aluminium	10	1 200	2 100	40–6 440	4
arsenic	1	9	–	–	0
barium	10	220	260	69–990	–
boron	8	11	13	5–45	–
cadmium	4	7	6	0.6–13	2
chromium	10	6	2	3–9	0
cobalt	6	6	9	1–26	–
copper	2	8	4	4–12	0
iron	10	750	1 000	38–3 410	5
lead	10	3	4	0.2–11	0
manganese	10	130	110	16–360	6
mercury	6	0.4	0.2	0.1–0.6	0
nickel	10	12	12	5–42	–
selenium	3	2	1	2–3	0
strontium	10	280	400	47–1 400	–
uranium	4	1	2	0.1–4.3	–
zinc	10	74	65	11–240	0

$N_{DETECTED}$: Number of sites with an elemental concentration beyond the detection limit (Table 3.1)
$c_{AVERAGE}$: average elemental concentration in samples exceeding the detection limit
S.D.: standard deviation in the average elemental concentration
c_{RANGE}: range of elemental concentrations in samples exceeding the detection limit
$N_{EXCEEDED}$: Number of sites with an elemental concentration beyond the WHO limit (Table 3.1)

Table 3.8 Trace metal levels determined by ICP-AES in groundwaters from the fractured bedrock aquifer at 87 sites in the Nyabisheki basin

Element	$N_{DETECTED}$ (of 87)	$c_{AVERAGE}$ (μg l^{-1})	S.D. (μg l^{-1})	c_{RANGE} (μg l^{-1})	$N_{EXCEEDED}$ (of 87)
aluminium	40	500	1 200	87–8 290	20
barium	83	63	46	1–340	–
boron	22	80	100	22–350	–
chromium	30	47	45	13–200	8
copper	35	30	60	11–380	0
iron	86	4 000	7 000	20–44 700	66
manganese	87	390	510	4–1 990	53
nickel	5	90	20	70–120	–
strontium	*				
zinc	87	10 000	9 000	38–44 700	53

$N_{DETECTED}$: Number of sites with an elemental concentration beyond the detection limit (Table 3.1)
$c_{AVERAGE}$: average elemental concentration in samples exceeding the detection limit
S.D.: standard deviation in the average elemental concentration
c_{RANGE}: range of elemental concentrations in samples exceeding the detection limit
$N_{EXCEEDED}$: Number of sites with an elemental concentration beyond the WHO limit (Table 3.1)
* analysis for strontium not performed

Table 3.9 Trace metal levels determined by ICP-AES in springwaters from 7 sites in the Nyabisheki basin

Element	$N_{DETECTED}$ (of 7)	$c_{AVERAGE}$ (μg l^{-1})	S.D. (μg l^{-1})	c_{RANGE} (μg l^{-1})	$N_{EXCEEDED}$ (of 7)
aluminium	2	140	30	110–160	0
barium	7	70	30	16–100	–
iron	7	60	80	8–220	0
manganese	5	200	500	5–1100	1
strontium	*				

$N_{DETECTED}$: Number of sites with an elemental concentration beyond the detection limit (Table 3.1)
$c_{AVERAGE}$: average elemental concentration in samples exceeding the detection limit
S.D.: standard deviation in the average elemental concentration
c_{RANGE}: range of elemental concentrations in samples exceeding the detection limit
$N_{EXCEEDED}$: Number of sites with an elemental concentration beyond the WHO limit (Table 3.1)
* analysis for strontium not performed

inorganic contaminants characteristically result from chronic ingestion of waters with levels not metabolized in the human body. Unlike biological contaminants, it is usually the bio-accumulation of inorganic constituents rather than the first contact that produces the ill effects. Nitrate has been linked to methaemoglobinaemia and the development of cancers. Methaemoglobinaemia is a condition where the ability of blood to absorb oxygen is impaired. It has been suggested that nitrate be reduced to less toxic forms through the introduction of soluble, biodegradable organic matter into groundwater sources (Custodio and Gurgui, 1989); however, prevention is usually more effective and less expensive than remediation. If localized sources of nitrate such as latrines are positioned sufficiently remote from water sources, denitrification (reduction of nitrates) will be allowed to occur naturally. In Malawi, a separation of at least 2 m from the bottom of a latrine to the water table and no less than 200 m from latrines to water sources has been recommended (Lewis and Chilton, 1984; Foster, 1985).

Levels of iron, manganese, aluminium, chromium and zinc exceed WHO standards over both study areas. The existence of Fe, Mn and Al in basement groundwaters is expected since they are produced from the alteration of minerals within granites and gneisses of the basement complex. However, the prevalence of corroded borehole rising mains throughout both catch-

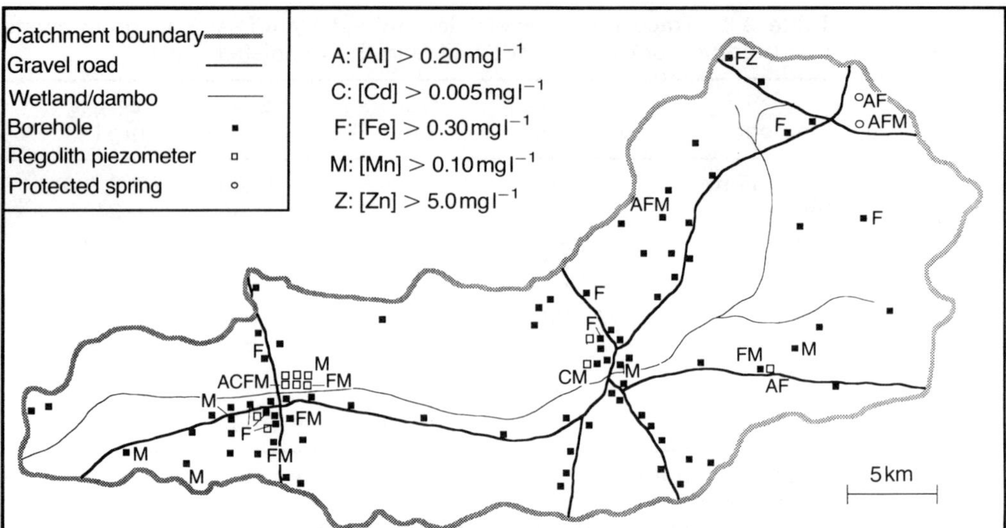

Figure 3.9 The distribution of sites across the Aroca catchment having trace metal concentrations in violation of aesthetic and health guidelines for drinking-water quality set by the World Health Organization.

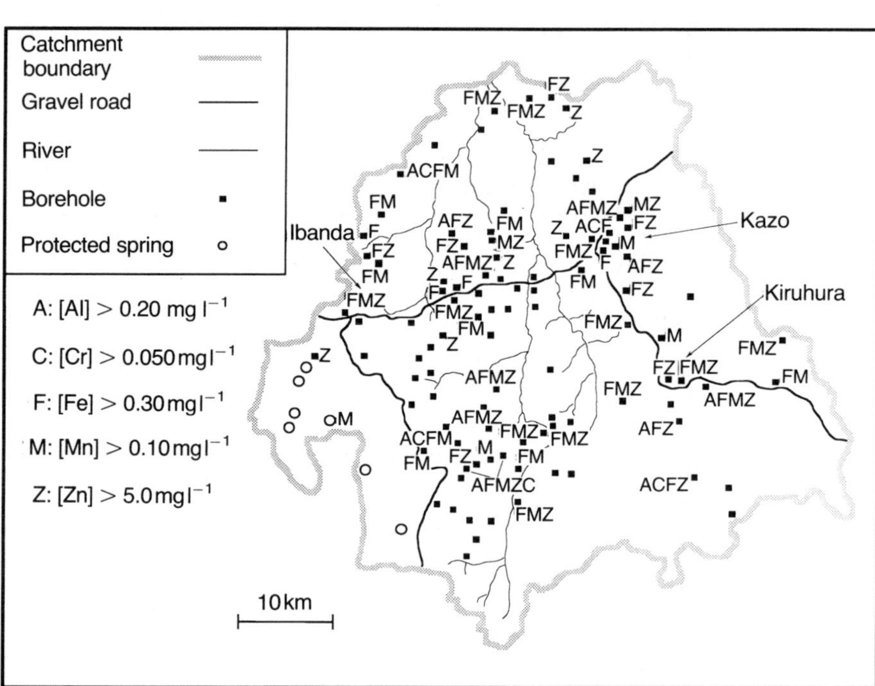

Figure 3.10 The distribution of sites across the Nyabisheki catchment having trace metal concentrations in violation of aesthetic and health guidelines for drinking-water quality set by the World Health Organization.

ments indicates that galvanized steel pipes in boreholes act as a secondary source of trace metals. Galvanized steel pipes have a zinc plating and are composed primarily of iron but may contain varying quantities of manganese, aluminium and chromium. The existence of significant quantities of zinc and the dramatic rise in iron recorded in groundwater samples from boreholes in relation to those from PVC-fitted springs and piezometers in the regolith provide an indication of the degree to which the galvanized pipes contaminate groundwater.

The hazards of utilizing waters enriched with Al, Fe, Mn or Zn are aesthetic, while consumption of waters with excessive levels of chromium constitutes a health risk. Both aluminium and zinc in concentrations beyond the WHO limits of 0.2 mg l^{-1} and 5.0 mg l^{-1} respectively, are capable of changing the appearance of water. Levels of iron and manganese in water beyond WHO regulations yield a salty taste and tend to stain clothes and pots. The formation of reddish-brown and black precipitates, hydroxides of iron and manganese, in 'jerry-cans' and pots transporting waters pumped out of boreholes is another common nuisance. The processes by which iron is precipitated are given in Figure 3.11. Treatment of waters with high iron and manganese typically involves the natural aeration and floccu-

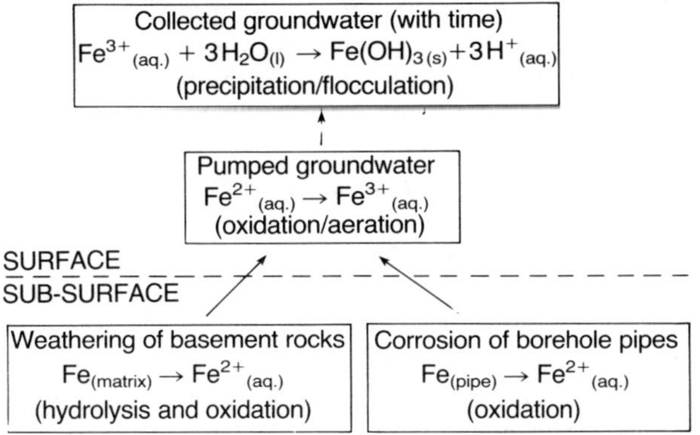

Figure 3.11 Mechanisms by which iron typically precipitates from basement groundwaters.

lation processes indicated in Figure 3.11 but includes a sedimentation component that enables precipitated species to settle out of solution prior to consumption.

Conclusions

Deterioration of groundwater quality through the corrosion of borehole rising mains and the seepage of sewage waste has been identified in the basement aquifers of two catchments in Uganda. Sewage wastes lead to elevated concentrations of nitrate while corroded pipe materials increase the content of principally iron, zinc and manganese. For the most part, these impacts are localized and can be minimized by careful borehole location and regular maintenance of pipe fixtures.

The degradation of groundwater quality through the natural weathering of metals from the basement matrix was also observed. In contrast to the localized sources, the effects of weathering are unavoidable and extend across both basins. It yields elevated iron, manganese and aluminium in groundwater, occasionally at levels exceeding WHO guidelines. In terms of the groundwater source, preliminary results suggest that the quality of groundwaters from the regolith aquifer is slightly superior to those in the fractured bedrock aquifer. The comparison is made difficult since lower levels of TDS, iron and manganese recorded in the regolith may simply reflect the sampling methodology which avoided the use of metallic pipes. Significantly, the quality of the regolith groundwaters is commonly compromised by an increase in aluminium content believed to stem from the dissolution of clay minerals.

Complaints from the rural communities over the aesthetic quality of their groundwater source were generally supported by chemical analysis. In the absence of routine groundwater quality monitoring programmes, perceptions of consumers may prove to be reliable indicators of aesthetic water quality. At the very least, opinions of the user community should be addressed. Without action in the form of remediation or education, communities may revert to biologically polluted, surface waters to meet their needs.

Acknowledgments

The collection and field analysis of groundwater were performed with the assistance of John Karundu, Nicholas Azza, Tindimugaya Callist and Rwarinda Edwardmartin of the Water Development Department of Uganda and Mark Hughes of the University of Toronto. Sociological data were supplied by research partners at the Makerere Institute for Social Research, Makerere University, Uganda. Funding for the collaborative research was provided by the International Development Research Centre (IDRC), Canada.

References

Acworth, R.I., 1987. The development of crystalline basement aquifers in a tropical environment. *Quarterly Journal of Engineering Geology*, vol. 20, pp. 265–272.

Angura, T.O., 1992. Socioeconomic considerations for groundwater development in Uganda: A comparative analysis of the Aroca and Nyabisheki catchments. A hydrogeological and socio-economic examination of groundwater resource management in Uganda: strategic issues for future development. *Proceedings of the International Development Research Centre's Hydrogeology Phase II Interim Workshop, December 10, 11, 1992* (Unpublished Report)

Callist, T., 1992. An investigation of the fractured-basement aquifer of the Aroca basin. A hydrogeological and socio-economic examination of groundwater resource management in Uganda: strategic issues for future development. *Proceedings of the International Development Research Centre's Hydrogeology Phase II Interim Workshop, December 10, 11, 1992* (Unpublished Report)

Chilton, P.J. and Smith-Carington, A.K., 1984. Characteristics of the weathered basement aquifer in Malawi in relation to rural water supplies. *Challenges in African Hydrology and Water Resources (Proceedings of the Harare Symposium, July 1984)*, International Association of Hydrogeologists'. Series Publication No. 144, pp. 15–23.

Custodio, E. and Gurgui, A., 1989. *Developments in Water Science*, vol. 38, p. 625.

Faillat, J.P., 1981. Physico-chemical characteristics of the ground water in the fractured rocks in the Ivory Coast. Quality of Groundwater (Proceedings of an International Symposium, Noordwijkerhout, The Netherlands, March 23–27, 1981), *Studies in Environmental Science*, vol. 17, pp. 671–674.

Foster, S.S.D., 1985. Groundwater pollution protection in developing countries. *Theoretical Background, Hydrogeology and Practice of Groundwater Protection Zones, International Association of Hydrogeologists*, vol. 6, pp. 167–200.

Howard, K.W.F., Hughes, M., Charlesworth, D. L. and Ngobi, G., 1992. Hydrogeologic evaluation of fracture permeability in crystalline basement aquifers of Uganda. *Journal of Applied Hydrogeology*, vol. 1, pp. 55–65.

Howard, K.W.F. and Karundu, J., 1992. Constraints on the exploitation of basement aquifers in East Africa – water balance implications and the role of the regolith. *Journal of Hydrology*, vol. 139, pp. 183–196.

Jones, M.J., 1985. The weathered zone aquifers of the basement complex areas of Africa. *Quarterly Journal of Engineering Geology*, vol. 18, pp. 35–46.

Lewis, W.J. and Chilton, P.J., 1984. Performance of sanitary completion measures of wells and boreholes used for rural water supplies in Malawi. *Challenges in African Hydrology and Water Resources (Proceedings of the Harare Symposium, July 1984)*, International Association of Hydrogeologists'. Series Publication No. 144, pp. 235–247.

Loehnert, E.P., 1981. Groundwater quality of dug wells in southern Nigeria. Quality of Groundwater (Proceedings of an International Symposium, Noordwijkerhout, The Netherlands, March 23–27, 1981), *Studies in Enviromental Science,* vol. 17, pp. 147–153.

McFarlane, M.J., 1992. Groundwater movement and water chemistry associated with weathering profiles of the African surface in parts of Malawi. In: *Hydrogeology of Crystalline Basement Aquifers in Africa*. (E.P. Wright and W.G. Burgess, eds) Geological Society (UK) Special Publication No. 66. 264 pp.

Malomo, S., Okufarasin, V.A., Olorunniwo M.A. and Omode, A.A., 1990. Groundwater chemistry of weathered zone aquifers of an area underlain by basement complex rocks. *Journal of African Earth Sciences*, vol. 11, No. 3/4, pp. 357–371.

Olivier, H., 1961. *Irrigation and Climate; New Aids to Engineering Planning and Development of Water Resources*. E. Arnold, London, 250 pp.

Omorinbola, E.O., 1984. Groundwater resources in tropical African regoliths. *Challenges in African Hydrology and Water Resources (Proceedings of the Harare Symposium, July 1984)*, International Association of Hydrogeologists' Series Publication No. 144, pp. 57–72.

Owoade, A. Hutton, L.G., Moffat W.S., and Bako, M.D., 1989. Hydrogeology and water chemistry in the weathered crystalline rocks of southwestern Nigeria. *Groundwater Management: Quantity and Quality (Proceedings of the Benidorm Symposium, October 1989)*, International Association of Hydrogeologists' Series Publication No. 188, pp. 201–214.

Taylor, R.G., Howard, K.W.F., Karundu J. and Callist, T., 1993. Distribution and seasonality of groundwater recharge to the regolith–basement aquifer system of Uganda. *GSA Abstracts* vol. 25, No. 6 (September, 1993).

World Health Organization, 1984. *Guidelines for Drinking-Water Quality*, vol. 1. *Recommendations*. World Health Organization, Geneva, 130 pp.

4 Controls on the occurrence of fluoride in groundwater in the Rift Valley of Ethiopia

R.P. Ashley and M.J. Burley

Abstract High concentrations of fluoride in groundwater used for drinking by 50 000 sugar estate workers in the Awash Valley of Ethiopia have resulted in bone and tooth deformities. As a part of a broader study, the distribution of fluoride in local groundwater and surface waters has been investigated, and compared with bedrock and drift geology and with hydrochemical parameters. The main chemical equilibrium control on the concentration of fluoride has been identified as the solubility of calcium fluoride, and variations in fluoride concentration can be related to changes in calcium concentration resulting from dissolution of calcium minerals, ion exchange and mixing with waters of different chemical composition. The results of the study have implications for the selection of well locations and design aimed at obtaining reduced-fluoride water supplies.

Introduction

The Ethiopian Sugar Corporation (ESC) owns the Wonji/Shoa and Metahara sugar estates in the valley of the River Awash, part of the northern Rift Valley. In 1984 the populations of the two estates were 30 000 and 25 000 respectively, comprising employees of the ESC and their families. They lived in small villages and camps distributed within and around the estates, and obtained almost all their water supply from wells, boreholes and springs. The high concentration of fluoride in the local groundwater was recognized in the early 1960s as being responsible for the symptoms of dental fluorosis (mottling and malformation of the teeth) and skeletal fluorosis (hardening of bone cartilage, causing the vertebrae to fuse and the rib cage to become rigid). The symptoms of skeletal fluorosis manifest themselves after 10–20 years of intake of high fluoride drinking water, and consequently the World Health Organization recommends a Guide Level maximum concentration of 1.5 mg l^{-1} fluoride in drinking water (World Health Organization, 1984).

In 1984 the ESC appointed Sir M. MacDonald & Partners Limited to investigate the feasibility of supplying reduced fluoride drinking water to the two estates. This chapter describes that part of the investigation which examined the origin and behaviour of fluoride in local groundwater.

The sites

Location and topography

The Wonji/Shoa and Metahara estates are in the Awash Valley, 100 km southeast and 120 km east of Addis Ababa respectively, at elevations of 1540 m and 960 m above sea level (Figure 4.1). At these locations the Rift floor is about 50 km wide and 1000 m below the level of the adjacent mountainous areas. In contrast to the rugged regional topography, the estates have flat or gently undulating terrain. The River Awash forms the northern boundary of the Wonji/Shoa estate and flows through the northern part of the Metahara estate. Figures 4.2 and 4.3 show the settings of the two estates,

Groundwater Quality Edited by H. Nash and G.J.H. McCall. Published in 1994 by Chapman & Hall. ISBN 0 412 58620 7

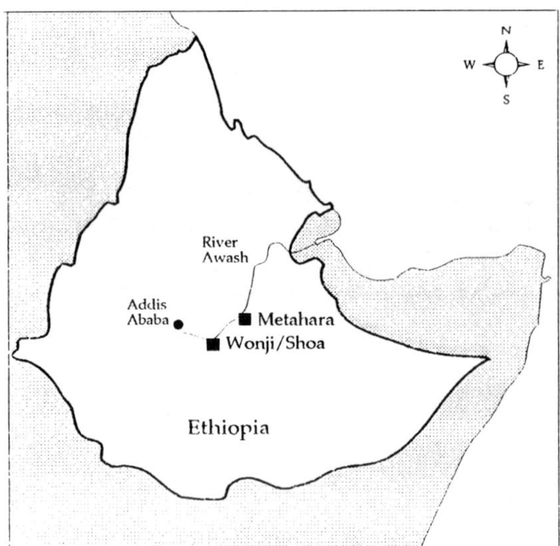

Figure 4.1 Location map.

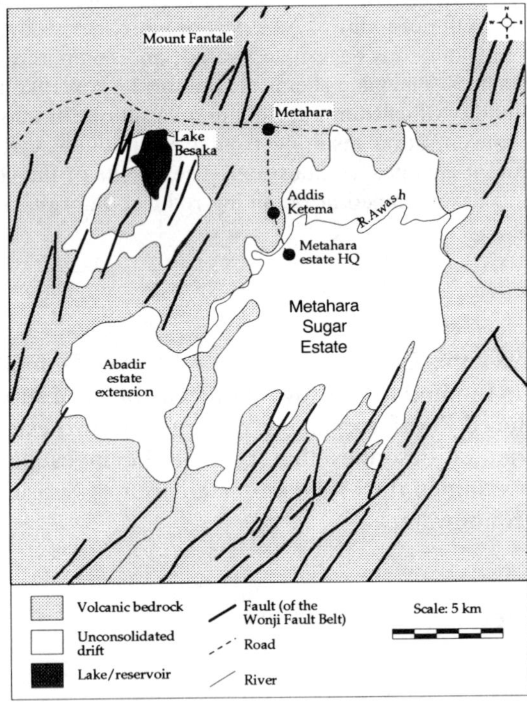

Figure 4.2 Geology of the Wonji/Shoa area (after Mohr, 1961).

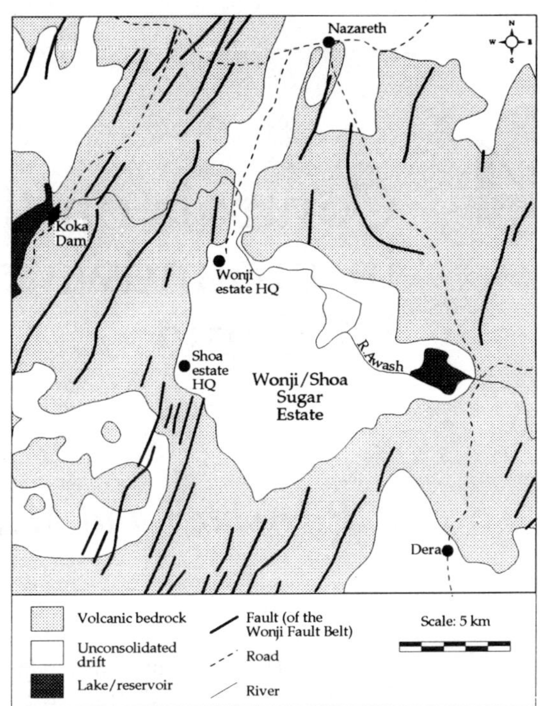

Figure 4.3 Geology of the Metahara area (after Mohr, 1961).

and Figures 4.4 and 4.5 show the main settlements and workers' encampments.

Geology

The bedrock of the Rift Valley at the two estates (Figures 4.2 and 4.3) is composed of volcanic rock of the Aden Volcanic series dating from ages between the Pliocene and Recent, after the initiation of faulting in the Rift (Mohr, 1961). The volcanic rocks of the Wonji/Shoa area include basaltic, intermediate and acidic types. All these lithologies tend to contain relatively high amounts of sodium/potassium compared to calcium, especially in the more acid lavas. The basaltic types tend to be scoriaceous, and the acidic rocks occur as rhyolitic ignimbrites, welded and unwelded tuffs, ash flow tuffs, pumice and obsidian-rich lavas. Thus all the volcanic rocks are potentially highly permeable, particularly at the upper and lower surfaces of lava flows. The acidic rocks are also chemically unstable and readily broken down to release constituents into groundwater. The bedrock lithologies at Metahara are similar to those at Wonji/Shoa, with the addition of Recent volcanic rocks from Mount Fantale to the north. The lavas tend to be high sodium/potassium acidic types, although the last eruption (in 1810) produced scoriaceous basalts which reached Lake Besaka. Visual inspection confirmed the high permeability and porosity of these rocks because of the high proportion of voids formed from solidified gas bubbles. No data were obtained on the fluorine content of the volcanic bedrock at either site.

The superficial deposits of the upper Awash Valley are of Pleistocene age and are concentrated in distinct basins, two of which support the sugar estates. The deposits in the Wonji/Shoa basin are more than 66 m thick and are lacustrine in origin. There is evidence, from some borehole records, that a 30 m thick layer of mainly clayey sediments overlies a lower layer of

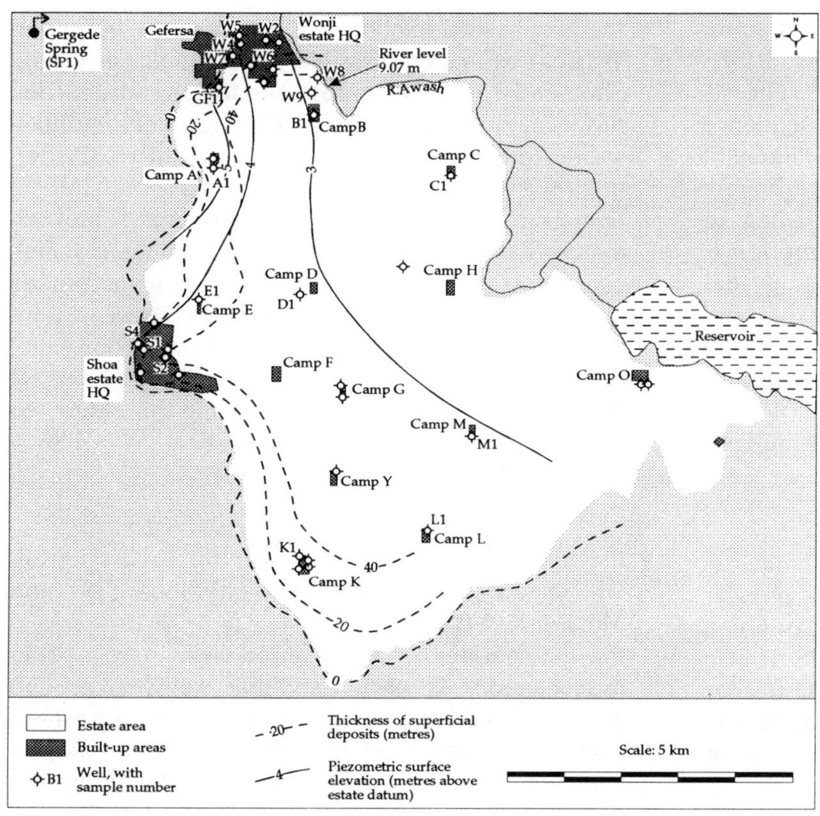

Figure 4.4 Hydrogeological data: Wonji/ Shoa.

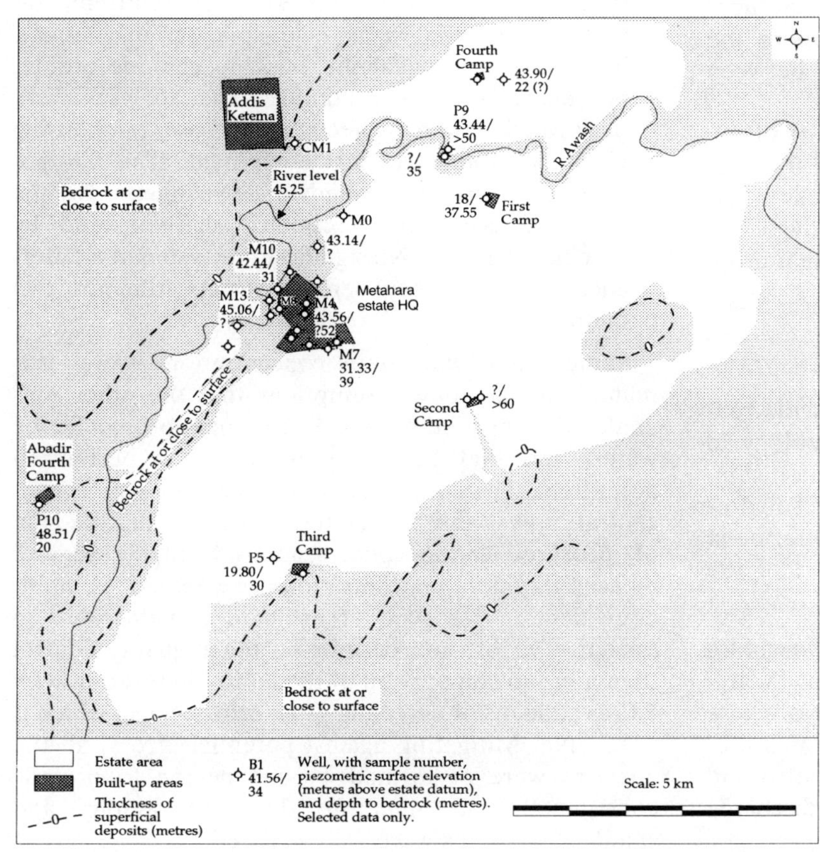

Figure 4.5 Hydrogeological data: Metahara.

medium-coarse sands. The Metahara basin is more than 50 m thick and is fluvial in origin (Mohr, 1961). The available information on lithological types suggests that the strata are more sandy throughout the sequence than at Wonji/Shoa. The difference in origin is likely to result in a greater concentration of calcium carbonate and sulphate in the Wonji/Shoa sediments than in those at Metahara (Hem, 1970). This presumption is supported by ESC technical staff, who stated that Metahara soils are deficient in available calcium and require liming, whereas Wonji/Shoa soils do not. However, no direct information was obtained on the nature of calcium minerals in the soils and sediments; for example, it is not known whether they occur as mineral grains or intergranular cement.

Hydrogeology

Groundwater is exploited for water supply from the bedrock, mainly from shallow hand-dug wells. A number of springs are also used for local supplies. As noted above, groundwater in the superficial deposits is also used for water supply, mainly by means of deeper boreholes. The boreholes tend to be deeper at Wonji/Shoa, where the shallow sediments are clayey, than at Metahara.

From the evidence of water levels measured in wells during the investigation, it appears that there is hydraulic continuity at Metahara between the River Awash, bedrock and superficial deposits, but that there is little or no continuity at Wonji/Shoa, where the shallow clays probably limit the connection between the Awash and the deeper sands.

The flow in the Awash is regulated by the Koka Dam and Reservoir upstream of Wonji/Shoa, and thus the river is a year-round hydraulic feature at both sites. At both sites there are river off-takes which provide the bulk of the water used for irrigation of the sugar crop.

Investigation methods

The investigation was designed partly to provide reliable data on the local geology and hydrogeology in order to supplement the pre-existing information summarized above. It was also designed to collect hydrochemical data relevant to the distribution of fluoride in waters around the two sites. The first aspect of the investigation comprised:

- the identification and level-surveying of all boreholes and wells in the vicinity of each site (39 at Wonji/Shoa and 31 at Metahara, although some of these had been abandoned and backfilled), the measurement of rest water levels and borehole depths and the collation of records of construction, discharge rates and specific capacities;
- the geophysical logging of as many of the boreholes as could be made accessible (16 boreholes at Wonji/Shoa and 15 at Metahara). The logging was carried out using a portable SIE T450 instrument, equipped with caliper, self potential/point resistance (SP/PR), natural gamma radiation and temperature/differential temperature (T/DT) sondes. There was some evidence that Rift Valley hot springs contained high concentrations of fluoride, and the last-named sonde was used to examine the correlation between temperature and fluoride concentration in a borehole; and
- the pump testing of three existing boreholes to measure aquifer permeability, one at Wonji/Shoa and two at Metahara.

The second aspect of the investigation comprised:

- the collection of water samples for analysis from all operating boreholes at Wonji/Shoa (18) and Metahara (14);
- the collection of water samples for analysis from miscellaneous springs, hand-dug wells, lakes and water holes, rivers and other boreholes near both sites. The river samples, which included samples from the Awash, were in addition to a large number of samples analysed as part of a separate investigation of river water treatability; and
- the analysis of water samples for a range of parameters including electrical conductivity (EC), pH and temperature (all measured in the field), calcium, magnesium, iron, manganese, alkalinity, sulphate, chloride, nitrate, phosphate, silica and fluoride. The analyses were performed at the project team's base camp using Hach portable analytical equipment. The fluoride analyses were performed by two methods: the SPADNS Reagent colorimetric method, and by specific ion probe. No significant or systematic difference was noted between the results from the two methods.

Sodium and potassium concentrations were estimated, on the joint assumptions that the other ions analysed comprised the main substances present in the water, and that the analyses had been performed accurately. The combined sodium and potassium concentration was thus calculated by subtracting the sum of measured cation concentrations from the sum of measured anion concentrations, expressed as milliequivalents. This type of estimate is vulnerable to analytical error, and permits no other quality control than a broad comparison of the ionic composition with EC as a means of detecting gross errors, a test which is unreliable. Mitigating against potential errors, all the analyses were performed by one person to provide consistency, and numerous duplicate samples and standard solutions were also analysed.

Table 4.1 Summarized results of analyses of samples from the Wonji/Shoa site

Sample	Group	pH	EC	T	Ca	Mg	Alk.	SO_4	Cl	NO_3	F
Plantation camp wells											
A1	C	6.8	1 830	29	124	35	645	285	4	1.8	2.7
B1	C	nm	968	25	146	14	566	89	2	26	3.1
C1	B	7.6	1 250	35	10	6	628	28	2	1.3	9.1
D1	C	7.1	1 725	28	66	30	526	nd	nd	nd	4.3
E1	B	7.5	1 200	30	25	3	676	32	1	1.8	8.9
K1	B	7.4	1 000	30	49	11	538	13	2	nd	6.8
L1	B	7.0	2 100	34	58	16	874	145	2	1.3	8.8
M1	C	6.8	1 750	26	186	46	935	130	<1	nd	2.1
Wonji area wells											
W2	B	7.4	1 400	31	43	14	630	48	6	11.8	9.9
W4	B	7.4	1 300	37	27	7	600	39	12	13.6	15
W5	B	nm	1 320	nm	27	5	586	31	66	20	14.5
W6	B	7.4	1 420	33	33	9	602	33	19	15.4	16.2
W7	B	7.5	1 500	35	20	7	586	30	11	21.1	11.8
W8	C	7.0	710	25	41	6	326	7.5	2	52.8	3.7
W9	B	nm	1 010	nm	18	8	572	40	12	1.8	12.6
Shoa area wells											
S1	B	7.1	1 190	nm	46	15	588	44	9	20.2	13.9
S2	C	7.3	1 490	nm	44	13	550	88	10	3.1	4.6
S4	B	7.1	1 270	nm	44	15	558	40	15	12.3	14.8
Others											
GF1 (Gefersa Hotel well)	A	8.1	1 550	38	6	0.5	610	59	62	48	20
SP1 (Gergede Spring)	A	nm	1 050	47	2	2	406	26	4	nm	17.6
AR2 (Awash River)	D	8.4	220	21	20	4	161	4	9	<1	0.9

Abbreviations: EC: Electrical conductivity ($\mu S\ cm^{-1}$)
T: Temperature (°C)
Ca: Calcium ($mg\ l^{-1}$)
Mg: Magnesium ($mg\ l^{-1}$)
Alk: Alkalinity ($mg\ l^{-1}$ as $CaCO_3$)
SO_4: Sulphate ($mg\ l^{-1}$)
Cl: Chloride ($mg\ l^{-1}$)
NO_3: Nitrate ($mg\ l^{-1}$)
F: Fluoride ($mg\ l^{-1}$)
nm: not measured
nd: not detected

Results

The investigation generated a considerable quantity of data, not all of which are relevant to the purpose of this chapter. The results presented below exclude detailed accounts of the well inventory, the geophysical logging and some other components of the hydrogeological studies; the less relevant hydrochemical parameters, such as manganese, silica and phosphate, are not reported.

Wonji/Shoa

Figure 4.4 summarizes the results of the hydrogeological investigations, showing the locations of all identified boreholes, contours of thickness of superficial sediments, and contours on the groundwater piezometric surface. The sediment thickness data were compiled partly from records of drilling, and partly from the natural gamma log, which showed an increase in gamma activity as the bedrock interface was approached.

Groundwater flow is from west and southwest to east, entering the lacustrine sedimentary basin from the volcanic bedrock. No data were available from the eastern edge of the basin, but it may be assumed that the groundwater flows into the bedrock again. The piezometric surface is about 6 m lower than the Awash in the northwestern part of the site, confirming that where the river crosses the basin there is limited hydraulic continuity between the groundwater and the river water.

The pumping test conducted for the investigation at well W8 took the form of a simple three hour constant discharge test at a rate of $3.0\ l\ s^{-1}$. In summary, a Theis analysis (Kruseman and de Ridder, 1970) of the recovery data concluded that the transmissivity was about $20\ m^2\ d^{-1}$. Specific capacity data from four other boreholes in the sedimentary aquifer ranged from 0.2 to $1.3\ l\ s^{-1}\ m^{-1}$; using the Logan approximation (Kruseman and de Ridder, 1970), the equivalent transmissivity range is estimated to be from 20 to $140\ m^2\ d^{-1}$. Specific capacity data from six wells in the bedrock ranged from 1 to $11\ l\ s^{-1}\ m^{-1}$, equivalent to a transmissivity range of $100–1200\ m^2\ d^{-1}$.

Table 4.1 records the results of the hydrochemical analyses of all the samples from Wonji/Shoa. The data

Table 4.2 Summarized results of analyses of samples from the Metahara site

Sample	Group	pH	EC	T	Ca	Mg	Alk.	SO_4	Cl	NO_3	F
Metahara estate area											
M0	C	7.2	1 220	27	57	20	642	125	50	7.9	15.3
M4	C	7	1 750	30	75	44	574	247	92	9.2	3.1
M7	C	7.2	1 510	28	50	13	364	150	184	29.9	3.7
M8	D	7.2	1 210	30	86	26	588	98	82	37	3.1
M9	D	7.4	870	28	61	16	546	20	16	4.4	3.8
M10	D	nm	680	30	52	13	364	3	8	1.8	2.8
M12	C	7.2	1 310	28	99	29	474	198	130	3.1	2.7
M13	D	7	1 090	28	74	19	524	63	18	5.7	6.8
Other wells											
P1	C	7.1	2 600	29	72	34	568	34	326	106	4.6
P2	C	7.1	6 700	nm	274	377	390	1 800	1 655	224	2.6
P5	C	7.4	3 400	32	118	89	382	770	270	30	3
P9	D	7.4	795	27	77	13	444	17	nd	1.3	2.8
P10	B	7.6	3 110	39	57	43	412	510	530	13.2	4.5
P13	B	nm	1 750	(hot)	26	20	346	450	242	9.2	3.5
CM1	B	nm	3 200	nm	nd	nd	1 476	240	15	8.5	26
Miscellaneous											
AR1 (Awash River)	E	8.4	250	nm	19	5	118	nd	nd	nd	1.3
LB1 (Lake Besaka)	A	9.2	6 100	nm	nd	nd	2 980	450	8	5	36

Abbreviations: See Table 4.1.

have been divided into groups, the first three of which represent successive stages along assumed flow paths:

- Group A: water samples from the volcanic bedrock, such as Gergede springs and Gefersa well water;
- Group B: water samples from wells in the sediments close to the bedrock or just penetrating into it;
- Group C: water samples from wells that abstract groundwater that has flowed a significant distance from the bedrock, or which are shallow compared to the thickness of sediment; and
- Group D: River Awash water samples.

The data are more conveniently considered in graphical form and are examined in the next section.

Metahara

Figure 4.5 summarizes the results of the hydrogeological investigations at Metahara, showing the locations of all identified boreholes, data on the thickness of the alluvial sediments and on the groundwater piezometric surface. The data are concentrated in the area around the sugar estate factory, and so cannot satisfactorily be contoured for aquifer thickness or piezometric surface. The sediment thickness data were compiled partly from records of drilling, but also from the natural gamma log, which showed an increase in gamma activity as the bedrock interface was approached. The topographic ridges which cross the site are fault-bounded slices of bedrock, and their subsurface extensions are responsible for the irregular sediment thickness.

Groundwater levels are similar to river levels, implying a higher degree of continuity than at Wonji/Shoa. It may thus be assumed that the overall direction of groundwater flow is parallel to the river flow direction from west to east, although there appears to be some flow away from the river. ESC staff reported that groundwater levels away from the river had risen at a rate of 0.6 m a^{-1} in the years prior to the investigation, probably as a result of infiltration of irrigation water. No data pertaining to hydraulic continuity between the bedrock and the alluvium were obtained, but good continuity probably exists.

Two pumping tests were conducted on wells M8 and M13, similar in conception to that at Wonji/Shoa. The test at M8 rapidly drew down the water level to the pump intake and, although the discharge eventually stabilized, the data could only be analysed by the Thiem steady-state method (Kruseman and de Ridder, 1970), which yielded a transmissivity estimate of 4.5 m^2 d^{-1}. At M13 the test was conducted for two hours of pumping, and a Theis analysis of both drawdown and recovery data gave a transmissivity estimate of 170 m^2 d^{-1}. Specific capacity data from three other boreholes in the alluvium ranged from 0.21 to 2.5 l s^{-1} m^{-1}; using the Logan approximation, it is estimated that the equivalent transmissivity range would thus be from 20 to 260 m^2 d^{-1}. Specific capacity data from three wells in the bedrock ranged from 0.4 to 11 l s^{-1} m^{-1}, equivalent to a transmissivity range from 42 to 1200 m^2 d^{-1}.

Table 4.2 records the results of the hydrochemical

analyses of all the samples from Metahara. The data have been divided into groups on the basis of sample origin:

- Group A: Lake Besaka sample;
- Group B: water samples from wells north of the river, but not close to it;
- Group C: water samples from wells south of the river, but not close to it;
- Group D: river water samples from estate wells close to the river; and
- Group E: River Awash water samples.

The data are more conveniently considered in graphical form and are examined in the next section.

Discussion

Occurrence of fluoride in groundwater

Fluorine occurs in igneous rocks world-wide, at an average concentration of 715 mg kg^{-1} compared, for example, to the average chlorine concentration of 305 mg kg^{-1} (Hem, 1970). In sea-water it occurs at typical concentrations of 1.3 mg l^{-1}, compared to chlorine concentrations of 19 000 mg l^{-1}. These figures lend perspective to the following discussion. Fluorine occurs as fluorite (CaF_2) in igneous as well as in sedimentary rocks. It is also common in apatite (as $(CaF)Ca_4(PO_4)_3$), and as a substitute for the hydroxide ion in amphibole and mica minerals. Rocks rich in alkali metals (such as sodium and potassium) and obsidian, both of which are common in the Rift Valley, tend to have relatively high amounts of fluorine. Thus at least a part of the fluoride in Rift Valley groundwater may be derived from the weathering of fluorine-containing minerals.

The association of high fluoride concentrations with hot springs in the Rift Valley is also known, and during the present investigation a sample from such a spring near Lake Shalla, 150 km southwest of Wonji/Shoa, was found to contain 90 mg l^{-1} fluoride. It is likely that fluoride can also enter groundwater either by mixture with juvenile waters and volcanic gases, or by deep percolation of groundwater to hot, high-fluorine rocks. In groundwater, fluorine occurs mainly as the simple fluoride ion. Although it is capable of forming complexes with silicon and aluminium, these are believed to exist only at a pH < 7.0, a condition which is not encountered in natural waters in this part of Ethiopia.

The solubility of fluoride is governed by the solubility of its least soluble compound, which in the present case is calcium fluoride (CaF_2). According to Hem (1970) the solubility of calcium fluoride may be expressed by the relationship:

$$\log[Ca] = 2 \log[F] - 10.57 \qquad (4.1)$$

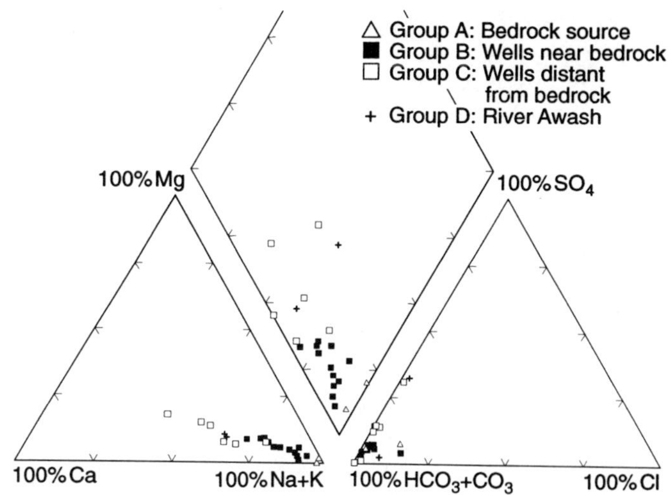

Figure 4.6 Piper Diagram of Wonji/Shoa analysis data.

where [Ca] and [F] are the activity of calcium and fluoride respectively. Where the ionic strength varies little from sample to sample, concentration data may conveniently be used in a relationship similar to equation 4.1. In the present case, however, the quantity of dissolved ions varies considerably and activities must be calculated and used.

The implication of equation 4.1 is that the higher the level of dissolved calcium in the groundwater, the lower the level of fluoride which can be supported. Hence, any increase in the calcium content must be compensated for by the precipitation of calcium fluoride. The data from this investigation are examined below in the context of this relationship.

Wonji/Shoa

Figure 4.6 is a Piper Diagram (Piper, 1944) representing the relative concentrations of major ions. Among the cations, the main variation is in the relative proportions of calcium plus magnesium (Ca + Mg) to sodium plus potassium (Na + K), with the ratio of the former to the latter increasing along the putative flow path. Among the anions, there is little chloride present in the samples, so the main variation is in the relative amounts of sulphate and carbonate plus bicarbonate ($CO_3 + HCO_3$). Figures 4.7 and 4.8 examine the same variations in absolute terms, expressing the concentrations as milli-equivalents per litre (meq l^{-1}). Following the putative flow path the processes that appear to operate successively are:

(1) Decrease in Na + K and increase in Ca + Mg at approximately equal rates as water moves from bedrock into the alluvium, presumably as a result of ion exchange between sodium and potassium in the water and calcium and magnesium in clays.
(2) Increase in Ca + Mg at a greater rate than the

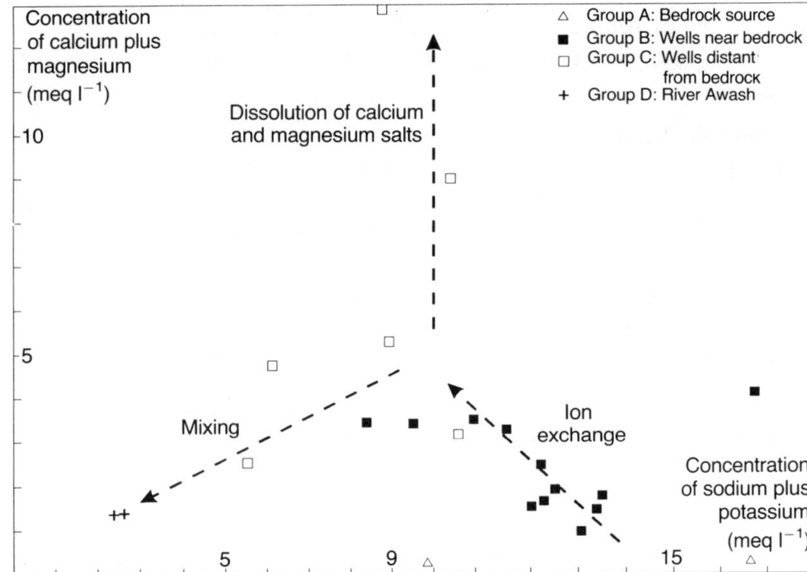

Figure 4.7 Variation in cation concentrations: Wonji/Shoa data.

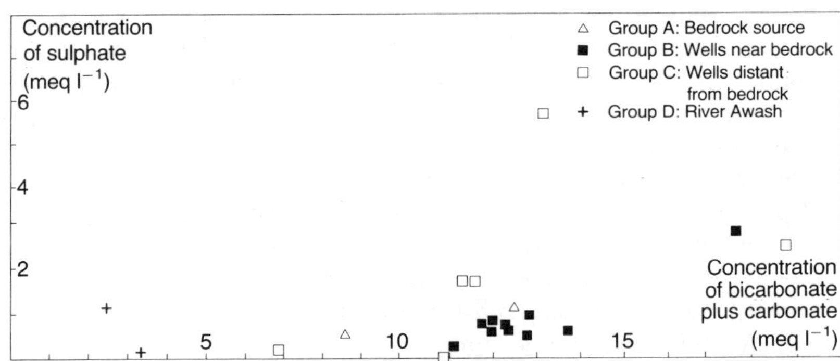

Figure 4.8 Variation in anion concentrations: Wonji/Shoa data.

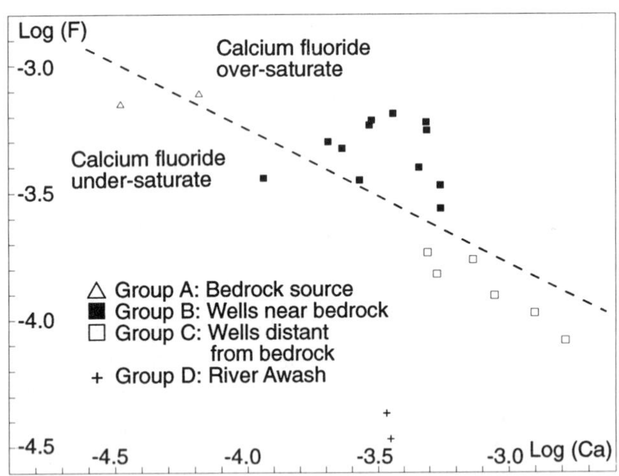

Figure 4.9 Variation in activities of calcium and fluoride: Wonji/Shoa data.

decrease in Na + K, together with some increase in sulphate and CO_3 + HCO_3, possibly as a result of dissolution of calcite and gypsum. A limited amount of mixing with Awash river water or, less likely, with infiltrating irrigation water (also from the Awash) may also be occurring.

Figure 4.9 is a graph of log[Ca] against log[F]: the activities were calculated from the concentrations of calcium and fluoride and the ionic strength of the water sample using the Davies relationship (Stumm and Morgan, 1981). Also marked on the graph is the theoretical solubility limit according to equation 4.1. There appears to be a clear trend from high-fluoride/low-calcium bedrock waters to low-fluoride/high calcium sedimentary aquifer waters, taking in an intermediate stage in waters close to a bedrock source. All these waters are at least close to saturation in calcium fluoride and the intermediate waters appear to be oversaturated. River Awash waters are clearly under-saturated in calcium fluoride.

Metahara

Figure 4.10 is a Piper Diagram of the Metahara major ion analysis data. The cation pattern may be interpreted as reflecting a mixing trend between high Ca + Mg water from the River Awash and high Na + K bedrock water, such as that from Lake Besaka. The Awash sample, however, has low concentrations of all constituents in absolute terms (Figure 4.11), and a better candidate for the high Ca + Mg end-member

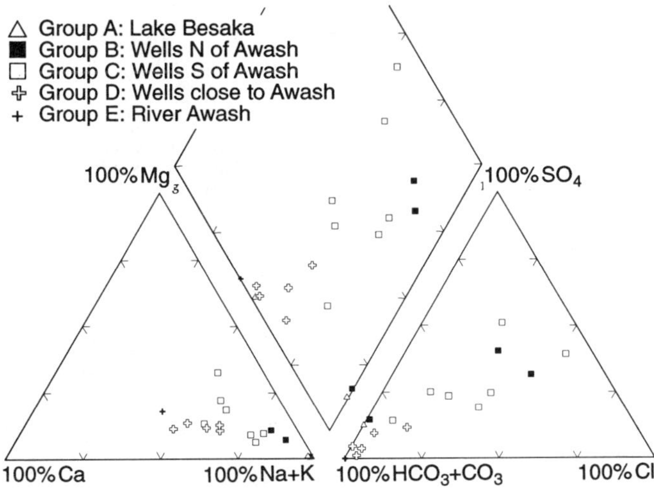

Figure 4.10 Piper Diagram of Metahara analysis data.

would be infiltrated Awash water used for irrigation and concentrated by evaporation. Figure 4.11 suggests the presence of three end-members in a mixing trend: Awash water; irrigation water; and a bedrock water. The existence of an irrigation water end-member is supported by the variation in sulphate concentration (Figure 4.12) and the very high nitrate concentrations in samples from beneath the sugar estate (Table 4.2), which may be caused by leaching of fertilizers. It should be noted that mixing with infiltrated irrigation water is to be expected at Metahara where the alluvial strata are believed to be sandy throughout, but is less likely at Wonji/Shoa where the shallow strata are clayey. The overall level of concentrations is greater than at Wonji/Shoa, probably due to the evaporative concentration of irrigation water. If other processes are taking place which increase the Ca + Mg concentration, such as calcite and gypsum dissolution or ion exchange, they cannot be distinguished from the mixing processes.

Figure 4.13 is a graph of log[Ca] against log[F]. Most of the samples plot on the undersaturated side of the calcium fluoride solubility limit, and it is noticeable that, compared to Wonji/Shoa, the degree of variation in the activity of calcium is much smaller: the variation resulting from mixing has not resulted in oversaturation and precipitation of calcium fluoride, thereby reducing the fluoride concentration. This suggests that one difference between the two sites may be the absence, at Metahara, of calcium for dissolution or ion exchange, which at Wonji/Shoa boosts the groundwater calcium concentration over the limit for calcium fluoride saturation. The other difference is that much of the groundwater sampled at Metahara originated, via one route or other, as River Awash water with a low fluoride concentration, diluting and masking high-fluoride bedrock groundwater.

Conclusions

The investigation has shown that high-fluoride groundwater at both Wonji/Shoa and Metahara sites originates in volcanic bedrock as a high Na + K, low Ca + Mg hydrochemical type. It flows from the bedrock into the sedimentary basins at both sites. At Wonji/Shoa the Ca + Mg content is increased, probably partly by ion exchange and partly by dissolution of calcite and gypsum in the lacustrine sediments. This causes the calcium fluoride solubility to be exceeded, and the fluoride concentration is consequently reduced by precipitation of calcium fluoride. Some mixing also occurs with groundwater entering the aquifer from the Awash, although this is limited by the low permeability of the shallower sediments in the basin.

At Metahara the sediments are alluvial, and more permeable throughout their thickness. Much more mixing is possible, both with Awash water entering the aquifer directly from the river, and with infiltrating irrigation water, originating from the river but concentrated by evaporation. There is little evidence that these processes have caused the solubility limit of calcium fluoride to be exceeded, or that the fluoride concentration has been thereby reduced. This is considered to be partly because there is less calcium available for dissolution in the alluvial sediments at this site and partly because the groundwater has a lesser proportion of high-fluoride bedrock groundwater than at Wonji/Shoa.

The investigations have shown that where there is a need to obtain groundwater supplies with low fluoride concentrations, the search should concentrate on strata where calcium concentrations in groundwater are high, as may result from the presence of calcite or gypsum. Although the resultant fluoride concentrations may not be reduced to completely acceptable levels, the groundwater would provide a more economical feed to a defluoridation treatment system than would otherwise be required.

The investigation has also demonstrated that by adopting appropriate quality assurance and control procedures it is feasible to conduct investigations of complex hydrochemical systems using field analytical equipment.

For completeness, it should be recorded that the project team recommended that future drinking and domestic water supplies at both estates should be obtained by treating water from the River Awash, piping it to the villages and camps from a central treatment works. The technical problems of treating water for the removal of fluoride will be the subject of a separate paper.

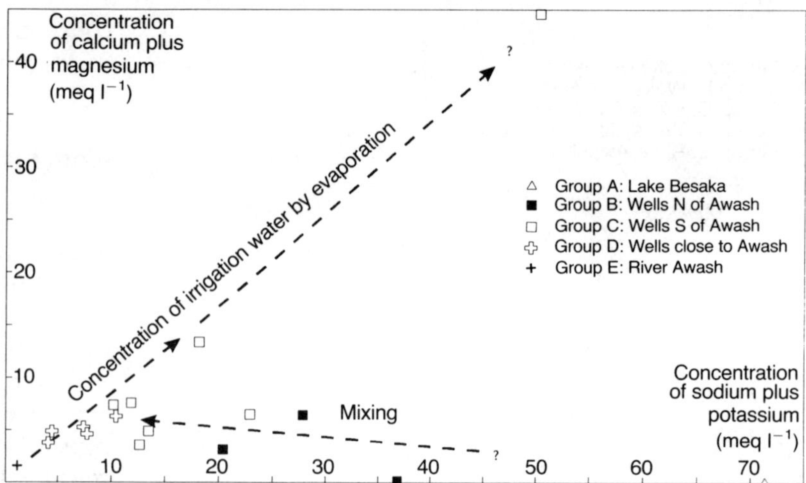

Figure 4.11 Variation in cation concentrations: Metahara data.

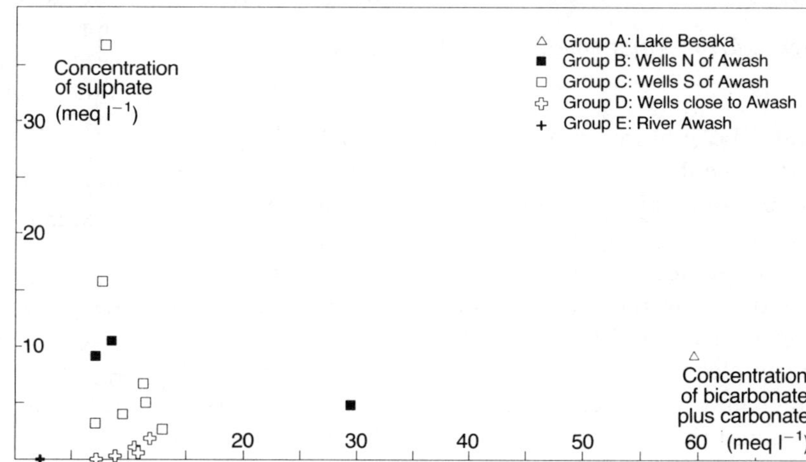

Figure 4.12 Variation in anion concentrations: Metahara data.

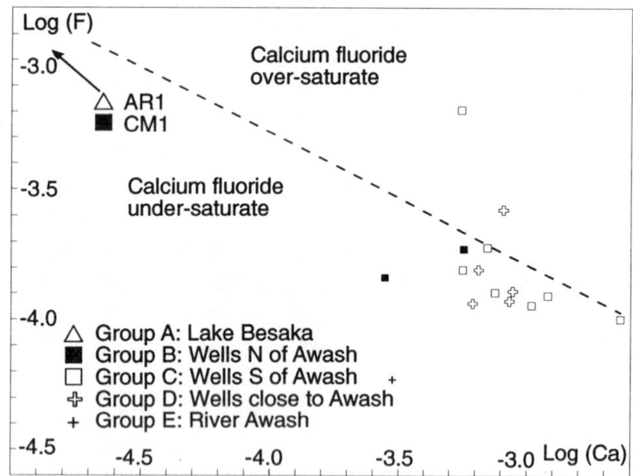

Figure 4.13 Variation in activities of calcium and fluoride: Metahara data.

Acknowledgments

The work described in this chapter was funded by the Ethiopian Sugar Corporation, which provided every necessary facility for the satisfactory completion of the investigation. The authors also wish to express their appreciation of the assistance and advice of many individual members of the engineering, medical, agricultural and administrative staff of the ESC at the Wonji/Shoa and Metahara Sugar Estates, and in Addis Ababa.

References

Hem, J.D., 1970. *Study and Interpretation of the Chemical Characteristics of Natural Water*. United States Geological Survey, Water Supply Paper 1473.

Kruseman, G.P. and de Ridder, N.A., 1970. Analysis and evaluation of pumping test data. *International Institute for Land Reclamation and Improvement, Bulletin 11*, Netherlands.

Mohr, P.A., 1961. *The Geology of Ethiopia*. University College of Addis Ababa Press, reprinted 1971.

Piper, A.M., 1944. A graphical procedure in the chemical interpretation of water analyses. *American Geophysical Union Transactions*, vol. 25, pp. 914–923.

Stumm, W. and Morgan J.J., 1981. *Aquatic Chemistry*. J. Wiley, New York, 2nd edn.

World Health Organization, 1984. *Guidelines for drinking Water Quality*. World Health Organisation.

5 Low-cost groundwater quality investigation methods – an example from the Bolivian Altiplano

P.L. Younger

Abstract It is feasible for Non-Governmental Organizations (NGOs) to develop an in-house capability in groundwater quality analysis, which will foster independence from expensive commercial laboratories. Simple low-cost analytical techniques are readily available for key parameters which determine the suitability of water for drinking and irrigation. The work of YUNTA, a Bolivian NGO specializing in water resources development, has benefited by application of these techniques during a hydrogeological survey. The results obtained have not only indicated the suitability of groundwaters for the above uses, but have also contributed to the understanding of recharge distribution and flow paths, thus aiding in the assessment of the total resource available for sustainable irrigation development.

Introduction

Development of reliable water supplies for rural communities in developing countries depends on the suitability of available water resources for human consumption and irrigation. For human consumption, the most important water quality parameters are the bacterial and viral purity; the surrogate parameter of total coliforms is the usual measure of such purity. For irrigation purposes, specific electrical conductance (or conductivity) and the Sodium Adsorption Ratio (SAR) are the most useful parameters. Conductivity gives an overall measure of salinity, while high values of SAR (in excess of $\sim$10) can lead to the dispersion of soil colloids, destroying the texture and drainage properties of the soil (Todd, 1980, p. 300). Given the definition of $SAR=(Na/\surd[(Ca+Mg)/2])$, where concentrations are in milli-equivalents per litre (meq l^{-1}) it is clear that a basic hydrochemical analysis must include determination of sodium, calcium and magnesium. Many other parameters are important for health (e.g. nitrate) and agriculture (e.g. boron), but the total coliforms, conductivity, Na^+, Ca^{2+} and Mg^{2+} may be regarded as the critical five determinants for most development studies.

As long as groundwater development is pursued at a very small scale (scattered dug wells with handpumps), there is little justification for determining further quality parameters. However, once groundwater development is expanded to include irrigation boreholes with motorized pumps, the question of regional groundwater resources becomes important. Sustainable groundwater development depends on effectively managing regional groundwater resources so that adverse environmental impacts are avoided. Management of regional groundwater resources depends upon a thorough understanding of both flow patterns and major-ion groundwater chemistry. If the cations which control SAR are already being determined, then only a few anions (Cl^-, SO_4^{2-} and HCO_3^-), temperature and pH need to be determined to allow a fairly complete assessment to be made of major-ion groundwater chemistry.

Groundwater Quality Edited by H. Nash and G.J.H. McCall. Published in 1994 by Chapman & Hall. ISBN 0 412 58620 7

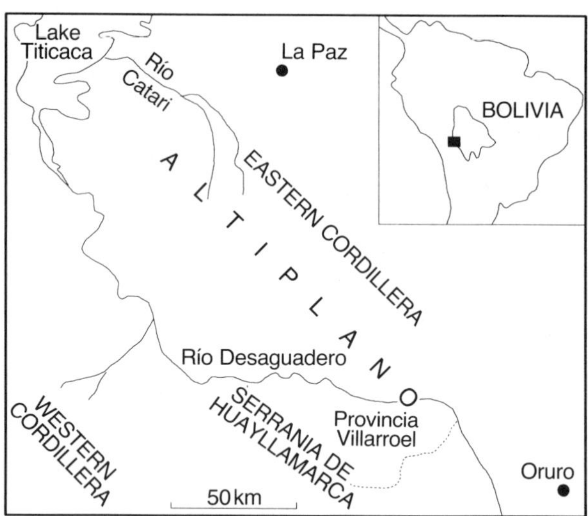

Figure 5.1 Location map of Provincia Villarroel, Altiplano, Bolivia.

In this chapter, recent developments in groundwater resource utilization by a Bolivian NGO are used to illustrate how thorough hydrogeochemical investigations can be undertaken (without recourse to commercial laboratories) by NGOs with restricted funding.

Previous hydrogeological work in the Bolivian Altiplano

Very little information exists on the hydrogeology and hydrogeochemistry of the Altiplano of Bolivia. The hydrogeology of the northernmost Altiplano, adjacent to Lake Titicaca (Figure 5.1), was investigated by the United Nations Development Programme (UNDP) and the Servicio Geológico de Bolivia (GEOBOL). This investigation involved drilling and test-pumping wells, sampling and analysis of groundwaters, definition of hydrochemical facies and calculations of water balances for sub-catchments (GEOBOL-UNDP, 1973).

Gumiel (1988) undertook a reconnaissance hydrogeological survey of an area of the Altiplano to the south of the area studied by GEOBOL-UNDP (1973). Field measurements of groundwater level, temperature and conductivity were made, along with full laboratory analyses of 115 samples. Some of this work was undertaken in the area described in this chapter, though the only wells sampled were traditional hand-dug holes, most of which are in shallow perched aquifers that are not part of the regional groundwater flow system (D. Cahuaya, Centro YUNTA, pers. comm., 1992). Subsequently, summaries of the work of Gumiel (1988) have been published in both Spanish (Quintanilla, Gumiel and Guyot, 1991) and French (Guyot and Gumiel, 1990). Further minor references to the hydrogeology of the Altiplano include Salm (1984), Aranyossy (1989) and Guyot *et al.* (1990).

Groundwater development in Provincia Villarroel

Figure 5.1 shows the location of Provincia Villarroel, which is the main work area for the Centro de Promoción y Cooperación YUNTA, an NGO with headquarters in La Paz. Provincia Villarroel covers some 2000 km^2 of the central Bolivian Altiplano. This is a complex Tertiary–Quaternary province of intermontane foreland basins, bounded to the west by the stratovolcanoes of the western Andean Cordillera and to the east by the folded and thrusted Palaeozoic sequence of the eastern Andean Cordillera (Richter, Ludington and Soria-Escalante, 1992). The western part of the Province is hilly, being underlain by a folded and thrusted Neogene sedimentary–volcanic sequence (Figure 5.2; Table 5.1). Sandstone units form thin aquifers in this area, and the dips ensure that they are mostly confined. The eastern half of the Province is remarkably flat, being underlain by a complex sequence of mainly lacustrine Quaternary sediments. Patterns of lacustrine sedimentation have resulted in most of the aquifers in this lowland zone being confined also, even though they may lie only 3–4 m below ground. Of particular interest in this lowland zone is the Pupusani Lineation (Younger, 1992; Figure 2), which is interpreted as an extensional reactivation of a southwestward-verging Miocene thrust front. The fault scarp operated as a focus for drainage during the Quaternary and sandy deposits of high permeability were preferentially stacked against the scarp, forming good aquifers located in the main modern recharge area of Villarroel.

Between 10 000 and 15 000 Aymaran people inhabit Provincia Villarroel, living mostly in scattered smallholdings; even the largest settlements (e.g. Mollebamba) do not exceed 40 families. Villarroel is effectively isolated from the surrounding areas of the Altiplano by the Río Desaguadero to the north, and by the rugged hills of the Serrania de Huayllamarca to the south and west (Figure 5.1). Because of this isolation, and the total lack of infrastructure, agricultural production is geared to local consumption rather than to supplying urban markets. Lying at around 3800 m above sea level, Provincia Villarroel is subject to extremes of heat and cold in the different seasons. Overall, the climate is semi-arid, with a marked wet season (October to March) and a long dry season (April to September) in which little or no rain falls. The mean annual rainfall over Villarroel is approximately 420 mm. Because of the elevation, solar radiation is intense and some 85% of the annual rainfall is lost to evaporation.

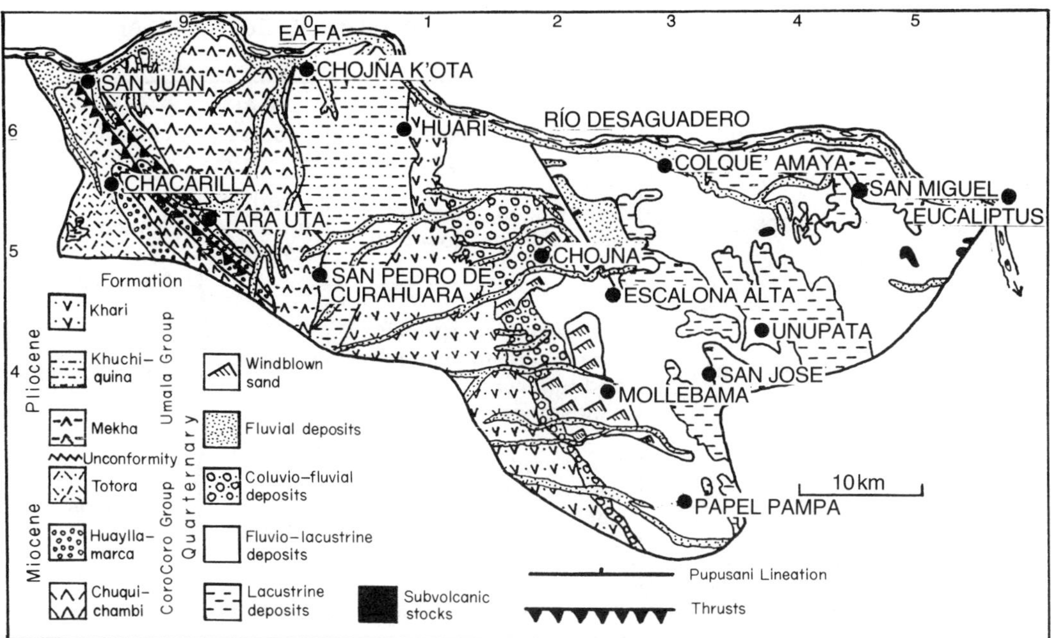

Figure 5.2 Geological map of Provincia Villarroel. Compilation partly based on published and unpublished maps of GEOBOL, and on aerial photography and field mapping by the author.

There are no perennial streams in the Province, with the sole exception of the Río Desaguadero which forms its northern boundary. Agricultural production in this environment is hindered by frequent droughts, frosts and floods.

After the exceptionally severe drought of 1983, YUNTA began to work with the local people on a programme of water resource development. With the exception of a few minor irrigation schemes along the Río Desaguadero, all of the development has depended on groundwater. By 1992, more than 90 wells with handpumps had been constructed in the Province, bringing dependable water to approximately 60% of the population for the first time. In the best traditions of participatory community water development (Kerr, 1989; Ball and Ball, 1991), YUNTA works only at the request of local communities who have themselves decided that they wish to change their water supply from a traditional hand-dug, unlined water hole to a properly constructed well with sanitary seal and handpump. The role of YUNTA is thus as a facilitator for the transfer of technology; the bulk of the labour, and at least some of the money for materials, come from the communities themselves. While hand-dug and hand-augered wells can be constructed in much of the lowland zone of Villarroel, a mechanical drilling rig must be used in areas with a deeper water table and harder rocks in the intermediate and upland zones of the Province. Even where a mechanical rig is used, community members work as driller's labourers. After a well has been completed, *dirigentes* see that it is maintained. *Dirigentes* are volunteers from within the local community, selected for their enthusiasm, who are given some relevant training and a small monthly honorarium. They encourage community participation in routine pump maintenance, and decide when to call for specialist help from YUNTA.

In the last few years, the scope of YUNTA's work in Villarroel has widened. It now includes programmes of health education and *concienciación* (consciousness-raising), in addition to improvements in agricultural production through the use of greenhouses and small-scale irrigation systems. The irrigation schemes require the construction of wells fitted with motorized pumps, each yielding up to 1200 $m^3\ d^{-1}$. By 1992, five such wells had been constructed. Since the discharge of an irrigation well at a single site exceeds the total combined discharge of all of the 90 manually pumped wells distributed throughout the Province, YUNTA has begun to consider the question of long-term sustainability of current and projected groundwater development in the Province. To address this issue, a hydrogeological study of Villarroel was initiated (Younger, 1993), and the author was enlisted to coordinate this study, providing training in hydrogeology for the permanent staff members of YUNTA in the process. This study has included a reconnaissance study of groundwater chemistry throughout the Province. For reasons of cost, speed of analysis, and independence from external support, it was felt that the development of an in-house capability in hydrochemical analysis was an important goal. To this end low-cost hydrochemical

Table 5.1 Tertiary stratigraphy of the Central Altiplano, Provincia Villarroel, Bolivia (K–Ar dates collated from literature cited by Richter *et al.* 1992)

Series	Group	Formation	Thickness (m)	Lithologies	K-Ar dates	Lateral equivalents
PLIOCENE	UMALA	Khari	3 000 m	Sandstones, mudstones and acidic ashflow tuffs	Overlying tuff: 2.2 Ma	Taraco Formation Mauri Formation Pérez Formation Remedios Formation
		Khuchiquiña	500–1 000 m	Laminated mudstones, rare thin distal acid tuffs and tuffaceous sandstones		
		Mekha	700–1 800 m	Laminated mudstones with gypsum beds, acid tuffs and rare sandstones	Basal tuff: 5.5 Ma	
		Unconformity				
MIOCENE	COROCORO	Totora	4 000 m	Muddy sandstones and sandy mudstones with acid tuffs	Ulloma tuff: 9.1 Ma	Caquiaviri Formation Kollu-Kollu Formation
		Huayllamarca	2 000–5 000 m	Arkosic sandstones with lenses of cupriferous mudstone	Tuff: 16.6 Ma Intrusion: 16.7 Ma Tuff: 18.4 Ma	Kollu-Kollu Formation Coniri Formation
		Chuquichambi	500 m	Gypsum and gypsiferous mudstones and sandstones		Coniri Formation? Tihuanacu Formation?

analysis methods were selected which had the potential to both yield accurate results and yet be sufficiently simple that non-specialist staff could learn to use the methods with reproducible reliability. Independent sampling and analysis undertaken in Villarroel by external organizations (SENAMHI, the national Bolivian meteorological and hydrological organization and CNRS, the French National Scientific Research Centre) provided an ideal opportunity to compare the results of YUNTA's work with those from commercial laboratories.

Low-cost hydrochemical analysis methods

Hutton (1983) compiled a guide to simple yet accurate analytical methods considered to have potential for use in developing countries, and many of the techniques mentioned below are detailed in that guide. In the hydrogeological study of Villarroel, a selection of such methods was chosen to cover the range of determinants required to assess the suitability of local groundwaters for human consumption, irrigation and hydrogeological modelling. Table 5.2 lists these determinants, along with an indication of their utility for these different purposes, and the low-cost analysis methods required.

Only six items of portable equipment are required for the analysis of all the determinants listed in Table 5.2. These are:

(1) conductivity meter;
(2) pH–Eh–temperature meter;
(3) multiple tube incubator (coliform determinations);
(4) portable spectrophotometer (DREL 5);
(5) hand-held digital titrator; and
(6) sodium ion-specific electrode (ISE) meter.

The choice of particular brands was based upon extensive cost–benefit analyses of alternative systems. The main considerations were:

- accuracy of results: e.g. titration for alkalinity yields far more accurate results than determination using a

Table 5.2 Groundwater quality parameters, their practical use and analytical techniques used in their determination in this study

Groundwater quality parameter (units)	Practical use[1]	Low-cost analytical technique
pH	H, G	Palintest pH meter
Eh (mV)	G	Palintest pH meter
Temperature (°C)	G	Palintest pH meter
Conductivity ($\mu S\ cm^{-1}$)	H, I, G	Palintest conductivity meter
Total hardness ($mg\ l^{-1}$ as $CaCO_3$)	H, I	Titration with EDTA
Alkalinity ($mg\ l^{-1}$ as $CaCO_3$)	G	Titration with concentrated H_2SO_4
Sodium ($mg\ l^{-1}\ Na^+$)	H, I, G	HACH ISE meter
Calcium ($mg\ l^{-1}\ Ca^{2+}$)	I, G	Titration with EDTA
Magnesium ($mg\ l^{-1}\ Mg^{2+}$)	I, G	Titration with EDTA
Sulphate ($mg\ l^{-1}\ SO_4^{2-}$)	H, G	Spectrophotometer
Chloride ($mg\ l^{-1}\ Cl^-$)	I, G	Titration with $AgNO_3$
Nitrate ($mg\ l^{-1}\ NO_3$–N)	H, G	Spectrophotometer
Bicarbonate ($mg\ l^{-1}\ HCO_3^-$)	G	Titration with concentrated H_2SO_4
Total coliforms	H	HACH multiple-tube incubator with MUG growth medium

[1] H = used to help determine suitability of water for human consumption; I = used to determine suitability of water for irrigation; G = used in hydrogeochemical interpretations.

colorimetric comparator, and titration was therefore selected;

- ease of use and calibration: this obviously has implications for the training of technicians to use and maintain the equipment with no outside help;
- costs of equipment and reactants: e.g. while battery-operated incubators using the membrane filtration method (CIIR, 1989) allow more flexibility in field schedules than the mains-operated multiple tube incubator manufactured by HACH, the latter is far less expensive to obtain and maintain than the former; and
- lack of alternatives: e.g. if recourse to a commercial laboratory offering atomic absorption spectrophotometry is to be avoided, the only feasible technique for sodium determinations is by use of a portable ISE meter.

Full details of the instruments mentioned above may be obtained from the manufacturers (or their agents), whose addresses are listed in the Acknowledgments at the end of this chapter.

To give an indication of costs, all the necessary equipment (with sufficient reactants for approximately 100 analyses) was purchased for less than US$4000 (1991 prices). The most expensive items were the ISE meter and the spectrophotometer, each costing about $1200, including reactants for more than 100 determinations. The pH and conductivity meters cost about $300 each, the incubator about $750, the digital titrator only $75, and the reactants approximately $350.

The hardware will remain serviceable for many hundreds of analyses, but the costs of even the initial 100 analyses, including hardware costs, amounts to less than a third of what a commercial laboratory would charge for the same work. These figures indicate that it is certainly cost-effective to establish an independent analysis capability within an NGO.

Field practices

The difficulties of obtaining representative samples of groundwater from boreholes have been discussed extensively in the North American literature over the last decade (for a review, see Barcelona *et al.*, 1990, pp. 108–127). It is generally acknowledged that certain unstable parameters must be determined on-site (and preferably in a flow-through cell; Barcelona *et al.*, 1990, pp. 118–119) if the results are to have any geochemical significance. The basic unstable parameters are temperature, pH and alkalinity, though Eh must also be measured on-site if it is to be measured at all. In addition, it is customary to measure conductivity on-site, even though it is not an unstable parameter, since it is an easy measurement to make, and gives a measure of the overall dissolved solids content. This information is a useful guide during well purging and for the alkalinity titration. Thus conductivity, pH, Eh, temperature and alkalinity were all measured on-site in Villarroel, the first four parameters in a flow-through bucket arrangement outlined below, and alkalinity by titration (Figure 5.3).

Perhaps the key difficulty in groundwater sampling lies in effective 'purging' of standing water from the well casing and gravel pack annulus so that the water which is finally sampled is unmodified by reaction

Figure 5.3 David Cahuaya, Chief Water Resources Technician of Centro YUNTA, performing an alkalinity titration with a HACH hand-held digital titrator during a pumping test of a new irrigation well at Chapi-Kkollu (UTM Grid Ref. FA 123393).

Figure 5.4 Arrangement for continuous monitoring of pH, temperature and conductivity during pre-sampling purging of a hand-pump well at Anat'ola (UTM Grid Ref. FA 267318). Probes are in a bucket receiving continuous discharge from the well.

within the well casing. While there is much debate over how best to purge a well for sampling, it is general practice to pump between 2 and 10 times the volume of water standing in the casing before taking the final sample. Best practice would involve the continuous monitoring of pH, temperature and conductivity in an on-line flow-through cell during the purging since the stabilization of these parameters may provide an indication of when true groundwater is produced. Using existing handpumps, it is generally not feasible to pump even two times the well casing volume before exhausting the strength and goodwill of both visiting technicians and co-operating villagers. In such cases in Villarroel, therefore, well purging was approached by placing a small bucket containing the temperature, pH and conductivity probes (to act as a crude flow-through cell) immediately beneath the handpump (Figure 5.4), and monitoring these parameters for several minutes of continuous pumping until they stabilized.

Results

A preliminary hydrogeological reconnaissance of Villarroel was undertaken from 17 to 20 March, 1992. It involved water level measurements, purging (as outlined above) and measurement of pH, Eh, conductivity and temperature in 50 wells with handpumps throughout the Province. The groundwater flow patterns (Figure 5.5) and conductivity distribution (Figure 5.6) yielded by this reconnaissance formed the basis for planning a series of eight pumping tests. The tests were planned principally to determine aquifer parameters and also to obtain representative samples of groundwater from different lithostratigraphic units. Pumping tests were conducted in July 1992 using a 4″ electric submersible pump run from a monophase portable generator. Discharges of about 120 $m^3\ d^{-1}$ were obtained, except at two sites with irrigation wells where discharges of 864 $m^3\ d^{-1}$ and 1037 $m^3\ d^{-1}$ were attained.

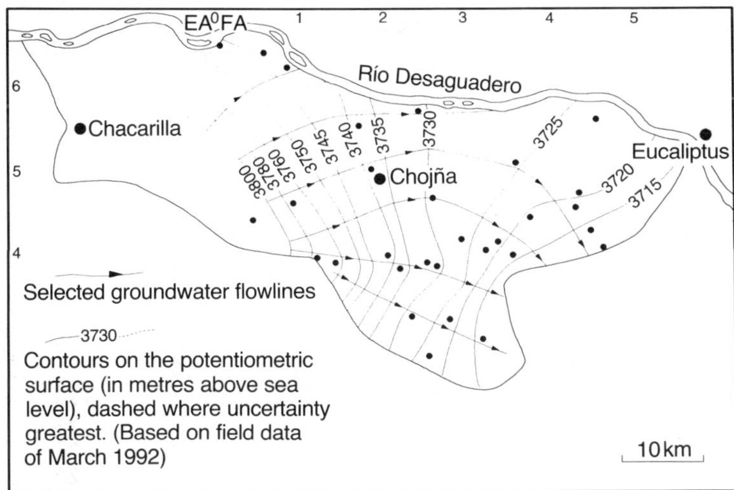

Figure 5.5 Groundwater head contours for Villarroel (with deduced groundwater flow directions), based on field measurements at the start of the dry season in March 1992.

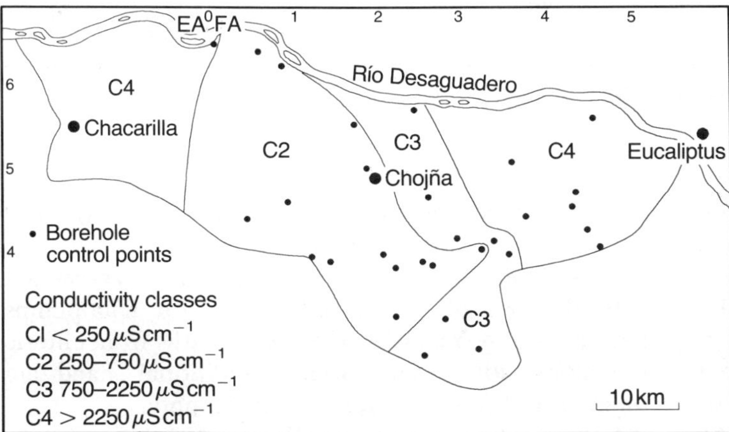

Figure 5.6 Distribution of conductivity classes in Villarroel.

Hydrochemical analyses of the samples collected during this work, and analysed using the simple low-cost methods outlined above, are presented in Table 5.3.

Data quality

A commercial laboratory in La Paz (SENAMHI) made full chemical analyses on water from some of the wells sampled in the YUNTA study. Comparison of these parallel data sets provides insight into the quality of data obtained using low-cost methods. The YUNTA field values for pH, temperature and conductivity are all less than those from the commercial laboratory, for SENAMHI did not determine pH and temperature in the field, but later in the laboratory. In the case of conductivity, the SENAMHI conductivity meter gave results on average 24% greater than those given by the YUNTA meter. Calibration tests on the two meters revealed that the SENAMHI meter was in error. The meter in question can only be calibrated by the manufacturers, and this would mean sending the meter to the USA; meters which can be calibrated easily by the user (such as that used by YUNTA) are clearly preferable.

According to the principle of electro-neutrality, bulk groundwater samples ought to yield an ionic balance (B_I) close to zero. The ionic balance can be written:

$$B_I = \frac{\Sigma\,(\text{Cations}) - \Sigma\,(\text{Anions}) \times 100\%}{\Sigma\,(\text{Cations} + \text{Anions})}$$

where Cations = concentration of each cation in meq l^{-1}, and Anions = concentration of each anion in meq l^{-1}.

Calculation of B_I for each analysis is a good check on data quality. If B_I is large then the analysis either contains inaccurate results, or some ion present at major concentrations has not been determined. The tolerance of B_I values for a given analysis depends to some extent on the use to which the data are to be put. For strict thermodynamic calculations, a B_I less than 5% is often preferred; for more general geochemical purposes (facies definitions, etc.) B_I up to 10% might be tolerated. However, when B_I begins to exceed 15% there is even doubt about the order-of-magnitude concentration of major ions, and it may be better to discard the analysis than to draw misleading conclusions from it.

The YUNTA data (Table 5.3) have a mean B_I of 9.1%, excluding the outlying data point for Yampara,

Table 5.3 Hydrochemical analysis results for eight wells in Villarroel, determined by YUNTA technicians in July 1992

Well No[1]	8-2	20-2	20-5	27-1	27-2	44-1	45-1	68-4
Grid	123	085	078	447	462	344	315	243
Reference	393	626	617	431	412	396	400	389
(FA)								
pH[2]	7.58	7.84	7.80	7.52	7.55	7.76	7.68	7.92
Eh (mV)[2]	+140	+30	+119	+43	−7	+90	+90	+107
Temperature (°C)[2]	15.2	15.0	13.4	14.1	14.0	13.8	14.4	13.6
Conductivity ($\mu S\ cm^{-1}$)[2]	647	561	586	4810	5060	1087	673	423
Total hardness ($mg\ l^{-1}\ CaCO_3$)	276	186	386	1973	940	393	306	500
Alkalinity ($mg\ l^{-1}\ CaCO_3$)[2]	302	189	135	221	203	144	114	209
Sodium ($mg\ l^{-1}\ Na^+$)	61	52	19	600	740	57	55	21
Calcium ($mg\ l^{-1}\ Ca^{2+}$)	64	70	113	297	251	160	90	138
Magnesium ($mg\ l^{-1}\ Mg^{2+}$)	28	2.9	25	295	75	bdl[3]	19	37
Sulphate ($mg\ l^{-1}\ SO_4^{2-}$)	140	53	120	400	500	160	98	32
Chloride ($mg\ l^{-1}\ Cl^-$)	28	72	21	1440	1493	205	91	22
Nitrate ($mg\ l^{-1}\ NO_3$–N)	3.7	2.9	3.3	2.2	0.7	1.6	1.7	4.2
Bicarbonate ($mg\ l^{-1}\ HCO_3^-$)	362	227	162	265	244	173	137	251
Ionic balance B_I %	8.3	3.3	19.5	10.5	5.2	6.6	10.5	33.6

[1] Well numbers in the YUNTA well inventory, available for inspection on request.
[2] Analysed in the field.
[3] bdl = below detection level

or 12.2% if the outlier is included. The range is from 3.3 to 33.6%. By comparison, new data for Villarroel produced by CNRS (A. Ribstein, pers. comm., 1992) give a B_I range of 0.1–47% with a mean of 11.34%. The YUNTA data are clearly on a par with these data from a commercial laboratory. The SENAMHI data have a B_I range 0.02–3.96%, with an approximate mean of 1.44%, and they are clearly superior to both the YUNTA and CNRS results. Nevertheless, the fact that the B_I values for the YUNTA data, obtained using simple low-cost methods, are comparable with those obtained from commercial laboratories is very encouraging. The approach outlined in this chapter would appear to be justified by these comparisons alone.

Data interpretation for Villarroel

In mapping and evaluating the groundwater chemistry of the Province, both the YUNTA data (Table 5.3) and the other analytical results have been pooled. The significance of the deduced patterns will be discussed from two viewpoints; immediate water-use issues, and hydrogeochemical interpretations.

Implications for water use

In terms of chemical quality, only the brackish groundwater found at Chojña K'ota (UTM Grid Ref. FA 001661) would be generally considered to be unsuitable for human or animal consumption in the long term. However the local population is accustomed to the water and does not appear to suffer any acute health problems attributable to the water (though a question mark over chronic health problems remains). The limited bacteriological information available to date suggests that there are no health problems associated with the use of the sealed wells with handpumps constructed by YUNTA. However deaths from enteric infections still occur where traditional hand-dug unlined wells are used for drinking supply.

For irrigation purposes, a classification scheme presented in detail by Todd (1980, pp. 296–304) has been adopted for use in this study. The risk of soil salinization by using groundwater of a given chemical composition is assessed by reference to sodium adsorption ratio (SAR) and conductivity. The classification scheme is summarized in Table 5.4. The map of conductivity classes (Figure 5.6) indicates that groundwater in approximately 60% of the Province would be classified as having a 'high' or 'very high salinity risk' (i.e. classes C3 and C4, Table 5.4 and Figure 5.6). Care must be taken to ensure that irrigation schemes using these waters are carefully planned to avoid dangerously elevated soil water concentrations; techniques to minimize evapotranspiration and ensure adequate soil drainage will be necessary. In terms of sodium adsorption ratio (SAR), a map for the entire province is not presented here, since only localized districts (around Laruta, UTM Grid Ref. FA 447431, and Unupata, FA 353507) in the eastern half of the Province have groundwaters with SARs in excess of 10 (Table 5.4). A few of the wells in the Unupata and Laruta area gave SARs in the 'high sodium risk' category. In these cases, irrigation schemes using groundwater would lead to irreparable soil deterioration, and any development would be unsustainable in the long term. The general

Table 5.4 Classification scheme of waters intended for irrigation

Conductivity class	Range of values ($\mu S\ cm^{-1}$)	Salinity hazard	Sodium Adsorption Ratio class	Range of values	Sodium (alkali) hazard
C1	< 250	low	S1	< 10	low
C2	250–750	medium	S2	10–18	medium
C3	750–2250	high	S3	18–26	high
C4	> 2250	very high	S4	> 26	very high

(Adapted from Todd, 1980, p. 304).

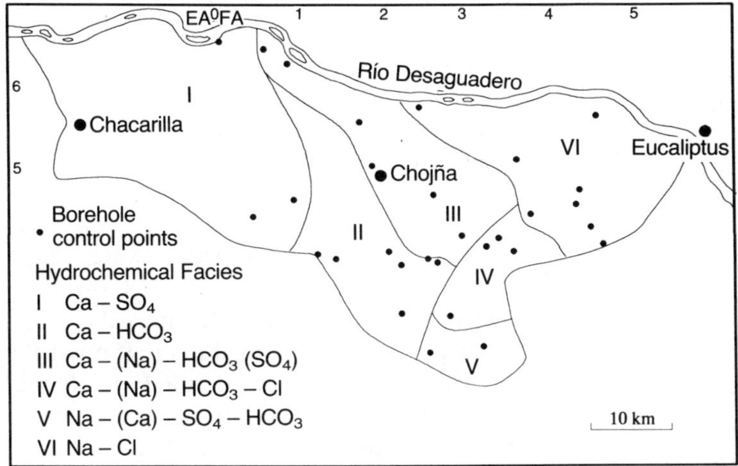

Figure 5.7 Hydrochemical facies in Villarroel.

prognosis is that groundwater development for irrigation purposes will be sustainable in most of the central area of the Province (areas with C2 and perhaps C3 waters on Figure 5.6), provided that adequate attention is paid to ensuring efficient soil drainage.

Hydrogeochemical interpretations

Without an undue prolongation of the discussion, only an outline assessment of regional hydrogeochemistry can be given here. Two complementary hydrogeochemical approaches have been used in this study – an assessment of aqueous speciation and mineral equilibria; and an assessment of hydrochemical facies. The first task was accomplished by entering all the data shown in Table 5.3 into the PC implementation of the well-known USGS WATEQF program (Truesdell and Jones, 1974). The significant results are that all the groundwater samples listed in Table 5.3 were found to be saturated with respect to calcite/aragonite ($CaCO_3$), but undersaturated with respect to halite (NaCl), gypsum ($CaSO_4.5H_2O$) and anhydrite ($CaSO_4$). The development of calcite saturation in recharging groundwaters will be discussed below with respect to different hydrochemical facies zones; the important point to note here is that calcite thermodynamics are unlikely to be controlling the overall geochemical evolution of the system at a distance from the recharge areas.

Comparing the concentrations of the various ions, expressed in meq l^{-1}, it is possible to define the dominant ionic species in each water sample. On this basis, hydrochemical 'facies' (Freeze and Cherry, 1979, pp. 250–254) for the entire groundwater system in Villarroel have been defined, and they are shown on Figure 5.7. This map must be interpreted in conjunction with the geological, piezometric and conductivity maps of the Province (Figures 5.2, 5.5 and 5.6). Three basic end-member hydrochemical facies can be readily identified, namely the Facies I (Ca–SO_4), II (Ca–HCO_3) and VI (Na–Cl). The remaining facies (III, IV and V) appear to represent 'mixtures' in varying proportions between the three end-members. Controls on the development of these distinct facies are discussed briefly below.

Facies I (Ca–SO_4) occurs in the western upland zone of the Province and reflects the dissolution of gypsum found in the Pliocene Mekha Formation and Miocene Chuquichambi Formation (Table 5.1) which crop out in that zone (Figure 5.2).

The second facies (Facies II; Ca–HCO_3) is of a type normally associated with relatively young recharge waters in many aquifer settings. In this case, it probably reflects the dissolution of calcite cements which occur in sandstones of the uppermost Khari Formation.

Facies III and V are fairly similar (they differ in the relative predominance of sodium and sulphate), and possibly share a common origin. The elevated sulphate concentrations are the most interesting feature of both facies. One possible interpretation is that recharge occurs in the areas characterized by Facies III and V by seasonal infiltration from ephemeral streams which drain the gypsiferous uplands to the west. Alternatively, the sulphate may be derived from dissolution of disseminated sulphate evaporite minerals within the local Quaternary deposits. There are geological grounds for supposing that such minerals may indeed be present. The sediments of the relevant zones are of fluvio-deltaic lake-marginal associations; in the Facies III area at least, the accumulation of these sediments was influenced by the Pupusani Lineation (Younger, 1992). The lowland zone to the east was lacustrine during the Quaternary cold stages (Clapperton, 1983). The modern saline lakes and salt pans of the southern Altiplano may be regarded as both remnants of, and partial modern analogues for, this former lake. The modern fluvio-deltaic sediments which fringe the Salar de Uyuni have been found to contain abundant microcrystalline gypsum (Risacher and Fritz, 1991, p. 214), whereas halite (NaCl) deposits are restricted to the central lake area. This zonal pattern of an outer gypsiferous belt and an inner halitic pan is repeated in saline lakes throughout the Andes, reflecting the higher solubility of halite compared to gypsum (Ericksen and Salas, 1989). Applying these analogues to the Quaternary of Villarroel, there is every reason to suppose that a gypsiferous belt might exist in the western margin of the lowland zone where groundwater is of Facies III and V. Because of the common ion effect, release of calcium by dissolution of sulphate minerals can be expected to result in calcite saturation in a water with an independent source of HCO_3 (such as dissolved atmospheric or soil zone CO_2). The calcite saturation of these waters is therefore not surprising.

Facies IV (Ca–(Na)–HCO_3–Cl) is a hybrid between Facies II (Ca–HCO_3) and VI (Na–Cl). The presence of a plume of low conductivity Facies II water intruding into the Facies IV zone around San José (UTM Grid Ref. FA 300400; compare Figures 5.6 and 5.7) strongly suggests that Facies IV is indeed a straightforward mixture of Facies II and VI. The mixing plume coincides with an extensive body of highly permeable sand deposits presumed to be associated with a buried southeastward extension of the Pupusani Lineation (Figure 5.2). Preferential groundwater flow through these high permeability sediments (further suggested by the piezometric contour pattern in this vicinity; Figure 5.5), would appear to have driven a wedge of Ca–HCO_3 water between the sulphate-bearing waters of Facies III and V, which might otherwise present a continuous facies belt. Mixing of this intruding water with the Na–Cl waters of Facies VI could then account for the evolution of Facies IV.

There are three hypotheses for the origin of Facies VI (Na–Cl):

(1) Applying the saline lake analogies outlined above, it is possible that the Na–Cl is derived from disseminated halite in the clastic lacustrine silts and clays of the Facies VI area.
(2) Periodic flooding of the lowland zone by the Río Desaguadero might lead to the infiltration of Na–Cl waters which have been further concentrated by evaporation (the Río Desaguadero has a conductivity of about 2300 μS cm^{-1}, whereas some of the Na–Cl waters of the Facies VI area reach 4800 μS cm^{-1}). This theory was first proffered by Quintanilla *et al.* (1991). However, residents of Villarroel do not support the suggestion of these authors that flooding is an annual event; widespread flooding generally occurs once a decade or less.
(3) Seasonal subsurface ingress of Na–Cl from the Río Desaguadero may occur when groundwater levels are low. The long-term pattern of stream–aquifer interaction is flow of groundwater into the river; however temporary reversal of this situation at the end of the dry season cannot be ruled out. Nonetheless, given that baseflow can be demonstrated by hydrograph separation, there is a danger of circular argument in pointing out the close similarity in hydrochemistry between the river and groundwater from wells near the river (e.g. Laymini; UTM Grid Ref. FA 233572). It is hoped that ongoing research on the distribution and movement of natural isotopes throughout the system might illustrate which processes are more important (A. Ribstein, CNRS, pers. comm., 1992).

These facies patterns and hydrogeochemical processes represent important constraints on the conceptualization of system behaviour. This has important consequences for the formulation of accurate conceptual and mathematical models to predict long-term system responses to various groundwater development alternatives for irrigation.

Conclusions

It is perfectly feasible for NGOs to become equipped to perform thorough hydrochemical studies without recourse to commercial laboratories. The results obtained using the methods outlined in this chapter are sufficiently accurate to allow judgments to be made on the suitability of water for human consumption and for agricultural use. Furthermore, major-ion hydrogeochemistry can be investigated in the same manner, shedding considerable light on flow and solute transport processes. This information is useful in formulating accurate predictive models for groundwater resource evaluation.

Acknowledgments

The author is most grateful to David Cahuaya, Mamerto Pérez and other colleagues at YUNTA (Casilla 14529, La Paz) for their support in the field and the office, and for their encouragement to publish the results. A contribution to subsistence from UNAIS is acknowledged.

The Hartlepools Water Company, UK, kindly donated the Palintest conductivity meter, and generous discounts for the purchase of equipment for YUNTA were granted by the following suppliers, which can provide technical details on all the equipment mentioned in the text:

(1) Palintest Ltd, Palintest House, Kingsway, Team Valley, Gateshead NE11 0NS, UK
(2) CAMLAB Ltd (for HACH equipment), Nuffield Road, Cambridge CB4 1TH, UK
(3) Rugged and simple groundwater sampling equipment was also donated by Waterra (UK) Ltd, Marlow House, 310 Haslucks Green Rd, Shirley, Solihull B90 2NE, UK.

I am grateful to Anne Coudrain-Ribstein of CNRS for providing helpful comments and sharing data. Thanks also to Trevor Elliot of the Water Resource Systems Research Unit (WRSRU) at Newcastle, who did the WATEQ analyses and contributed significantly to the discussion of hydrogeochemistry. Suzanne McLean of the WRSRU drafted the figures.

References

Aranyossy, J-F., 1989. Quelques exemples pratiques d'application des isotopes de l'environnement aux études hydrogéologiques. *Hydrogologie*, 1990 (3), 159–166.

Ball, C. and Ball, M., 1991. *Water Supplies for Rural Communities*. Intermediate Technology Publications, London. 56 pp.

Barcelona, M., Wehrmann, A., Keely, J.F. and Pettyjohn, W.A., 1990. *Contamination of Groundwater: Prevention, Assessment, Restoration*. Noyes Data Corporation, New Jersey. 213 pp.

CIIR, 1989. *Nicaragua: Testing the Water – From Village Wells to National Plan*. Catholic Institute for International Relations, London. 60 pp.

Clapperton, C.M., 1983. The glaciation of the Andes. *Quaternary Science Reviews*, 2, 83–155.

Ericksen, G.E. and Salas, R., 1989. Geology and resources of salars in the central Andes. In: G.E. Ericksen, M.T. Cañas Pinochet and J.A. Reinemund (eds), *Geology of the Andes and its Relation to Hydrocarbon and Mineral Resources*. Houston, Texas. Circum-Pacific Council for Energy and Mineral Resources, Earth Science Series 11, 151–164.

Freeze, R.A. and Cherry, J.A., 1979. *Groundwater*. Prentice- Hall, New Jersey. 604 pp.

GEOBOL-UNDP, 1973. *Los recursos de agua del Altiplano norte y del Area de Oruro. Proyecto de desarrollo de los recursos de aguas subterraneas en el Altiplano. (The water resources of the northern Altiplano and of the Oruro area. Project of Groundwater Resources Development in the Altiplano)*. Servicio Geológico de Bolivia, United Nations Development Programme, La Paz. 215 pp. + appendices. (In Spanish.)

Gumiel, D., 1988. *Prospección hidrogeológica del àrea altiplanica del PMPR II*. CEE-CORDEPAZ, La Paz, Bolivia. 92 pp.

Guyot, J.L., and Gumiel, D., 1990. Premières données sur l'hydrogéologie et l'hydrogéochimie du Nord de l'Altiplano bolivien. *Hydrogéologie* 1989 (3), 159–164.

Guyot, J.L., Roche, M.A., Noriega, L., *et al.*, 1990. Salinities and sediment transport in the Bolivian highlands. *Journal of Hydrology*, 113, 147–162.

Hutton, L.G., 1983. *Field Testing of Water in Developing Countries*. Water Research Centre, Medmenham, UK. 125 pp.

Kerr, C. (Editor), 1989. *Community Water Development*. Intermediate Technology Publications, London. 279 pp.

Quintanilla, J., Gumiel, D., and Guyot, J.L., 1991. Evaluación preliminar de la hidrogeología e hidrogeoquímica del norte del Altiplano Boliviano. *Revista Boliviana de Química*, 10, 59–64.

Richter, D.H., Ludington, S. and Soria-Escalante, E., 1992. Geologic Setting. In US Geological Survey and Servicio Geológico de Bolivia, Geology and Mineral Resources of the Altiplano and Cordillera Occidental, Bolivia. *US Geological Survey Bulletin* 1975, 14–24.

Risacher, F. and Fritz, B., 1991. Quaternary geochemical evolution of the salars of Uyuni and Coipasa, Central Altiplano, Bolivia. *Chemical Geology*, 90, 211–231.

Salm, H.R., 1984. Algunas consideraciones sobre el uso de agua subterranea en una comunidad típica del Altiplano central de Bolivia. *Ecología en Bolivia*, 5, 29–37.

Todd, D.K., 1980. *Groundwater Hydrology*. (Second edition). Wiley, New York, 535 pp.

Truesdell, A.H., and Jones, B.F., 1974. WATEQ: A computer program for calculating chemical equilibria in natural waters. *Journal of Research of the US Geological Survey*, 2, 233–248.

Younger, P.L., 1992. Quaternary geology and hydrogeology: The value of an interdisciplinary approach. *Geoscientist*, 2 (5), 24–27.

Younger, P.L., 1993. Simple generalised methods for estimating aquifer storage parameters. *Quarterly Journal of Engineering Geology*, 26, 127–135.

Part Two

Groundwater and Salinity

J.W. Lloyd

Introduction

Groundwater with total dissolved solid concentrations that are too high for potable, agricultural or normal industrial uses are more prevalent than those available for use. Such waters are generally classified as saline, although there is a range of concentration classifications covering brackish and saline groundwaters and brines. To some extent the term 'saline' is like the term 'aquifer' – it is relative and dependent on the potential end-use or particular hydrogeological environment being studied.

Irrespective of classification, the recognition of groundwaters with total dissolved solids (TDS) in excess of say 1000 mg l^{-1} in an aquifer where fresher groundwater occurs and is to be exploited, gives cause for concern and will normally be vigorously investigated and hydrogeologically assessed.

In undertaking an assessment, it is obviously important to recognize the various salinity environments that can exist and the governing hydrogeological and hydrochemical processes. Some of the more important of these are discussed below.

Natural salinity environments

Although much of the hydrogeological literature relates to **saline intrusion**, which is seen, often erroneously, as sea-water entering an aquifer, by far the bulk of saline groundwater occurs in those parts of the system that have not been flushed by relatively modern recharge. The bulk of saline groundwater therefore occurs at depths below fresh groundwater or lateral to fresh groundwater, often in low flow zones.

Three major natural factors can control groundwater salinity:

(1) the geological distribution and hydraulic properties; and the rock-mass chemistry;
(2) the driving head controls; and
(3) the hydrogeological evolution and long-term transient conditions.

In semi-arid or arid areas a fourth natural salinity factor can be recharge. In a very mature landscape, as for example in central Australia, recharge through run-off can acquire a sodium chloride signature and rapidly become saline, at least in excess of 1000 mg l^{-1}.

Inland, in most environments of reasonable sized landmass, saline groundwaters predominate at depth. As most sedimentary rocks are of marine origin, the initial saline influence of the matrix and groundwater is marine, although from purely *in situ* water–rock reaction processes, the groundwater chemistry can change radically with time. Much groundwater salinity can also be attributed to non-marine deposits with ancient sabkha environments which provide readily reactive minerals. An additional source of salinity is hydrothermal convective cells in geologically recent intrusives which can move saline groundwaters from depth within igneous rocks.

In most hydrogeological systems the driving heads controlling the distribution of saline groundwaters in balance with fresh waters are those related to modern recharge. In large arid basins it is possible that the present-day balance is in fact a transient condition

Groundwater Quality Edited by H. Nash and G.J.H. McCall. Published in 1994 by Chapman & Hall. ISBN 0 412 58620 7

reflecting past recharge events. However, other head controls can be important, such as the hydrothermal convective cell noted above, and, more importantly, the density of the saline groundwater and especially of any brine that is present. For any study of major groundwater salinity, the density control, as reflected by the environmental head rather than the equivalent freshwater head, has to be understood for interpretation of the vertical gradient.

The supposition of transient conditions at the scale of geological time is often invoked in groundwater studies when head controls are not readily explained, such as in the case of regional arid basin groundwater gradients. In saline groundwater studies possible transient effects bear more consideration, since salinity in an aquifer can often be derived through natural upward and downward leakage from juxtaposed non-aquifer materials. In such materials, because of their low permeability, head adjustments can be extremely slow and salinity redistribution even slower.

In many situations, such as coastal or small island environments, the relationship between fresh and saline groundwater is conceptually straightforward. Groundwater derived from modern sea-water **interfaces** with fresh groundwater. However, this relationship can be complicated in coastal areas (as opposed to most small islands) by lack of fresh groundwater flushing during the more recent hydrogeological evolution. Good examples of this are seen in such diverse geological environments as the Netherlands and Java (Indonesia). In coastal areas sea-level will normally provide the base level for groundwater flow from inland. In consequence, head gradients in saline groundwaters moving at depth from inland will be upwards.

Anthropogenic salinity environments

The most usual case where salinity poses particular hydrogeological problems, such as threatening a fresh groundwater resource, is the disturbance of the balance between fresh and saline groundwaters due to extraction. This may be seen in the form of ingress of groundwater derived from modern seawater, a well-known problem in various Mediterranean countries. Another case is the upconing or lateral movement of old saline groundwaters. In such circumstances the dominant issues are the salinity source, which can usually be intensified by the hydrochemistry, the head controls, and the resource management response.

Man-made salinity environments can be created where salinity does not necessarily occur naturally. One of the most extreme cases is in the Indus Valley ot Pakistan where surface water irrigation losses over decades have caused a substantial rise in the groundwater level, to such an extent that evaporation can occur from the ground surface. With the passage of time, the evaporation has resulted in extensive salinization of the land and high salinity concentrations in the groundwaters.

A similar effect is seen in semi-arid parts of southern Australia, where widespread tree felling within recharge areas some hundred years ago led to rising groundwater levels. The ensuing salinization of the ground and the groundwater has devastated vast areas of land.

However, the most frequent anthropogenic source of salinity is probably from irrigation field losses, where soil zone evaporation concentrates salts that are subsequently leached into the groundwater table.

Study and management

The study of salinity is fraught with difficulty. Taking the most widely studied scenario of coastal groundwater salinity, only very approximate and bulk effects can really be assessed for eventual management purposes.

Two aspects have to be considered: hydraulics and hydrochemistry. In most cases the control on these will be hydraulically layered or heterogeneous. Stressing of the aquifer due to abstraction will create a varied response geometry so that a mixed fresh–saline groundwater zone will ensue. Because of natural fresh–saline groundwater relationships it is rare for a true interface to exist between the two types of water, rather a mixing zone will be present, which will be variably extended through abstraction. For example, studies in the Caribbean islands of Cayman and Bermuda indicate that the calculated Ghyben-Herzberg interface ratio of 1 : 40 falls within the mixed salinity zone, the base of which is at about 1 : 50. The base of the fresh groundwater is about 1 : 20. The fact that the Ghyben-Herzberg ratio does not readily apply, and indeed can overestimate fresh groundwater resources, reflects the difficulty in using a steady-state relationship to represent a highly dynamic scenario.

Although salinity distributions can probably only be defined in the broadest sense, it is often salinity rather than hydraulics that provides the best understanding of long-term salinity movement. This is because most of the measurements of head are normally made in the fresh groundwater areas, where variable abstraction does not easily allow the definition of small **reversed** groundwater gradients that can allow saline ingress.

As only a broad definition of salinity distribution is normally possible, its representation, for example in terms of numerical modelling, is also difficult. In most cases a distributed 2-dimensional approach is used to represent a 3-dimensional relationship, which can be very complex, particularly in considering the shape of mixing zones. A very considerable database is essential for such resource studies.

Where non-density coupled flow models are used, saline groundwater movement can be numerically estimated by determining reversed groundwater gradients, with a classical analytical approach adopted to locate an interface position at specific time steps. For groundwater modelling of islands a transient non-density coupled model can be used, defining the base of fresh water using the modified Ghyben-Herzberg ratio noted above. The use of density coupled flow models is increasing and where adequate data are available, these should provide better results than the above models, as they account for the density head drive. Such models are, however, difficult to apply to transient conditions and require further mathematical development.

In the management of coastal groundwater salinity the obvious option is manipulation of salinity. A good example is seen in the chalk aquifer in the south of England, where groundwater is abstracted close to the coast during periods of high head and away from the coast during periods of low head. Similar procedures are being used in Israel.

The creation of a fresh groundwater mound or barrier on the landward side of the ingressing sea-derived groundwater is often proposed as a good management procedure. It has proved effective, e.g. in Long Island, USA, but care has to be taken to ensure that salinity is derived laterally and not vertically from beneath the abstraction area.

In some islands, the head of fresh groundwater is conserved and the saline head depressed by using a caisson and abstracting from the inside of it. Although such structures can work they are expensive and, without protection, are susceptible to pollution.

In non-coastal areas, with both fresh and saline groundwater, management of abstraction is essential to avoid lateral saline movement or upconing. Frequently, salinity problems occur because the units underlying the aquifer can provide upward leakage of poor quality water when the head differential with the aquifer is disturbed. Structural dislocations often provide important pathways for such leakage and so need to be investigated. Unfortunately, the time required for leakage to be established may prove to be much longer than can be identified by any testing techniques, so that countering problems may ultimately depend on pragmatic manipulation of abstraction.

Where anthropogenic groundwater salinity has been created, the difficulty arises of removing the saline groundwaters so that the soil salinity and groundwater salinity are not enhanced further. Gravity drainage and pumping of saline groundwater to a river course, as practised in Pakistan, can be a solution where groundwater gradients are suitable. In some instances fresh groundwater can be scavenged from above the saline groundwater being removed. If gradients are unsuitable, where effective drainage cannot be installed and where there are limitations on river water quality, the reduction of salinity is extremely difficult. This is partly the problem in the Murray River Basin, Australia, where some minor relief is provided by drainage to saline evaporation ponds, but where some ground has to be sacrificed to salinity to benefit other ground.

Although saline groundwater poses problems for most water supply and irrigation engineers, it can be used directly in some industries. In northern Chile, for example, copper flotation is carried out in water with 73 000 mg l^{-1} TDS; ale is brewed with waters normally in excess of 1000 mg l^{-1}; and many spa towns depend solely upon saline groundwaters for their reputations.

Saline groundwater systems are also seen in beneficial terms as disposal facilities. In the Bahamas deep well waste disposal has been carried out for some time. In many countries landfill disposal sites are located in saline groundwater areas. The benefits are difficult to gauge and can only be decided in the local context on environmental grounds.

Increasingly, saline groundwater is being viewed as a resource. In Bahrain and Saudi Arabia, saline groundwaters are abstracted and desalinized for potable purposes, and the same approach is being investigated in Jordan. The limited experience of such processes shows that stable long-term chemistries are being achieved, which is important for desalination processes, and with sea off-takes, the well installations are far less susceptible to pollution. In the USA some brackish groundwaters are being protected as a possible future resource, a practice that could with advantage be extended to other countries.

Conclusions

The salinities that are found in groundwaters arise from a variety of natural sources and a variety of anthropogenic causes. Their control is obviously dependent on good resource management, but in order to exercise such management, a comprehensive hydrochemical and hydraulic assessment is required together with careful monitoring. In most environments, a total hydrogeological system assessment is necessary.

6 Mixing phenomena due to sea-water intrusion in the limestone Zarakas Mountain Range

C.Tavitian

Abstract The Zarakas Mountain Range in the southeastern Peloponnese, Greece, is composed of thick and highly karstified Tripolitsa limestones, which can be divided into three catchment basins. They discharge their groundwaters through numerous onshore and offshore springs on the eastern coast of the eastern (Akra Malea) peninsula. Part of the groundwater (25%) is also discharged on the western coast (where these limestones do not crop out widely) through a system of onshore and offshore springs, concentrated in one area south of Elea and fed by a fault controlled karst conduit. All spring waters are of sodium-chloride type. The process of ion exchange between the fresh groundwater and the underlying heavier sea-water takes place in the limestone aquifer, inland, due to infiltrating sea-water through faults and fractures. The chemistry of the karst groundwater varies with depth and with distance from the sea. Near the water table it is calcium-magnesium bicarbonate type whereas at greater depths or closer to the coast it is of sodium-chloride type. Isotope analysis of the karst and contact spring waters reveals a linear relationship between oxygen-18 depletion and spring basin mean recharge elevation.

Introduction

Permeable rocks (e.g. limestones of the Zarakas Mountain Range) in coastal regions act as aquifers and provide a hydraulic connection between the inland fresh groundwater and the sea-water. The salinization of the near shore inland aquifers by infiltrating sea-water represents a worldwide phenomenon. It is the result of disturbance of the equilibrium between inland fresh groundwater and denser sea-water due to fluctuations of the sea level and/or groundwater level. In the case of the Zarakas Mountain Range, as in any other coastal aquifer, the fresh water is in contact with the underlying salt water which intrudes landward along faults, fractures and other karst channels. The change-over from fresh water to salt water is marked by the presence of a transition (mixing) zone. The two hydrodynamic processes of dispersion and diffusion have created this zone where the chloride content gradually changes upward from that of salt water via brackish water to fresh water. The thickness of this zone varies considerably at different sites. Here, it is greater in the inland area and drops down to a few cm towards the coast. This zone is subject to changes vertically as well as horizontally which can be mainly due to the mode of groundwater recharge as well as to the withdrawals which are a common practice in the region, particularly for irrigation purposes.

The removal of sodium ions from sea-water which has infiltrated into freshwater aquifers has been ascribed by a number of workers to the action of ion exchange (Revelle, 1941; Piper and Garrett, 1953). Sodium ions present in sea-water will be exchanged against Ca ions. Sea-water thus becomes $CaCl_2$-type water (Appelo and Geirnaert, 1991). The conversion of calcium bicarbonate water to sodium bicarbonate water in many aquifers is also undoubtedly due to ion exchange (Foster, 1942; Back, 1960). Thus the process

Groundwater Quality Edited by H. Nash and G.J.H. McCall. Published in 1994 by Chapman & Hall. ISBN 0 412 58620 7

is reversible, the direction of exchange depending on the relative concentrations of ions (Davis and De Wiest, 1966). The fresh water will then change into $NaHCO_3$-type water (Appelo and Geirnaert, 1991).

The Revelle coefficient is defined as the ratio Cl/(CO_3 + HCO_3) ion concentration in meq l^{-1} (Revelle, 1941). Values of more than 10 are considered to be indicative of the presence of sea-water in the aquifer whereas values less than 1 imply good quality fresh water (i.e. with no sea-water contamination). The use of this coefficient however, requires a thorough knowledge of the hydrogeological conditions of the region that is being studied.

The Zarakas Mountain Range has an areal extent of over 450 000 km^2. It provides annually some 240 $\times$ 10^6 m^3 of high quality water received through infiltration. However, the discharged water through the numerous onshore and offshore springs (Figures 6.1, 6.3, and 6.5a) is contaminated with NaCl especially in the eastern coast of the peninsula which dips into the Myrtoon Sea (Figure 6.1). On the western coast, at the Laconian Gulf, the only discharge takes place through the Glyfada-Elea complex springs (Figures 6.1, 6.3 and 6.5a) which are fed by a conduit with karst groundwater from the Western Zarakas Mountains, and are also affected by internal aquifer contamination by mixing with intruded sea-water.

Chemical analysis of the waters of the coastal and submarine Glyfada–Elea springs and the seven boreholes drilled in the Molai area was first carried out by German Water Engineering during 1970–1972 for the preliminary hydrological study of the area. A second and more detailed study by FAO-UNDP during 1978–1980 continued this study; including also the springs of Monemvasia Bay, most of the inland contact freshwater springs and six more wells drilled in these limestones together with some private boreholes. During this period, isotope analysis of all these points was carried out by Democritos Nuclear Research Center of Greece, in order to correlate isotope content and the mean elevation of the recharge area (Figure 6.4).

After completion of this study the sampling area was extended further north from Monemvasia Bay to Kyparissi Bay. This time the additional meteorological information of the whole area up to 1990 was used in order to establish a better understanding of the hydrogeochemistry of the karst groundwater of the Zarakas Mountains and the numerous springs all along its coastlines. This chapter describes the karst groundwater movement within the Zarakas Mountain Range and its significance to regional water resources, the salinity study and its results, and the value of those results to aquifer management and to regional water resources planning.

The water management in karstic coastal lands, as here, has a profound influence on the position of the interface between salt and fresh water. The subtle equilibrium established over a long period of time may be suddenly disrupted by human activities. The withdrawal of large quantities of fresh water in such semi-arid areas of limited recharge may induce a disastrous change in the groundwater regime.

Water use in the area

The main occupation of the people here (total 10 000) is agriculture on the Molai Plain and animal rearing, mainly goats and sheep in the hilly and mountainous areas (from 100 to 1000 m elevation). Since antiquity, the traditional cultivations have been olive and fig trees, which consumed limited amounts of water, supplied entirely from rainwater; and needed no artificial irrigation: surface water is rare and streamflow on the plain occurs only in wintertime (every 4–5 years) and even then lasts only for a few weeks. At that time the water budget was thus balanced.

For the last three decades, however, there has been a sudden increase in the demand for irrigation water, due to a change in type and intensification of cultivation, and to citrus and vegetables replacing olive and fig trees. There has been a simultaneous increase in the area of arable land under cultivation from 2.5 km^2 in 1963 to nearly 18 km^2 today and this has forced farmers to look for new sources of water. Also, the considerable rise of tourism has caused increased demand for fresh water. Deep wells were drilled in the Upper Pliocene fine grained, unconsolidated aquitard in the 1960s and 1970s, but these measures proved to be inadequate to meet the rapidly growing water demands, so people in the 1980s turned their attention to the karst aquifer northeast of the Plain. This, however, proved to be of limited capacity, and the quality was poor. Due to the lack of large amounts of fresh water, the people here have now overpumped both aquifers in order to satisfy the demands of irrigation, resulting in the intrusion of sea-water further inland, both horizontally and vertically and thus its mixing with the thin freshwater layer. Thus, groundwater has been tapped in ever increasing quantities, overdeveloping the local resources. As a consequence, the quality of irrigation and drinking water is greatly affected by chloride ions from the intruding sea-water.

Stratigraphy

The oldest rocks (Carboniferous–Permian) which crop out in the area belong to the Phyllite-Quartzite Series

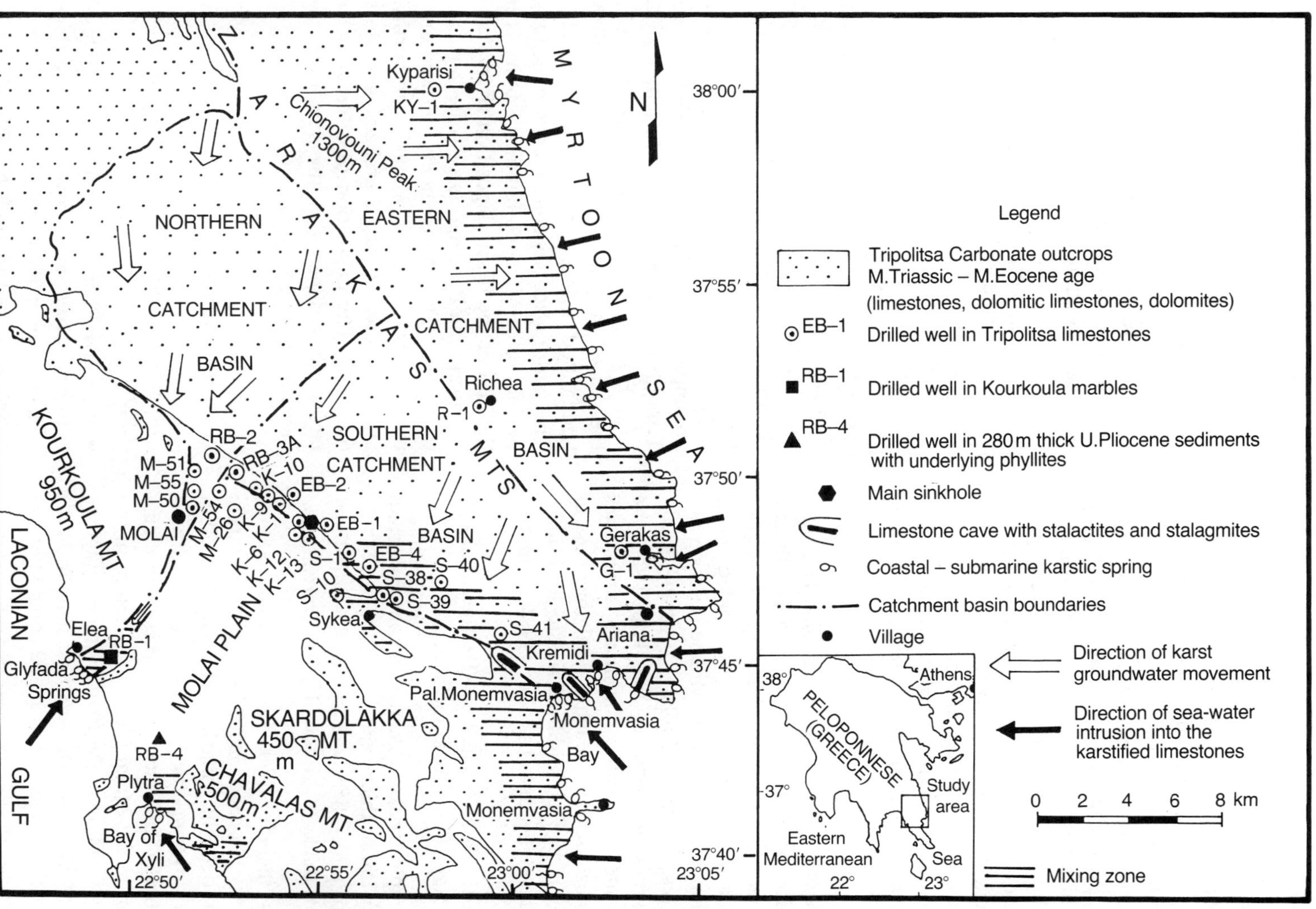

Figure 6.1 Map showing the outcrops of the Tripolitsa limestone in the Epidaurus Limira District, SE Peloponnese, Greece.

which includes mainly phyllites (mica schists) and to a lesser extent quartzites, greywackes, metaconglomerates and lenses of metabasalts (Jacobshagen, 1986). The base of this unit (Platy limestones) does not crop out in the study area. The visible thickness of this formation in outcrop just south of Molai is nearly 100 m (Tavitian, 1990).

The Tyros Beds (Ktenas, 1924, 1926) of Carboniferous–Middle Triassic age (total thickness 900 m) overlie the Phyllite-Quartzite Series. The Lower Tyros Beds include clastic, partly metamorphosed, thin, alternating beds of schist, quartzite, marble, phyllite and graywacke. Their thickness is nearly 300 m. This group of folded beds alternates with thick, white–light grey, slightly folded, massive Kourkoula Marbles which are also 300 m thick. The Upper Tyros Beds in this area crop out mainly to the north and northeast of the Kourkoula Mountain and include andesites and some tuffs with pyroclastic rocks of a total thickness of 300 m (Pe Piper, 1982; Pe Piper *et al.*, 1981; Papazeti, 1984, 1986).

The Tyros Beds are succeeded by the thick Tripolitsa carbonates which cover most of the area (north and east of the map, Figure 6.1). Between these two formations lies a thin transition zone of folded clastic 30 m-thick sedimentary rocks of Middle Triassic age cropping out just north of Molai (Tavitian, 1990). These thick carbonate rocks of the Tripolitsa zone comprise dolomites, dolomitic limestones, limestones and breccia limestones and reach a total thickness of 1850 m and their age is Middle Triassic to Middle Eocene (Exindavelonis and Taktikos, 1984; Stamatis, 1984).

The rocks of the Pelagonian zone (southeast part of the geological map, Figure 6.1) include:

- ophiolitic complex rocks (volcanic, metamorphic and sedimentary) 150 m thick; and
- thick bedded, pink, slightly crystallized oolitic limestones, 250 m thick, of Middle–Upper Triassic age.

Tertiary rocks include thick (greater than 600 m) granular Upper Pliocene lagoonal–shallow marine unconsolidated deposits (marls, clays, fine–medium grained sands) covering the total area of the Molai Plain (Figure 6.4). The thickness of these sediments varies: > 600 m in the Molai Graben; < 150 m in the Assopos Horst; > 300 m in the Plytra Graben (Figures 6.3, 6.4). Upper Pliocene deposits also include hard, well cemented, 150 m thick intercalated conglomerates in the northern part of the Plain.

Quaternary superficial sediments include marine terraces, terra rossa, talus cones, coastal dunes, and thin stream channel deposits. The main stratigraphic units of the area are shown in Figure 6.2.

Tectonics

The folded Phyllite-Quartzite Series is thrust onto the Platy limestone series and the folded Tyros Beds are thrust onto the Phyllite-Quartzite Series. The ophiolites of the Pelagonian zone are thrust onto the dolomites of the Tripolitsa zone. The overlying oolitic limestones are seen to be thrust onto the ophiolitic rocks of the same zone. The direction of the main thrusting is from northeast to southwest.

Up to the Middle Triassic, there was synsedimentational injection of andesitic magma up northwest and northeast trending faults (Doutsos and Koukouvelas, 1986). During the Middle Triassic the area emerged and the sea became shallower, thus creating a platform for the deposition of the carbonate rocks of the Tripolitsa zone, that lasted into the Middle Eocene.

The main orogenic stage can be placed in the Middle Miocene during the tangential movements of the Alpine orogeny (Kowalczyk, 1977) when extensive folding of the phyllites, the Tyros Beds and the transition zone occurred due to the high plasticity of their rock types. The hard (low plasticity) Tripolitsa limestones were less affected by this process showing open folds.

The regression of the sea during the Eocene was accompanied by faults that ended the deposition of Tripolitsa limestones (Jacobshagen, 1978). During the Late Eocene, the Tripolitsa zone was submerged and the sea of the flysch basin transgressed the karstified and faulted surface of these limestones (Richter, 1978). This faulting activity continued into the Oligocene. The beginning of Miocene faulting activity is indicated by the end of the flysch deposition (Kowalczyk, 1977).

During the Pliocene, large scale vertical movements took place. These movements followed the trend of the older faults (northeast primary, and northwest secondary) and created large grabens and horsts (Figures 6.3, 6.4). The deposition of alternating lagoonal and lacustrine sediments is the result of these vertical movements.

Geomorphology

The area comprises the following morphological units (Figure 6.1):

- the Molai Plain sloping from north (elevation 100 m) to south (sea-level) covered by the thick unconsolidated Upper Pliocene sediments;
- the Zarakas Mountain Range, north and east of the Plain, consisting entirely of thick bedded karstified Tripolitsa (> 1800 m thick) limestones;
- the Kourkoula Mountain west of the Plain, consisting of the Upper and Lower Tyros Beds; and

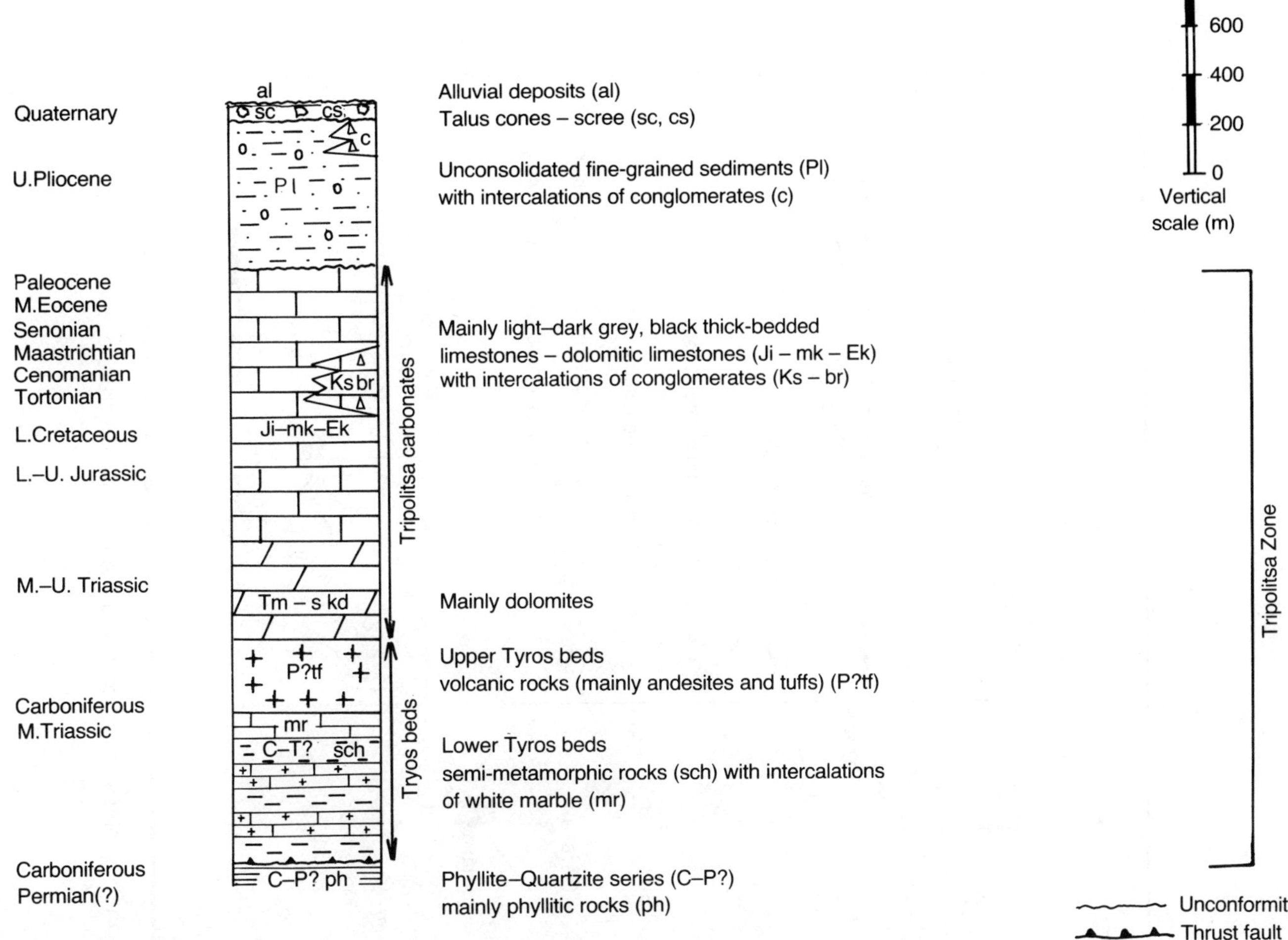

Figure 6.2 Generalized stratigraphic column in the Molai–Richea area.

- Small hills, south-east of the Plain, consisting of Upper Tyros Beds and oolitic limestones of the Pelagonian zone.

The main characteristic features of each morphological unit are as follows (Figure 6.1).

Molai Plain

The Molai Plain is a structural valley composed of the three major structures; the Molai Graben (600 m deep), the Assopos Horst (150 m deep) and the Plytra Graben (300 m deep). All these structures are filled with thick Upper Pliocene sediments. The Plain in the north is delineated by the 100 m contour line, whereas in the south it is bordered by the Gulf of Xyli.

Zarakas Mountain Range

The northern and eastern parts of the Plain are bordered by a hilly area (1300 m highest point) which forms the west flank of the Zarakas Mountains, composed of slightly folded Tripolitsa limestones: the range trends northwest and slopes eastwards to the Myrtoon Sea. There are four karst erosion surfaces observed in this mountain range (Tavitian, 1993):

(1) Chionovouni Surface (altitude 600–1200 m) which is characteristic of the higher elevations having cone shaped dolines with a sinkhole at their topographically lowest point. The dolines alternate with conical hills.
(2) The Koupia Surface (altitude 400–600 m) is characterized by closed basins (poljes).
(3) The Niata Surface (altitude 100–400 m) is characterized by shallow and very wide dolines, each having its own sinkhole.
(4) The Palea Monemvasia Surface (altitude < 100 m) is characterized by karst caves (with stalactites and stalagmites) near Kremidi close to sea level, and a large sinkhole system near Metamorphosis (north-east of the Plain).

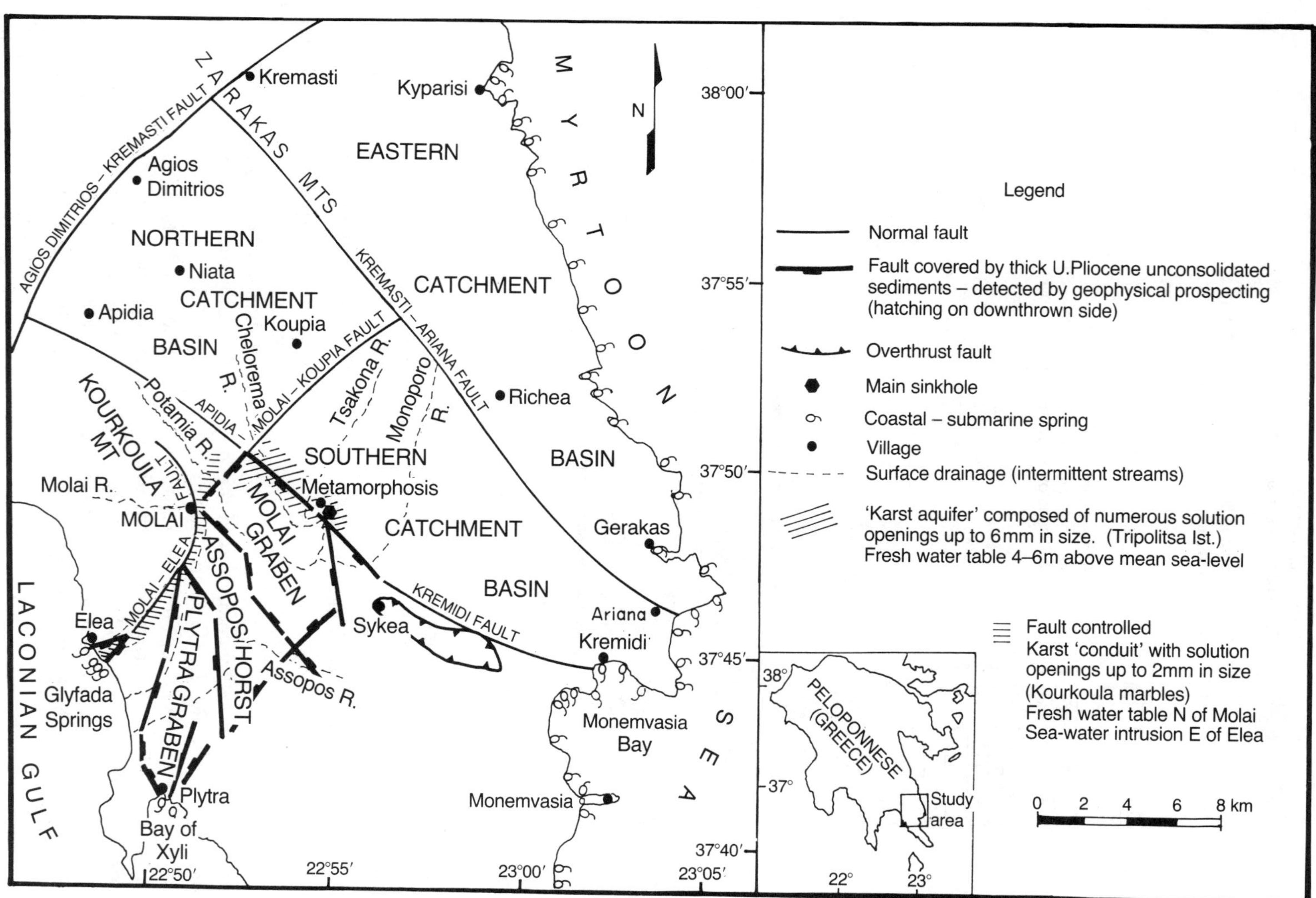

Figure 6.3 Map showing the main fault pattern in the Tripolitsa carbonates and the tectonic structure of the Molai plain.

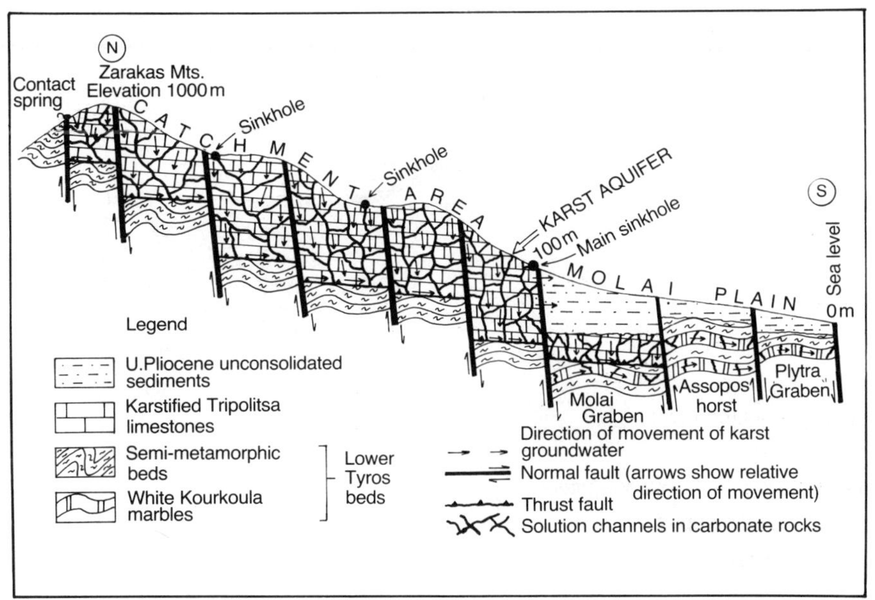

Figure 6.4 Schematic north–south geological section west of the Kremasti–Ariana fault.

Kourkoula Mountain

This is an elevated mountain system. The major Molai–Elea fault delineates its eastern borders. The highest elevation is close to 1000 m and its area is approximately 30 km^2. The mountain is composed totally of the Lower and Upper Tyros Beds. The white marbles here show distinct karst features (though less than the Tripolitsa limestones): there are a few large solution openings (~ 0.5 m wide). One of them drains a small doline near the peak.

Small hills southeast of the Plain

The oolitic limestones that make up the two hills here (Chavalas and Skardolaka, Figure 6.1) show a small degree of karstification mainly in the fractured and faulted zones.

Hydrology

There are three main hydrologic surface flow divisions (Figure 6.3). The first is the inland drainage complex of the Tripolitsa highly karstified mountains which drain to numerous closed depressions (dolines) (Figure 6.4). They have a high infiltration rate and little or no runoff. Infiltration takes place by water percolating underground through solution channels, mainly to the eastern coast (from Kyparisi to Monemvasia), in the form of numerous coastal and submarine springs. On the other hand, the major Glyfada–Elea karstic spring system (both coastal and submarine) on the west coast is fed by rainwater falling over the western Zarakas Mountain Range (Ghikas *et al.*, 1983). Although this spring complex is located within two small independent catchments, its discharge suggests that it is interconnected with the main underground drainage system of the northern basin following the Molai–Elea fault north of the Plain (Figures 6.1 and 6.3).

The second complex is the northern Molai Plain drainage system which has an integrated surface flow of four small intermittent streams (Potamia, Chelorema, Tsakona and Monoporo) to a central drainage point at the main sinkhole just south of Metamorphosis village, northeast of the Plain (Figure 6.3).

The third complex is the southern section of the plain which has two intermittent outflow stream channels (Molai and Assopos) that carry surface flow to the sea (Laconian Gulf), but only under exceptional rainfall conditions (once every five years).

Hydrogeology

The Tripolitsa carbonates of the Zarakas Mountains are seen to have three different catchment areas (Figure 6.1). The first (eastern) lies east of the Kremasti–Ariana fault. It discharges through the numerous coastal and submarine springs between Kyparisi and Ariana on the eastern coastline, mainly through the east–west trending faults (Figures 6.3 and 6.5a).

The second (northern) catchment basin is delineated by the four major faults in the Tripolitsa carbonates (the Agios Dimitrios fault in the northwest, Kremasti–Ariana fault in the northeast, Koupia–Molai fault in the south, and Apidia–Kremidi fault in the southwest). It discharges through the Glyfada springs via a karst conduit within the Molai–Elea fault (Ghikas *et al.*, 1983). This catchment area is limited in the north by the impervious semi-metamorphic beds of the Lower Tyros north of Kremasti. In the northwest part of the Plain, the Molai–Elea fault brings the buried Tripolitsa

Zarakas Mts
1
1
2
6
m.s.l.
4
3
5
Offshore spring
Zarakas Mts
1
1
2
6
3
5
4
m.s.l.
Onshore spring
Legend
Karstified limestones
1. Piezometric surface
2. Fresh water
3. Narrowing of solution channel (conduit)
4. Suction conduit of sea-water
5. Saline water – zone of mixing fresh with sea-water
6. Offshore – onshore karstic saline spring
Direction of movement of water
Eastern Zarakas karstic springs

Zarakas Mts
Offshore spring
m.s.l.
6
1
2
3
4
5
Zarakas Mts
Onshore spring
m.s.l.
6
1
2
3
4
5
Legend
U.Pliocene sediments
Karstified limestones
White Kourkoula marble
Phyllitic rocks
Glyfada – Elea karstic springs

Chavalas Mt
1
Onshore spring
m.s.l.
6
5
3
4
2
Chavalas Mt
Offshore spring
m.s.l.
6
1
2
4
3
5
Legend
U.Pliocene unconsolidated sediments
Karstified dolomites of Tripolitsa zone
Andesites of Upper Tyros Beds
Plytra karstic springs

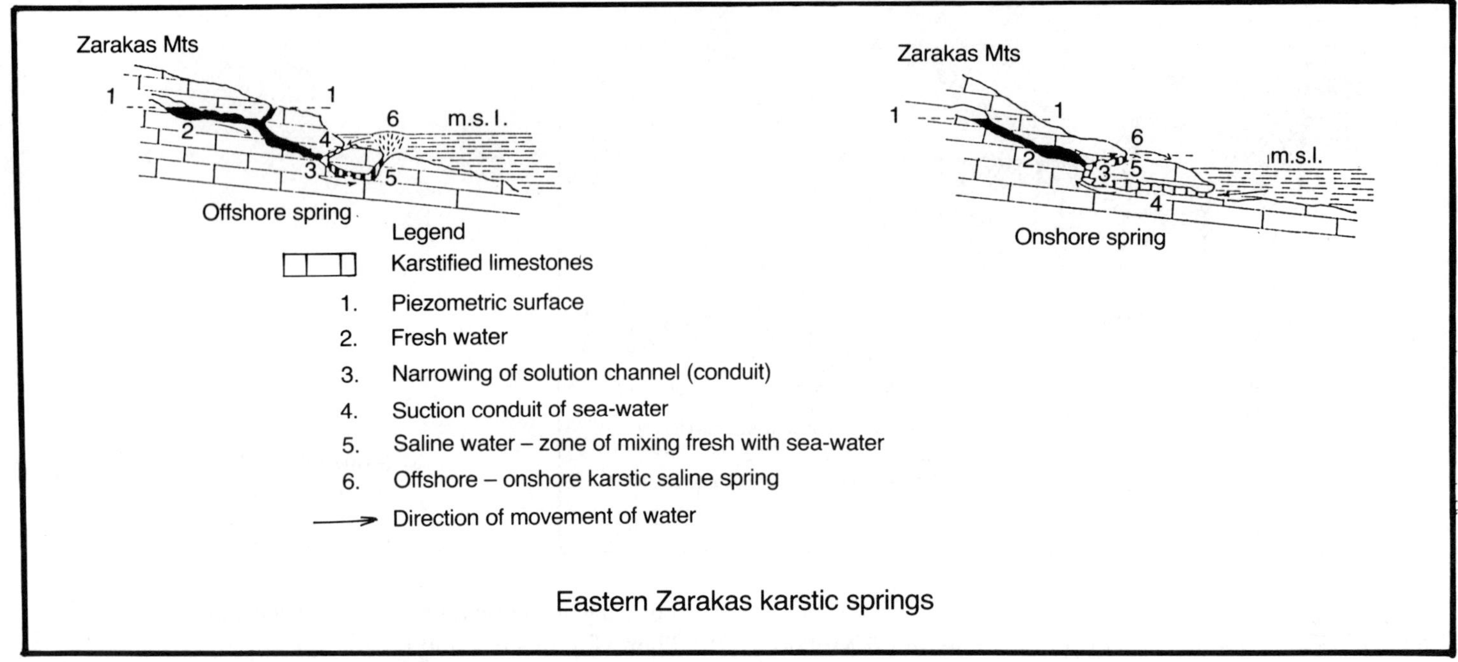

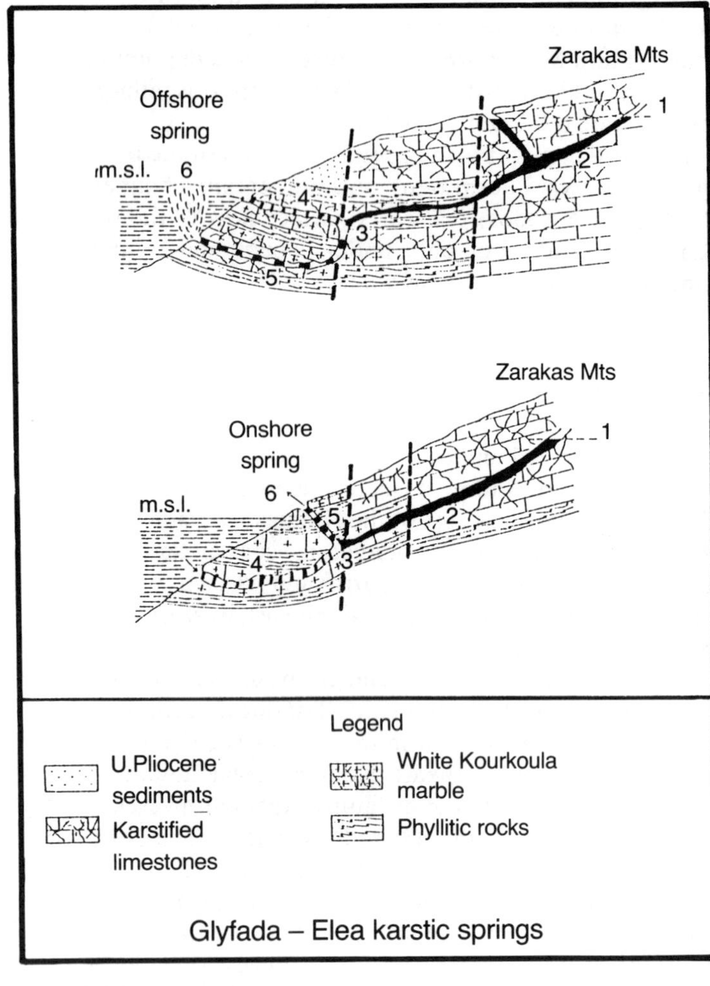

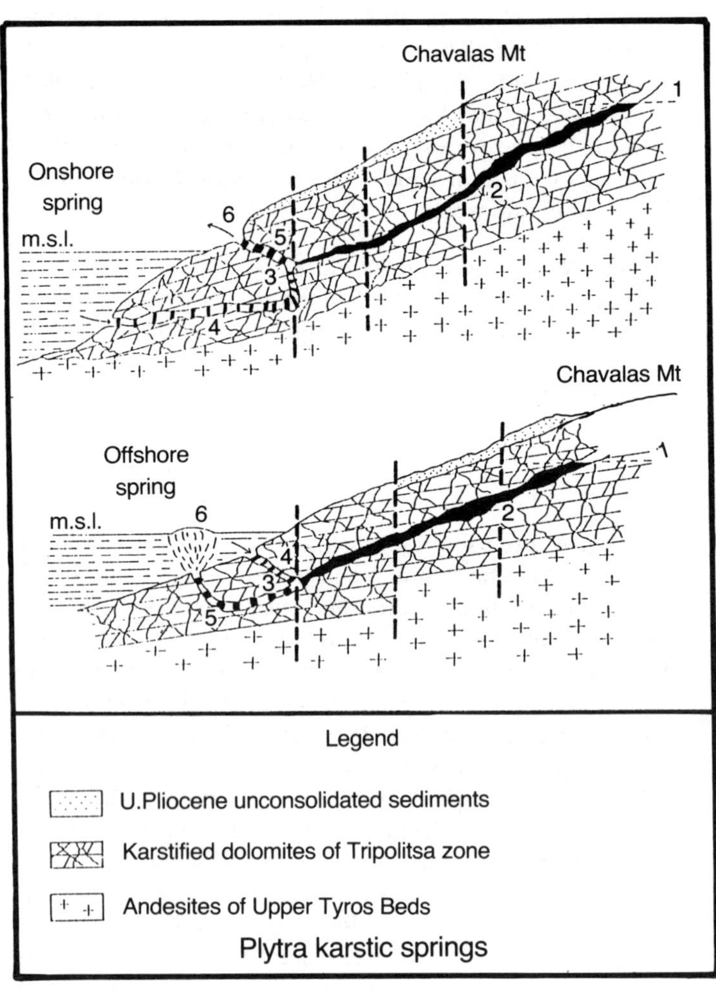

(a)

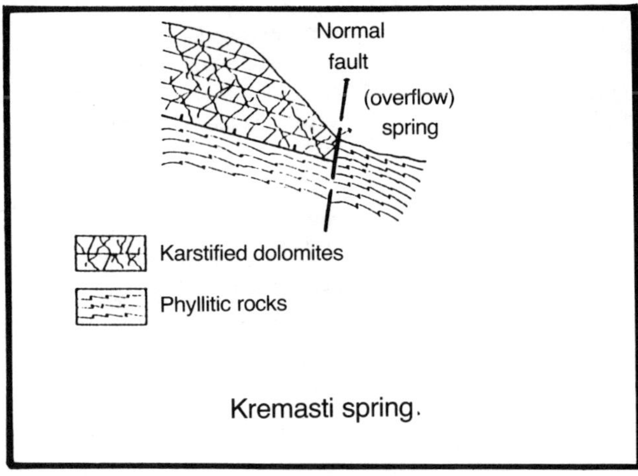

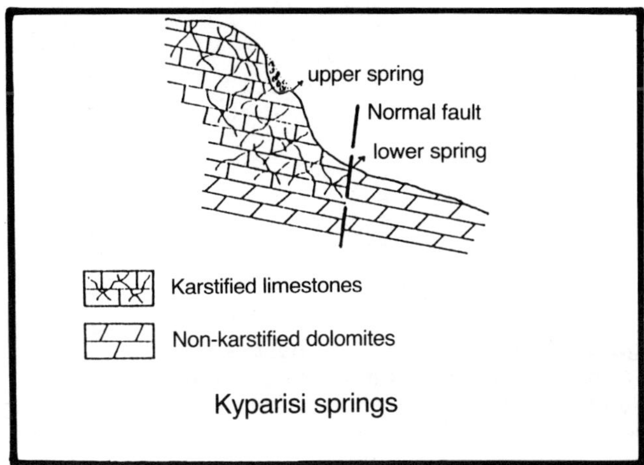

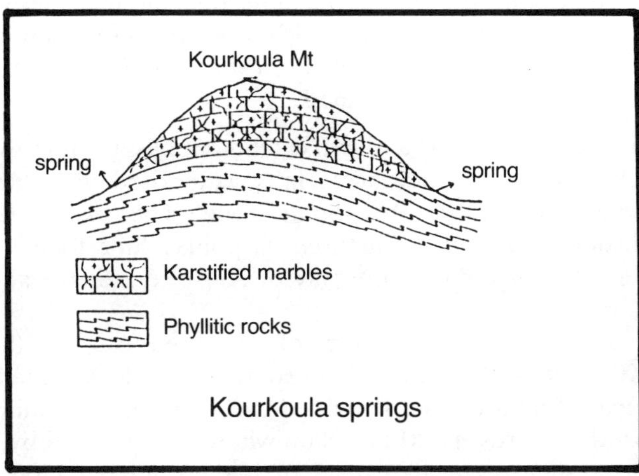

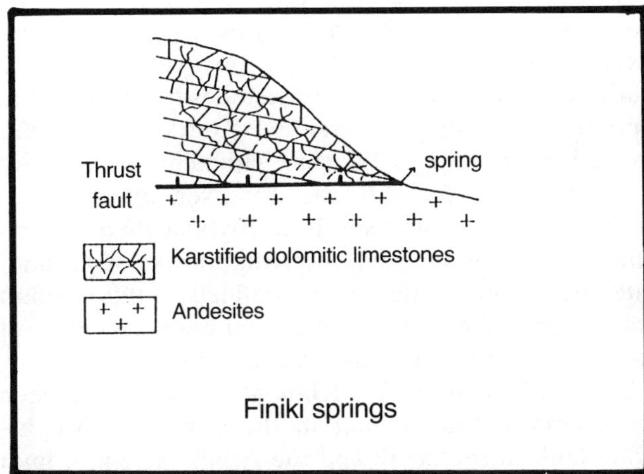

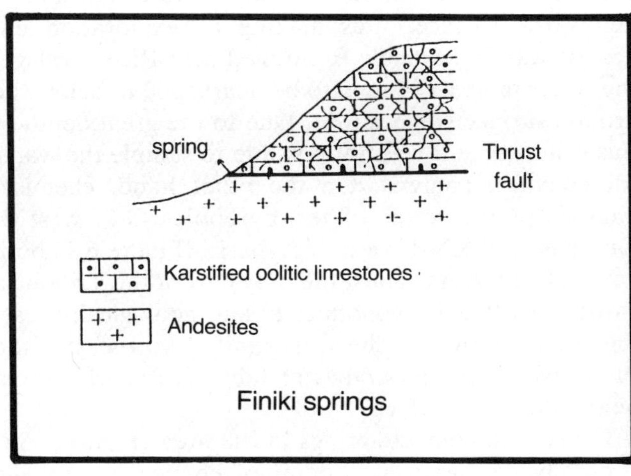

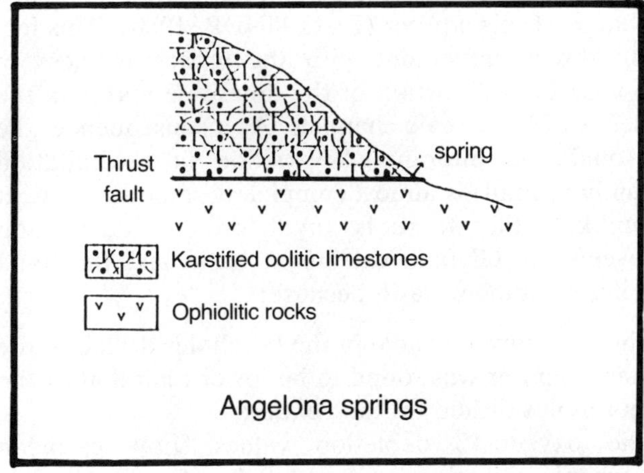

(b)

Figure 6.5 (a) Flow mechanism of onshore and offshore karstic springs in the Myrtoon Sea and Laconian Gulf. (after Bogli, 1980) (b) Flow mechanism of various contact springs.

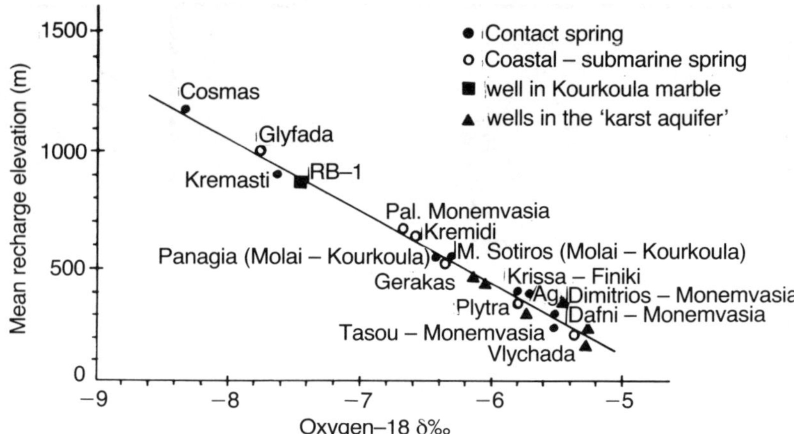

Figure 6.6 Correlation curve between spring basin mean recharge elevation and relative isotope composition. (after Leontiadis, 1981; Ghikas *et al*., 1983)

carbonates (encountered in boreholes M-50, M-51, M-54 and M-55) into direct contact with the marbles of the Lower Tyros Beds (encountered in borehole RB-1). Thus the karst system that forms an underground passage for the groundwater to move from north to south through the Tripolitsa limestones north of Molai, continues through the karstified Kourkoula marbles covered by the Upper Pliocene sediments just east of Molai (Figures 6.3 and 6.5a). Moreover, on the coastline north of Elea and south of Glyfada there are no other springs discharging, implying that karst groundwater movement is definitely confined to this conduit that follows the eastern and southern foothills of Kourkoula Mountain (Figures 6.1 and 6.3).

The third (southern) catchment basin lies between the Kremasti–Ariana fault in the east, the Koupia–Molai fault in the north and the Apidia–Kremidi fault in the west (Figure 6.3). This catchment area discharges through the springs of the Monemvasia Bay and its northwestern part is known as the 'karst aquifer'. The flat groundwater hydrographs indicate a very slow recharge of this aquifer (FAO-UNDP, 1980). This fact is in direct agreement with the great thickness of the unsaturated portion of the limestone and with the absence of large-scale channels. As a consequence, the seasonal and interannual variation in precipitation (mainly rainfall) is almost completely smoothed out. It is unlikely that there is any hydrologic connection between the Glyfada catchment basin and the karst aquifer catchment basin because:

- the groundwater table in the boreholes drilled in the karst aquifer was found to be lower than that of the boreholes drilled on the conduit;
- the oxygen-18 depletion values show a mean recharge elevation basin of 1 000 m for the Glyfada springs and borehole RB-1, whereas for the karst aquifer the elevation given lies between 250–500 m (boreholes RB-3A, EB-2, EB-1, EB-4, S-1, and S-10, Figure 6.6); and
- Considering the type of karst features in the several boreholes drilled it can be concluded that unconfined conditions prevail in the limestone aquifer. However, as the Kourkoula marble aquifer is overlain by impervious layers of phyllites (borehole RB-1), it can be considered as confined.

In addition there is a fourth catchment basin within the Chavalas and Skardolakas hills, composed of oolitic limestones of the Pelagonian zone. This discharges through the highly fractured Tripolitsa limestones at the Plytra springs in the Bay of Xyli (Figures 6.3 and 6.5a).

Thus the general form of the area east of the Kremasti–Ariana fault is seen to be a thick, tilted, highly fractured and karstified carbonate mass sloping south towards the Molai Plain where it is blocked by a thick semi-permeable mass of unconsolidated Upper Pliocene sediments (Figure 6.4).

In the central part of the Zarakas Mountains, the depth of the water table is several hundreds of metres below the surface, thus making its exploration uneconomical. In borehole R-1 drilled near Richea village, the water table was found to be nearly 500 m below the ground surface (Figure 6.1). Due to the great depth of this water table it was not possible to sample the water for chemical analysis. On the other hand, chemical analysis of the water of the boreholes G-1, west of Gerakas, and KY-1, west of Kyparisi (Figure 6.1) both gave saline waters. Thus, the only part of the Tripolitsa carbonates that is economically and geologically feasible for exploration is the karst aquifer, northeast of the Plain, where the groundwater table lies nearly 100 m below the land surface.

Finally, all contact springs in the area (Figure 6.5b) generally discharge at a point of change in geologic formation, e.g. overlaying limestones or marbles and underlying phyllites, andesites or ophiolites. The exception to this is the Kyparisi contact spring which emerges at the contact between the overlying highly fractured and karstified limestones and the underlying less fractured dolomites of the Tripolitsa zone. In the case of contact springs, spring flow responds rapidly to rainfall, decreasing drastically or ceasing entirely during the dry season. Although these springs are used for local domestic supply, hydrogeologically they are unimportant.

Hydrochemistry

The northern basin discharges its groundwater through the Glyfada–Elea spring system. Its water quality is of the sodium-chloride type (Figure 6.7a). The springs, though varying in flow, are very constant in water quality. During the period 1970–1985 their conductivity varied between 3.3 and 3.4 μS cm^{-1}. Isotope analyses gave values of oxygen-18 depletion as high as – 7.9 δ‰ which corresponds to a mean recharge elevation of 1000 m (Figure 6.6). The water of the northern catchment area, that was sampled in borehole RB-1, 1.5 km east of the Glyfada springs, was found to be sodium-chloride type (Figure 6.7a) with an electrical conductivity value of 2.8 μS cm^{-1}. Chemically, this water is similar to that of the Glyfada coastal springs. This well is situated on the lower part of the karst conduit and the upper part was encountered in the wells RB-2, M-51, M-54 and M-55. However, the water of these wells is of calcium-magnesium bicarbonate type and the water conductivity ranges between 1.0 and 1.1 μS cm^{-1} (Figure 6.7a). The Revelle coefficients of these vary from <1 in the northeast, to >14 in the southwest of Molai (well RB-1), indicating sea-water intrusion in the conduit up to at least 1.5 km inland.

The southern basin which includes the 'karst aquifer' has water quality that varies with depth. The calcium-magnesium bicarbonate type water in the upper portion of the reservoir boreholes changes to a sodium-chloride water at greater depths (Figure 6.7b). Isotope values of oxygen-18 depletion, range from – 5.3 to – 6.3 δ‰ which correspond to a recharge basin with a mean elevation from 250 m to 500 m (Figure 6.6). The groundwater of the southern recharge basin in the vicinity of the karst aquifer shows a variable salinity. The electrical conductivity values range from 1.0 to 6.6 μS cm^{-1}. The water type also varies from calcium bicarbonate and calcium-magnesium bicarbonate to sodium chloride (Figure 6.7b). This increase in salinity is observed with an increase in depth in all boreholes as well as in a horizontal southeastwards direction (i.e. towards Monemvasia Bay). Thus, boreholes drilled just east of Sykea have sodium-chloride type waters irrespective of the depth. Their water type is close to that of the Kremidi coastal springs (Figure 6.7c).

The Revelle coefficients of the water samples close to the water table of the boreholes near Metamorphosis are <1. At greater depths these values are close to 14. Also, water samples from the boreholes near Sykea (EB-4, S-38, S-39, S-40, S-41: Figure 6.1) gave Revelle coefficients of 14.0–14.5 and electrical conductivity values between 6.5 and 7.5 μS cm^{-1}. This indicates sea-water intrusion here also, and this is also supported by the fact that the Cl-/Br-ratio of a water sample from borehole EB-2 drilled in the 'karst aquifer' is 217 (FAO-UNDP, 1980; Ghikas *et al.*, 1983). The same ratio in the case of a water sample from borehole S-40 is 231. Comparing these values with that of sea-water, which is 300, leads to the conclusion that the salinity of the groundwater deep below the water table in the case of borehole EB-2 and at the water table in the case of borehole S-40 is due to mixing with sea-water.

The onshore and offshore springs of this basin at Kremidi and Palea Monemvasia have oxygen-18 depletion values ranging between −6.6 and −6.8 δ‰ indicating a mean recharge basin elevation of just over 500 m (Figure 6.6). The electrical conductivity values of the onshore spring waters vary between 6.7 and 7.6 μS cm^{-1} and of the offshore springs (sea bottom) 17.0–17.5 μS cm^{-1} and to 44.5 μS cm^{-1} at the surface and the Revelle coefficients range between 10.1 and 20.5, respectively.

Using a Stiff diagram method (Figure 6.8) to define individual patterns of chemical character, a 3% sea-water contamination of a limestone aquifer water in sampled inland wells, such as S-1, EB-4, S-38, S-39, S-40 or S-41 produces a chemical pattern similar in character to that of the Glyfada onshore springs. The pattern of the water from the boreholes just east of the Glyfada springs is also similar to that of these springs. The chemical content of these spring waters is, in general, due to salt-water intrusion from the coast. The same pattern is also observed in the case of the other seven springs. However, these have a greater percentage of sea-water contamination. The pattern of the contact freshwater spring compared to that of the sea-water shows a striking contrast.

The eastern catchment basin discharges through the numerous onshore and offshore springs of the Myrtoon Sea (Figure 6.1). The waters here are all of the sodium-chloride type (Figure 6.7c) and the Revelle coefficients range between 13.5 and 14.5. Isotope values (oxygen-18) range between −5.4 and −6.4 δ‰ and correspond to a recharge basin with mean elevation from 200 to 650 m (Figure 6.6). The electrical conductivity values of the onshore spring waters vary between 6.9 to 7.8 μS cm^{-1}. This explains why all villages of the eastern slopes of the Zarakas Mountains suffer from the absence of fresh water for domestic supply.

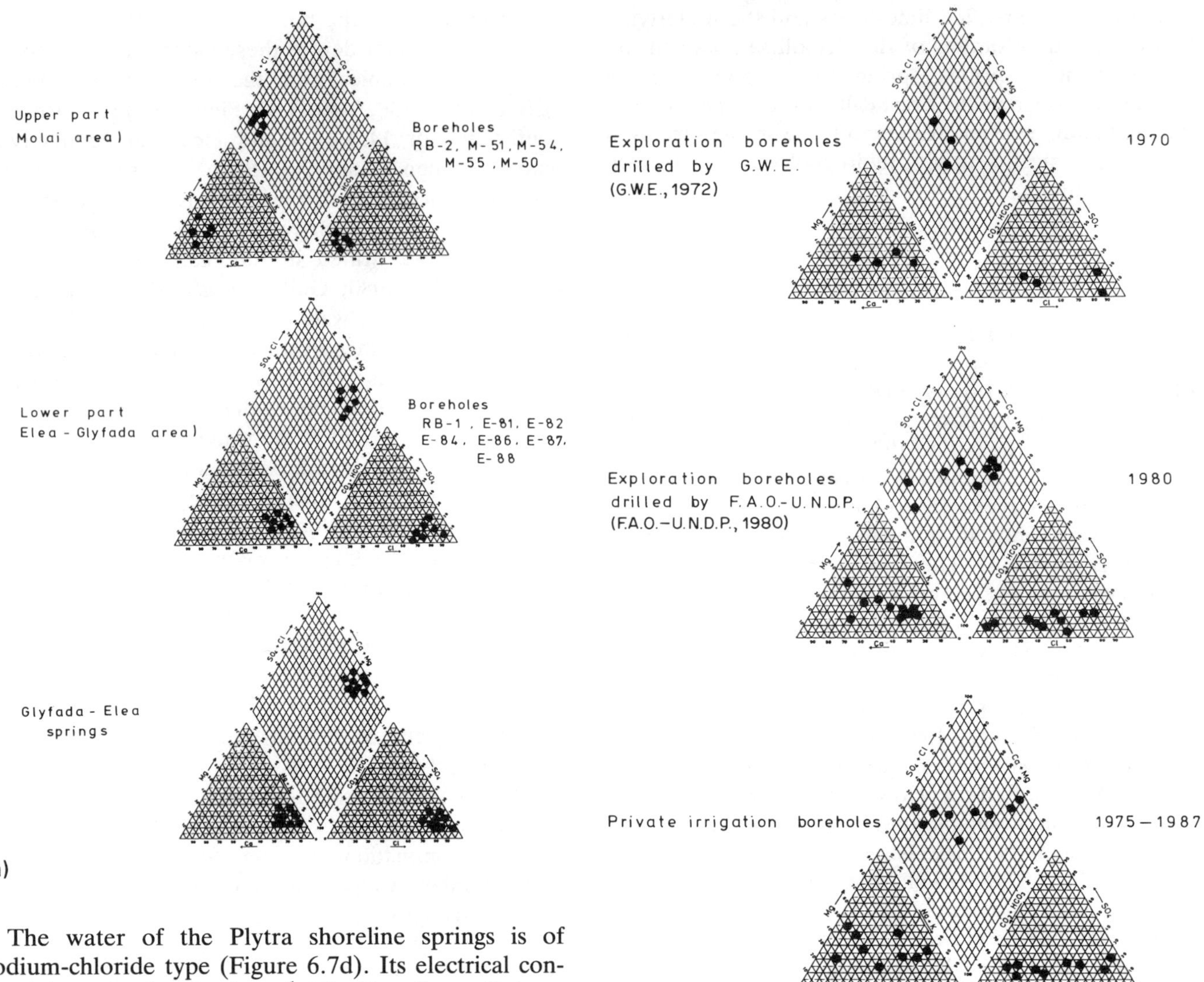

The water of the Plytra shoreline springs is of sodium-chloride type (Figure 6.7d). Its electrical conductivity value is 8.0 μS cm^{-1}. The Revelle coefficients of these springs range between 10 and 11. The offshore springs are difficult to sample but the electroconductivity values at spring source and at sea surface range between 50.1 and 50.4 μS cm^{-1}. The Plytra springs can be related isotopically to the Chavala Mountain just northeast of them, as the waters originate from a source area with an elevation of 400 m (Figure 6.6).

The electrical conductivity of the freshwater inland contact springs all over the area ranges between 0.26 and 1.0 μS cm^{-1} and the water types are generally calcium bicarbonate waters (Figure 6.7d). Where dolomites are present (e.g. Kyparisi spring), the magnesium content is also relatively large. The Revelle coefficients are usually <1. In order to calculate the water budget of the three catchment basins, data from nine meteorological stations were used (Tavitian, 1990). The results for each basin are listed in Table 6.1. Table 6.2 lists the chemical properties of all the waters discharged on the eastern as well as the western coast of the Zarakas Mountain Range.

Conclusions

(1) The Zarakas Mountain Range occupying an area 450 000 km^2 and composed of 1850 m thick, highly karstified limestones can be divided into three catchment basins. The average amount of water infiltrating from precipitation is 290×10^6 m^3 yr^{-1}. The southern and eastern basins discharge their groundwaters (75% of the total) through the numerous springs all along the eastern coastline. The third basin (northern) discharges its groundwaters (25% of the total) at the western coastline through a system of springs focused at Glyfada and Elea which are fed by a fault controlled karst conduit.

(2) All onshore and offshore springs have sodium-chloride water types and pick up salts due to sea-

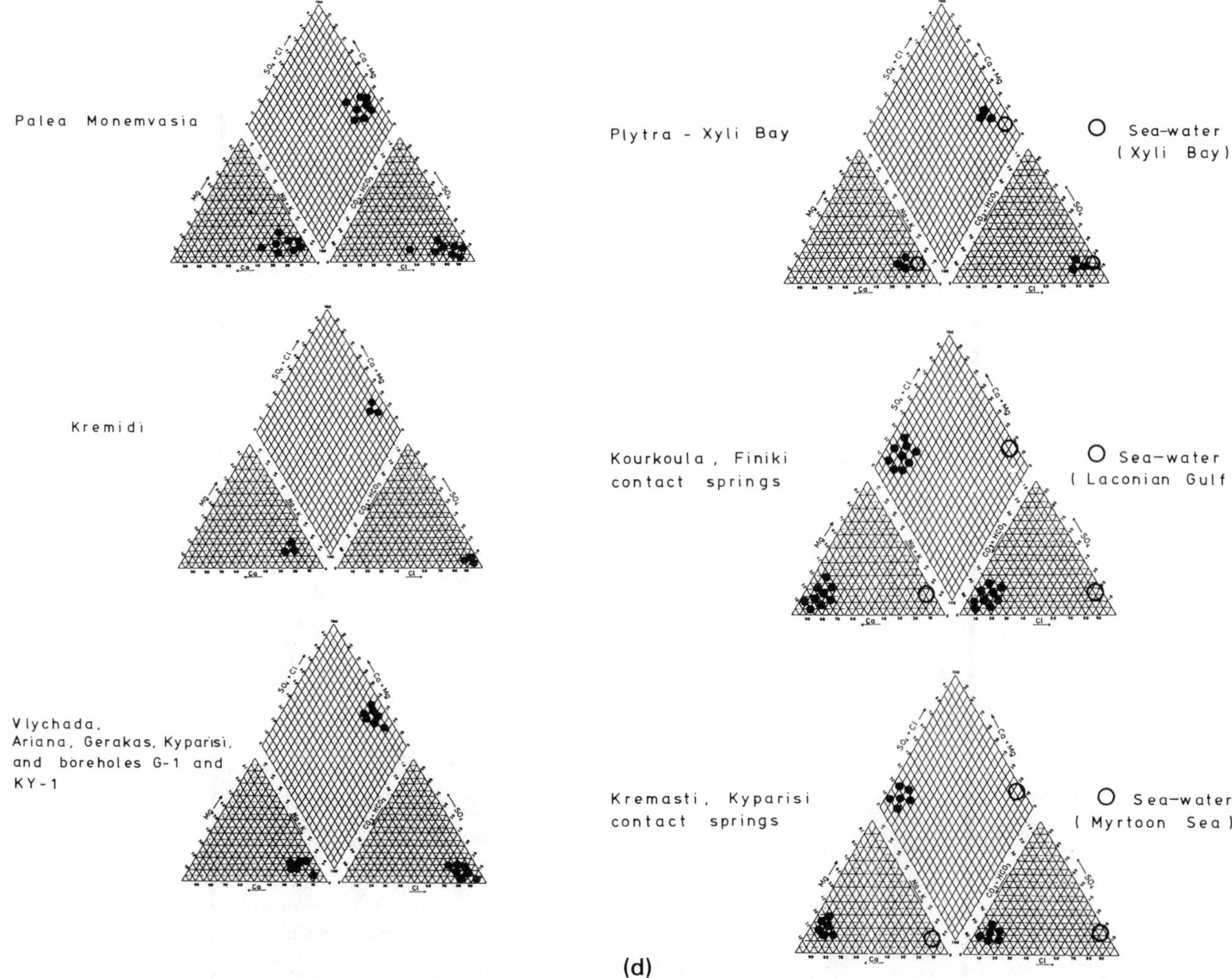

Figure 6.7 Trilinear diagrams: (a) the 'conduit' system of water; (b) the 'limestone reservoir' system of water; (c) the onshore and offshore springs of the 'eastern coastline' system of water; and (d) the 'Plytra Bay (Xyli Bay) onshore and offshore springs' and the 'Contact Springs' systems of waters.

water intrusion through fractures and faults. The total amount of water discharged through these springs is 240 × 10^6 m^3 yr^{-1} which is not suitable for irrigation, however. In contrast, the contact springs have calcium-bicarbonate type waters. The amount of fresh water discharged through these springs is limited (3 × 10^6 m^3 yr^{-1}) and can satisfy neither the irrigation nor the domestic demands of the area. This explains the acute freshwater shortage here.

(3) Piezometric heads in the boreholes of the 'karst aquifer' are usually low, about 4–6 m above mean sea-level at most, assuming a lens-like shape. Heads lessen southeastwards until they become negligible along the eastern shoreline. Hence, saline water is encountered underlying fresh groundwater at comparatively shallow depths. This strongly limits the possibility of using the aquifer as a major resource for irrigating the whole Molai Plain since there is always a danger that the equilibrium between fresh and saline groundwater will be disturbed, resulting in salt contamination.

(4) The karst aquifer lies northeast of the Molai Plain and has a calcium-magnesium bicarbonate water type near the water table, the quality changing to sodium-chloride type at depth. The water of this aquifer is saline just east of Sykea, at a distance of 10 km inland from the sea coast, indicating sea-water intrusion due to low piezometric heads. The conditions of equilibrium between the Zarakas aquifer's limited fresh groundwater (the only currently available karstic water resource here) and sea-water can be severely curtailed due to over-pumping. The amount of fresh water that can be extracted safely from this aquifer is limited: 5 × 10^6 m^3 yr^{-1}, without any danger of NaCl contamination. Obviously the major problem is to find ways and means of implementing a coordinated policy for the management of groundwater of this aquifer

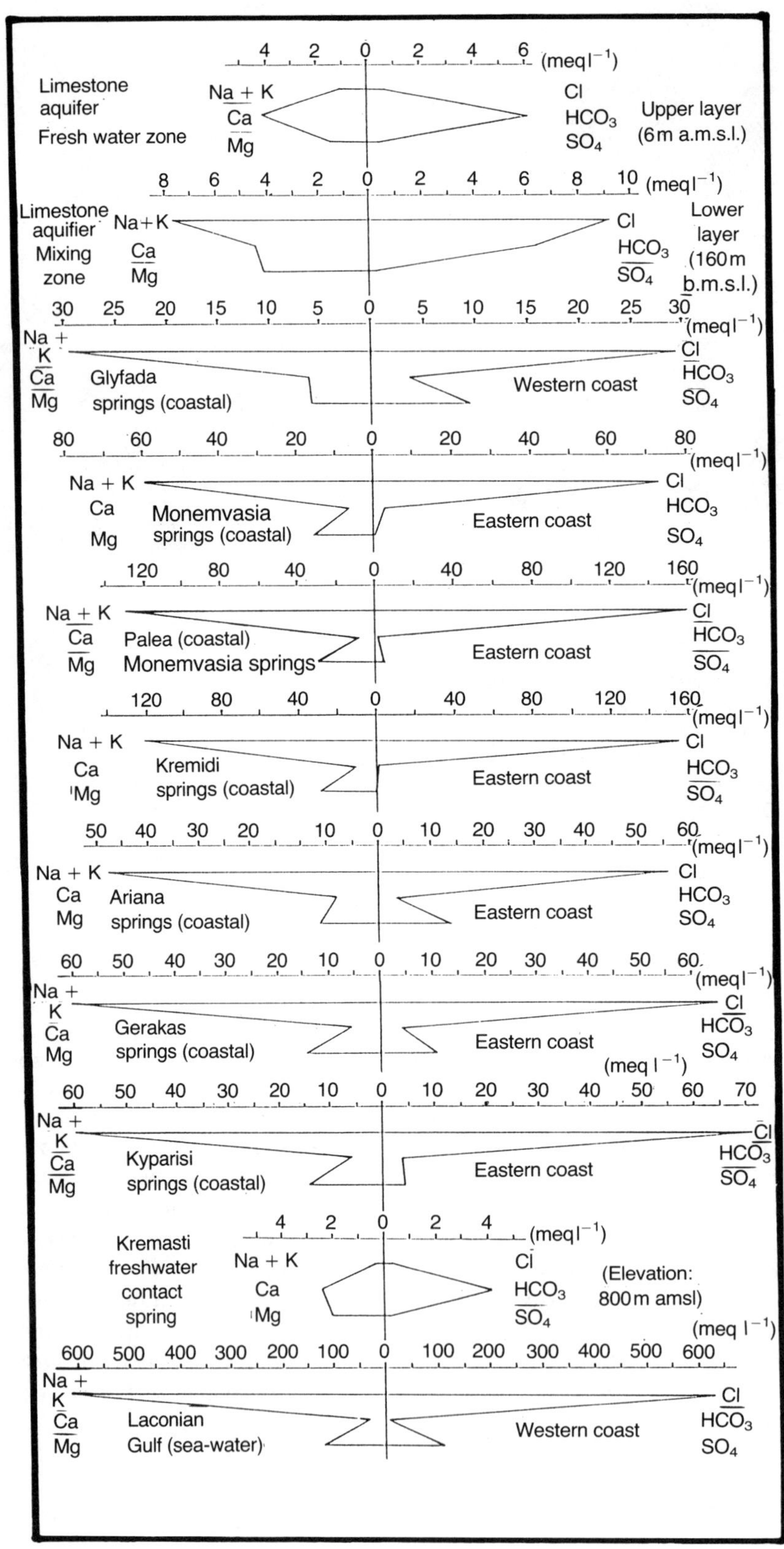

Figure 6.8 Stiff diagrams showing the various water patterns of the Molai–Zarakas area.

Table 6.1 Water balance of the Zarakas Mt Range (Tripolitsa limestones)

Catchment basin	Area (km^2)	Precipitation (mm)	Precipitation ($\times 10^6$ m^3)	Evapotranspiration ($\times 10^6$ m^3)	Discharge ($\times 10^6 m^3$)	Location of discharge (on–offshore springs)
Northern	132.8	852	113	53	60	Glyfada—Elea
Southern	111.1	749	83	39	44	Monemvasia Bay
Eastern	212.5	1198	254	119	135	Kyparisi–Ariana coastline
Total	456.4		450	211	239	

Table 6.2 Chemical properties of the discharged waters of the limestone Zarakas Mountain Range

Catchment basin	Discharge spring	Water type	Electrical conductivity (μS cm^{-1})	Oxygen-18 depletion value	Corresponding recharge basin elevation (m.)	Revelle coefficient	Per cent (%) sea-water mixing (estimated)
Northern	Upper conduit Glyfada–Elea	Ca–Mg–HCO_3	1.0–1.1			<1	
	(lower conduit)	NaCl	3.3–3.4	−7.9	1000	>14	3
Southern	Monemvasia Bay	NaCl	6.6–14.5	−5.3 to −6.3	250–350	14.0–14.5	3–6
	Kremidi Bay	NaCl	6.6–14.5	−6.6 to −6.8	400–500	14.0–14.5	5–6
	Ariana Bay	NaCl	6.6–14.5				3–6
	Karst Aquifer						
	(1) G.W. Level	Ca–Mg–HCO_3	1.0–1.1			<1	<1
	(2) Mixing Zone	$NaHCO_3$ to NaCl	2.5–7.5			5.0–14	2–3
Eastern	Myrtoon Sea springs	NaCl	13.5–14.5	−5.4 to −6.4	200–600	13.5–14.5	3–6
Small Hills NE of Plytra	Plytra Bay	NaCl	8.0	−5.8 to −6.0	350–400	10.0–11.3	3–6

so that the total amount of annual discharge from the private boreholes drawn from it, will be in equilibrium with the hydrological conditions and balance the recharge.

(5) The karst conduit has a calcium-magnesium bicarbonate water type, north of Molai. The amount that can be extracted is limited to not more than 2×10^6 m^3 yr^{-1}. South of Molai, towards the spring discharging point, the quality changes to sodium-chloride type due to mixing with sea-water at least 1.5 km inland.

(6) The detailed picture in ever changing hydrogeologic situations, especially in karst areas like this one, can only be understood by thorough hydrogeological field investigations accompanied by extensive hydrochemical analyses and meteorological observations.

References

Appelo, C.A.J. and Geirnaert, W., 1991. Processes accompanying the intrusion of salt water. In: W. De Breuk (ed.) *Hydrogeology of Salt Water Intrusion, A Selection of SWIM Papers*. I.A.H. Vol. 11/1991, Verlag Heinz Heise.

Back, W., 1960. Origin of hydrochemical facies of groundwater in the Atlantic Coastal Plain: *Internat. Geol. Cong., 21 Session, Part 1, Geochemical Cycles*, pp. 87–95.

Bogli, A., 1980. *Karst Hydrology and Physical Speleology*. (Translated by June C. Schmid) Springer-Verlag, New York, 284pp.

Davis, S.N. and De Wiest, R.J.M., 1966. *Hydrogeology*, John Wiley and Sons, Inc., New York, 463pp.

Doutsos, T. and Koukouvelas, J., 1986. Structural analysis of the Molai ore, *Min. Wealth*, 40, 7–15, (in Greek).

Exindavelonis, P. and Taktikos, S., 1984. *Geological Map of Greece, Molai Sheet*, Scale 1:50,000, I.G.M.E. Athens.

FAO-UNDP, 1980. *Water Resources Development in the Molai Area, Greece*. Final Technical Report.

Foster, M.D., 1942. Base exchange and sulfate reduction in salty ground waters along Atlantic and Gulf coasts: *Am. Assoc. Petrol. Geologists. Bull.*, vol. 26, pp. 838–851.

Ghikas I.C., Kruseman, G.P. Leontiadis, I.L. and Wozab, D.H., 1983. Flow pattern of a karst aquifer in the Molai area, Greece, *Ground Water*, vol. 21, No 4, pp. 445–455.

G.W.E. (German Water Engineering) GmbH, 1972. *Hydrological Study and Feasibility Report for the Lower Basin of the Evrotas River and Molai Region, Greece*. vols. I–VIII, OECD/Govmt. of Greece.

Jacobshagen, V., 1978. Geology of the Hellenides. In: H. Closs, D. Roeder and K. Schmidt (eds), *Alps, Apennines and Hellenides*. Inter-Union Commission on Geodynamics, Scientific Report, part 2, No 38, p. 415–423, Stuttgart.

Jacobshagen, V., 1986. *Geologie von Griechenland Beitrage zur Regionalen Geologie der Erde Gerbruder Boentraeger*, Borntrager, Berlin, 1986, pp. 1–363.

Kowalczyk, G., 1977. Jungquartare Strandterassen in S.E.-Lakonien, Peloponnes. In: G. Kallergis (ed.) *VI Coll. on the Geol. of the Aegean Region*. vol. 1, pp. 435–445, I.G.M.R. Athens.

Ktenas C., 1924. Formations primaires sedimentamorphiques en Peloponnese central. *CR. somm. Soc. geol. France*, 24, 61–63.

Ktenas, C., 1926. The development of the Paleozoic in central Peloponnese. *Prak. Acad. Athens*, vol. 1, pp. 53–59.

Leontiadis, I.L., 1981. Isotope hydrology study of Molai area in Lakonia. *Nuclear Research Center Demokritos, report Demo 81/4*, 27pp.

Papazeti, H., 1984. Petrological–mineralogical study of the rocks that host the mixed sulfide ores at Molai–Lakonia (Unpublished report in Greek) I.G.M.E., 36pp., Athens.

Papazeti, H., 1986. Petrographic research in the region of Tyros–Leonidion. Unpublished report, IGME, Tripolis Section, 53pp.

Pe Piper, G., Panagos, A.G., Piper, A.M. and Kotopouli, C.N., 1981. The (?) mid-Triassic volcanic rocks of Lakonia, Greece. *Geol. Mag.*, vol. 118, p. 118–127.

Pe Piper, G., 1982. Geochemistry, tectonic setting and metamorphism of mid-Triassic volcanic rocks of Greece. *Tectonophysics*, vol. 85, 252–272.

Piper, A.M. and Garrett, A.A., 1953. Native and contaminated waters in the Long Beach–Santa Ana area, California. *U.S. Geol. Survey Water-Supply paper 1136*, 320pp.

Revelle, R., 1941. Criteria for recognition of sea water in ground water. *Am. Geophys. Union Trans.*, vol. 22, pp. 593–597.

Richter, D., 1978. The main flysch stages of the Hellenides. In: H. Closs, D. Roeder, and K. Schmidt (eds) *Alps, Apennines, Hellenides*. Inter-Union Commission on Geodynamics, Scientific Report, No 38, pp. 434–438, Stuttgart.

Stamatis, A., 1984. *Geological Map of Greece, Richea Sheet*, Scale 1:50,000, I.G.M.E., Athens.

Tavitian, C., 1990. The hydrogeologic conditions of the karst and Neogene systems of the Molai Basin (Laconia). Unpublished PhD thesis (in Greek), Univ. of Patras, 483pp.

Tavitian, C., 1993. Karst morphology of the Tripolitsa carbonates in Epidaurus Limira district, S.E. Peloponnese, Greece, *Min. Wealth*, (in press).

7 Hydrogeologic implications of a cyclonic rainfall event in central Oman

P.G. Macumber, Barghash bin Ghalib Al-Said, G.A. Kew and T.B. Tennakoon

Abstract During a cyclonic event in late 1992, rain fell widely across the otherwise hot, arid Al Wusta Region of central Oman. This led to widespread run-off, wadi flow and groundwater recharge, with the Wadi Rawnab in places being several hundred metres wide and 2 m deep. Rainfall and run-off were, at different times, either enriched or depleted in the heavy stable isotopes, deuterium and oxygen-18, situations which are normally associated with monsoonal and frontal storm systems, respectively. Infiltration was rapid and evaporation was insignificant with recharge water being fresh with total dissolved solids (TDS) being <350 mg l^{-1}. This was also shown by isotopic and chemical data from the fresh, locally derived groundwater flow system which overlies a saline regional system found throughout Al Wusta. Prior to recharge, the run-off rapidly evolved from being a bicarbonate water to a sulphate-rich water by dissolution of gypsum. Solution of bicarbonate occurs later within the aquifer.

Water is extracted at a rate of 200 000 l d^{-1} from a 50 m thick, fresh (TDS 250 mg l^{-1}) groundwater lens commencing at 100 m depth beneath the Wadi Rawnab. Its presence shows that, despite the extreme aridity, sporadic cyclonic/storm events can produce important fresh groundwater resources, given the right physical conditions. However, these resources can be sustained only with proper management.

The objective of the present study of local flow systems in Al Wusta is to determine the distribution and water quality of the freshwater lenses, and the processes and rates of groundwater recharge. This information is essential in order to develop models for sustainable groundwater management.

Introduction

Physical setting

The Al Wusta Region covers about 25% of Oman, centred midway between Muscat and Salalah, and extending from the Saudi Arabian border to the Arabian Sea (Figure 7.1). To the south of Al Wusta is the Nejd area of the Dhofar Region of Oman.

The core of Al Wusta consists of a broad low plateau – the Central Plateau – which commonly lies at an elevation of between 120 m and 240 m (Figure 7.1). Major features of the plateau are a number of elongated depressions up to 30 km in length (Ma'abar, Hayma, Mukhayzanah etc.) which may represent structural features and the remnants of ancient drainage courses. To the southeast, the plateau is dissected by a number of coastward trending, normally dry, wadis which are deeply incised, with cliffed valley walls up to 50 m high – this is the Coastal Wadi Region (Figure 7.1). Alluvial infill is thin in the middle and upper reaches of the wadis, and limestone bedrock is often visible in the valley bottoms. Westwards of the Central Plateau is the A'Rub Al Khali, with its predominantly aeolian landscape superimposed on an otherwise flat relict marine or coastal plain. The elevation of the plain is about 70–100 m with dunes rising up to 100 m above the plain.

Beneath the Central Plateau, Tertiary limestones and marls form an essentially unconfined aquifer with the water table generally occurring 100–150 m below the ground surface. By contrast, in the eastern A'Rub

Groundwater Quality Edited by H. Nash and G.J.H. McCall. Published in 1994 by Chapman & Hall. ISBN 0 412 58620 7

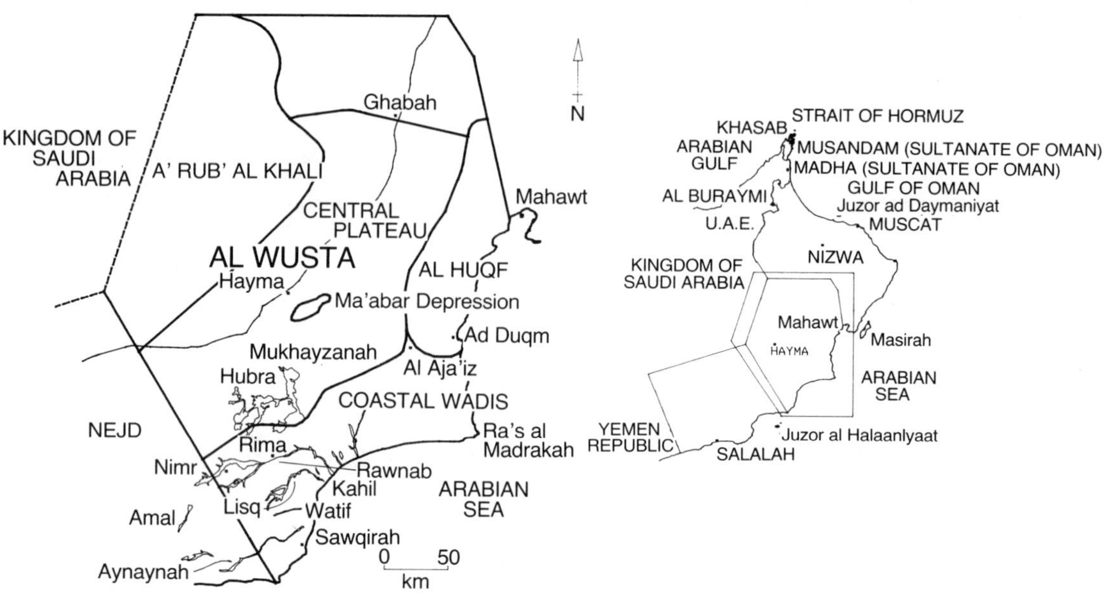

Figure 7.1 Locality map of Al Wusta Region, central Oman.

Table 7.1 Frequency of tropical cyclones

Period	Number entering Oman region	Number with associated precipitation
1950–59	4	2
1960–69	11	6
1970–79	19	10
1980–86	3	1

(from Taylor, Jones and Wigley, 1990)

Al Khali, groundwater is often close to the surface resulting in large areas of active sabkha.

The climate of the Al Wusta region is dry and arid with an annual average rainfall estimated at between 10 and 30 mm. The average rainfall is, however, misleading in that high intensity rainfall events capable of producing significant run-off and recharge occur infrequently. They are associated with low pressure systems moving across Oman (Barrett, Beaumont and Richards, 1990) or tropical storm/cyclone events tracking westwards across the Arabian Sea (Power, Barrett and Beaumont, 1989). An analysis of the frequency of tropical cyclones over Oman from 1950 to 1986 is given in Table 7.1. This shows that rainfall associated with tropical cyclones occurs about once every second year, with a steady increase from 1950 to 1979, then a fall in frequency. While many cyclones develop, they often dissipate before reaching the Oman coastline. Once they make landfall they often dissipate within 24–26 h (Taylor, Jones and Wigley, 1990).

This was not the case from 30th September to 5th October, 1992 when tropical cyclone O6-A passed over Al Wusta, causing rain to fall both as intermittent heavy showers and more sustained rain over a period of six days. This provided the first samples of cyclonic storm rainfall and run-off for chemical and isotopic analysis from the interior of southern Oman.

Hydrogeology

Regional groundwater flow

Much of Al Wusta is underlain by the Tertiary Hadhramaut and Fars Groups, which consist predominantly of marine limestones, marls and thin interbedded evaporites but with terrestrial sequences in the uppermost Fars Group. The limestones, marls and evaporites of the Hadhramaut and Fars groups are subdivided (Rippon, 1986) into a number of stratigraphic units – the Umm er Radhuma, Rus, Dammam and Fars Formations – which provide the framework for the regional aquifer system.

Most of central and southern Oman lies within a single groundwater basin stretching from the Northern Oman Mountains southwards to the Dhofar Mountains, and westwards into Saudi Arabia and the United Arab Emirates. Regional groundwater flow passes inland, from recharge zones occurring in and adjacent to the Northern Oman Mountains, and in the vicinity of the Dhofar Mountains in the south (Parker, 1985). Flow is towards zones of regional groundwater discharge in A'Rub Al Khali, and along the eastern coastline of Oman.

Groundwater quality deteriorates on passing inland. It is generally good in the Nejd, although locally degraded by dissolution of evaporites from the Rus Formation. On approaching Al Wusta, the overall groundwater salinity rapidly increases as it gains salts from the Tertiary sequences, and consequently the regional system is brackish to saline throughout Al Wusta (Parker, 1985), where it can only be used as a potable water supply after desalination.

The basic information on flow patterns, and the chemistry of the regional groundwater system in south central Oman was published by Parker (1985). These data arose from the water demands for oil exploration. The isotope study by the Public Authority for Water Resources, Oman (PAWR, 1986) provides an important insight into the nature and origins of distinct hydrofacies present. Such isotope studies (PAWR, 1986; Clark *et al.*, 1987; Wushiki, 1991) provided much information on the recharge processes across southern Oman. For instance, Nejd water is strongly depleted in oxygen-18 and deuterium and shows little deviation from the global meteoric water line (GMWL) indicating that rapid infiltration occurs, with only minor evaporation at the surface prior to recharge (Clark *et al.*, 1987). In contrast, groundwater within the Salalah Plain and the Dhofar Mountains is strongly enriched in stable isotopes, reflecting a predominantly monsoonal (meteoric) source and perhaps some sea-water intrusion nearer the coast. The regional groundwater passing northwards into Al Wusta from the Nejd is therefore clearly not derived from any simple monsoonal source such as that which currently recharges the Dhofar Mountains. Instead it is thought to derive from cyclonic or depression induced storm precipitation in which the stable isotopes are depleted as the storms rain themselves out over central Oman (Clark *et al.*, 1987).

On the basis of stable isotopes, two quite different groundwaters have been distinguished in southern Oman – the O-18 and D-enriched **monsoonal** type occurring nearer the coast and a depleted **cyclonic storm** or Nejd type, found further inland.

Tritium analyses from the uppermost groundwaters indicate that modern day recharge occurs to the Dammam and upper Rus Formations (Zone A) in the Nejd (Clark *et al.*, 1987). This is supported by modern C-14 dates (Clark *et al.*, 1987, Table 2, p. 82).

Local flow system

Despite the saline regional groundwater system, small yields of potable to marginally potable groundwater have been obtained from a number of places in Al Wusta. Such water normally underlies large surface depressions such as at Hayma and nearby Ma'abar (Figure 7.1) or beneath the major wadi systems draining eastwards towards the coast. The fresh water is mostly found in the Fars Formation (Parker, 1985), which crops out extensively throughout Al Wusta.

Groundwater salinity generally ranges from a TDS of 250 mg l^{-1} to > 2000 mg l^{-1} in the local flow systems (the upper limit given is an arbitrary limit for the local flow system: the actual value for the local flow system at any one place will depend on the salinities of both the local flow system and the underlying regional flow system, which are separated by a boundary within a transition zone). At Ma'abar, small scale irrigation is carried out on a few farms using transitional water with a salinity of 3000–4000 mg l^{-1} TDS. The fresher water occurs as lenses directly overlying and merging downwards and laterally into the saline regional groundwater. The fresh water is from a locally derived groundwater flow system of recent to sub-recent age, recharged from the overlying depressions and wadis during periods of intense rainfall, when there is significant run-off and stream flow.

These freshwater lenses are a valuable but limited resource, being used in a number of instances as potable water supplies and for minor irrigated agriculture. For instance, it has been estimated that 42% of families in Al Wusta use groundwater, the figure rising locally to 98% in the Wilayat of Jazir (Sultanate of Oman, Regional Plan of Al Wusta Area, 1992).

An indication of the salinity differences between the overlying locally derived groundwater and the underlying regional groundwater can be seen in a number of bores scattered throughout Al Wusta. For instance at Hayma (Figure 7.1), a thin lens of fresh water provides small quantities of groundwater for local usage (Kirkpatrick, 1978; Parker, 1985). The fresher water at Hayma ranges from about 1000 mg l^{-1} TDS up to about 5000 mg l^{-1}. The underlying saline water has salinities in excess of 16 000 mg l^{-1}. The salinity variation in the local and regional groundwaters can be seen in the groundwater exploration bore Hayma W-16 (Table 7.2) in which a fresher water lens overlies saline water with a 3 m thick transition zone. The bore is wholly in the Fars Formation.

Because of their limited thickness and the presence of underlying saline regional groundwater, the water quality of the lenses will gradually deteriorate on extraction unless managed properly. Indeed, there is commonly a gradual increase in salinity once extraction commences. The rate of water quality deterioration depends on the size and thickness of the freshwater lens, on extraction rates, on recharge rates, and on well design. For sustainability, the extracted potable water must be replaced by locally derived groundwater recharge. The rate of this recharge is therefore the limiting factor in groundwater usage.

The amount of recharge in turn depends upon the periodicity of more intense rainfalls, and the extent

Table 7.2 Salinity variations in the Hayma Bore W-16

Depth (m)	EC*
92	3 000
103	3 000
106	15 000
123	25 000
140	35 000

* EC = units in μmhos cm^{-1} electric conductivity. EC is analogous to salinity– the relationship for Al Wusta groundwater with salinities less than 10 000 mg l^{-1} being approximately 0.65 mg l^{-1} TDS = 1 EC. (from Kirkpatrick *et al.*, 1978)

to which the resulting run-off is concentrated in depressions and wadis from which it may infiltrate. The local flow system is important as a source of sustainable potable water only where the run-off is concentrated into such localized recharge zones capable of producing exploitable freshwater lenses. Therefore, apart from rainfall, the drainage pattern (i.e. geomorphology) is the most important factor which influences the rate and areal distribution of groundwater recharge in the Al Wusta Region and elsewhere in central Oman, and hence the availability and sustainability of potable groundwater resources.

The salinity and extent of the groundwater lens, and its sustainability upon exploitation, is different in the cases of the closed depressions of the Central Plateau, and the coastal flowing wadis. The major wadis have a coordinated drainage system, whereby run-off is derived from an extended catchment area. Furthermore, they are deeply incised permitting the development of extensive tributary networks. During the occasional periods of flow, the wadis are an important line source of recharge to the underlying limestone (or occasionally alluvial) aquifers. Although as yet largely unexplored, groundwater systems underlying the larger wadis with their extensive catchments are likely to yield the largest, best quality and perhaps most sustainable freshwater lenses in Al Wusta. This is already the case in the Wadi Rawnab, which flows from the vicinity of Nimr to the coast – a distance of about 100 km (Figure 7.1). In the Wadi Rawnab near Rima, groundwater with salinity of less than 250 mg l^{-1} is extracted daily at rates of about 20 000 l h^{-1} from a 50 m thick freshwater lens. The water table is 100 m below the wadi floor in Fars limestone.

In contrast, the relatively shallow closed basins of the Central Plateau normally have a limited catchment, which often extends no further than a few hundred metres beyond gently sloping perimeters. Groundwater extraction rates are normally low and salinity levels are higher, often being only marginally potable or brackish. For instance, at Ma'abar, the water quality of the freshwater lens varies from <1000 mg l^{-1} TDS upwards, however the water quality on small scale irrigation farms is about 4000 mg l^{-1} TDS, reflecting the high extraction rates needed. As at Wadi Rawnab, the water table is about 100 m below ground surface.

Cyclone O6-A

Rainfall and run-off

During late September and early October, 1992, tropical cyclone O6-A passed over Al Wusta causing extremely heavy rainfall over a wide area. Two rain gauges established by the Yalooni Oryx Project in central Al Wusta (at Yalooni and Wadi Faysr) received 41.25 and 54.5 mm of rain respectively over the 24 h period on the 3rd October.

In the Rima area 120 km to the south (Figure 7.1), rain fell over a six-day period from 30th September to 5th October. Precipitation was initially in the form of isolated sporadic rain storms, followed several days later by more sustained general rain commencing on the evening of 4th October and lasting for longer than 10 h. Run-off from the plains led to flow in small tributaries feeding the larger wadis. The rain caused widespread local flooding and the normally dry Wadi Rawnab flowed for about five days, being several hundred metres wide in places, and over 2 m deep (Figure 7.2). The surface water had virtually disappeared within a few days after the rains ceased, being lost into the groundwater systems as recharge. Only a few surface pools remained after two weeks. Despite the high potential evaporation rates, evaporative losses at the surface were negligible as water remained there for such a short period before infiltrating and recharging the groundwater.

Sampling

During and after the rainfall period, rainfall, run-off and wadi flow waters were sampled for chemical and isotopic analysis, and elsewhere spot electrical conductivity (EC) measurements were taken during the rainfall period. Rainfall samples were obtained from the Rima district at the beginning of the storms on 30th September and later at the beginning and during the sustained rain period on 4th to 5th October. In addition, a dew sample was obtained at Rima several days before the rain started. Isotopic data from the cyclone are shown in Table 7.3 and in Figure 7.3. Run-off and wadi flow samples were taken over the period 30th

Figure 7.2 The Wadi Rawnab near Rima, flowing during the passage of cyclonic storms – early October, 1992.

September to 16th October, many being taken during and soon after the first rainfall event, and therefore closely reflecting the rainfall properties.

A second set of samples was collected on 16th October, by which time virtually all of the surface water had gone. The only water present on the surface was found in what had been especially deep pools or in rock pools sheltered from direct sunlight. The results from the surface systems were then compared with groundwater data obtained prior to the cyclonic event.

Run-off-recharge salinity

Run-off salinities rarely exceeded 3000 EC, and were normally < 500 EC. Higher salinities were associated with areas of ground disturbed by earth-moving equipment. Thus, during the short period water remained on the surface, dissolution and evaporation were minimal, and the resulting recharge water was virtually fresh (< 500 EC). This agrees with the occurrence of locally recharged groundwater underlying Wadi Rawnab at a depth of 100 m, which has a salinity of about 390 EC (250 mg l^{-1} TDS). It suggests that similar quality groundwater exists along wadi systems elsewhere in the region.

Stable isotopes

Waters sampled during the episode fall into two isotopically distinct groups (Figure 7.3 and Table 7.3). In the case of the initial rainfall event on 30th September, the rain, run-off and wadi flow waters were all enriched in oxygen-18 and deuterium (as was the dew sample obtained prior to the initial rains), with oxygen-18 values ranging from −1.66 to −0.36 δ‰ and deuterium from −4.4 to 5.7 δ‰. These waters are therefore isotopically similar to monsoonal rainfall and groundwater recorded from the Salalah/Jebel region (Clark *et al.*, 1987; Wushiki, 1991). They are significantly different from the O-18 and D-depleted groundwater previously noted in the Nejd (Clark *et al.*, 1987) and from Al Wusta groundwater obtained during this study (see below).

In contrast, the later rains of 4th and 5th October and the resulting run-off and wadi flow were depleted in O-18 and D, with the rains (oxygen-18 ranging from −8.15 to −7.67 δ‰, and deuterium from −68.1 to −57.5 δ‰) being the most depleted waters yet observed in Oman. Run-off was likewise depleted, but less so than the rain; run-off also showed a range of isotopic values falling into two distinct groups 'A' and 'B'. Group A was obtained soon after the rainfall event. It lies close to the meteoric water line, but is significantly less depleted than the contributing rain. This reflects a mixing of the earlier (enriched) and the later (depleted) rainfall types once on the ground, since run-off from the later event mixed with remnant pools created during the earlier storms. It could also contain a rainfall input lying between the two extreme rain isotopic compositions sampled during the cyclone. However, this is only conjecture.

Group B was collected on 16th September, that is, 11 days after the rains had ceased; it represented the last vestige of surface water. This latter group was obtained to examine the effect of prolonged evaporation on the isotopic composition.

The isotopic character of rains and resultant run-off from Cyclone O6-A therefore encompasses both the O-

Table 7.3 Oxygen-18 and deuterium from precipitation and run-off

Sample No. Precipitation	Oxygen-18	Deuterium	Location
96	−1.26	1.7	Dew sample – 29th Sept.
119	−1.45	5.7	Asad – 30th Sept.
148	−7.67	−68.1	Rima – 4th Oct.
150	−7.79	−57.5	Rima – 4th Oct.
151	−8.15	−58.8	Rima – 4th Oct.
Run-off			
Enriched run-off from earlier rainfall event			
115	−1.52	−2.0	Asad[1]
116	−1.63	−0.6	Asad[1]
117	−1.61	−1.0	Asad[1]
118	−1.45	−3.9	Asad[1]
122	−0.50	0.7	Thayfut
123	−1.40	1.1	4 km south of Thayfut
124	−1.43	0.7	5 km south of Thayfut
128	−0.36	2.0	South Rima
132	−0.50	−2.2	South Rima
133	−1.66	−3.9	South Rima
135	−1.22	−4.4	South Rima
136	−1.56	−4.2	South Rima
137	−1.41	−3.0	South Rima
138	−1.30	−3.3	South Rima
139	−0.76	−1.9	Wadi Rawnab
142	−1.37	−3.7	Wadi Rawnab tributary
145	−0.73	−0.7	South Rima – resample 136
Depleted run-off following rains of 4–5th Oct.–Group A			
156	−6.10	−42.2	Wadi Rawnab
157	−6.35	−46.5	South Rima
158	−4.33	−28.7	Rawnab gathering station
160	−3.34	−25.8	South Rima
161	−5.55	−37.4	Near Maha 1 oil bore
Evaporated run-off, sampled on 16th Oct.–Group B			
168	2.04	−3.1	Near Maha 1 – resample 161
169	6.12	22.6	South Rima – resample 133
170	2.39	0.2	South Rima – resample 157
172	5.74	12.6	Wadi Rawnab tributary
173	2.57	15.7	Wadi Rawnab tributary

O-18 and D are expressed as δ‰ relative to Standard Mean Ocean Water (SMOW) in Tables 7.3 and 7.4 and all diagrams showing O-18 and D values.

[1] The Asad samples were taken from the surface during the initial storm; they can be considered as either rain or run-off.

18 and D enriched monsoonal type previously recognized from the Salalah Plain and Jebel, and the depleted cyclonic type observed in the Nejd groundwaters. However, the extent of O-18 and D depletion of the later rains is greater than found in either the Nejd or Al Wusta groundwater systems, and only the 'mixed' run-off of Group A resembles the isotopic composition of the inland groundwater.

Effect of evaporation on isotope enrichment

A cyclonic water line (Figure 7.3) determined for the overall event based on rainfall and early run-off (Table 7.3, excluding group B evaporated samples) is:

$$\delta D = 8.7\ \delta^{18}O + 9.13 \qquad (7.1)$$

This does not differ greatly from the Global Meteoric Water Line,

$$\delta D = 8\ \delta^{18}O + 10 \qquad (7.2)$$

and the two lines converge in the zone of isotopic composition encompassing the Salalah/Jebel groundwaters. The gradient reflects conditions under which evaporation takes place (temperature and humidity), and the deuterium excess is dependent upon the initial isotopic character of the surface water undergoing evaporation. While too much emphasis cannot be placed on a water line derived from a single cyclonic event, surface, rain and groundwater data from throughout central and southern Oman (Al Wusta and Dhofar) fit the line closely (Figure 7.4), and this water

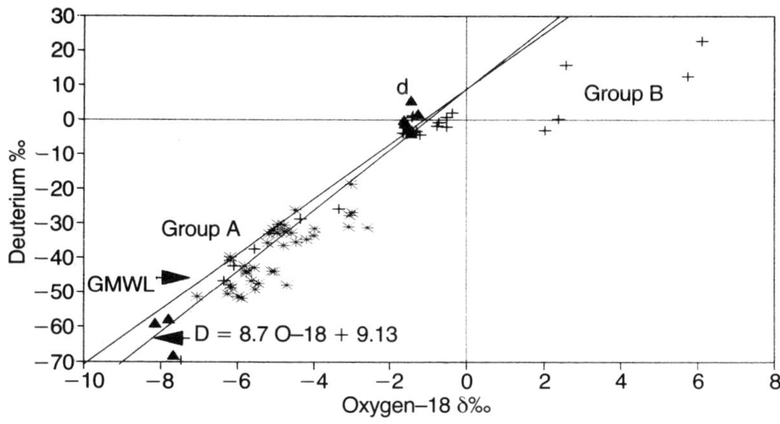

Figure 7.3 Oxygen-18 vs deuterium data from Al Wusta. All run-off and rainfall data are from the cyclonic event. The dew sample (d) was collected several days prior to the cyclone.

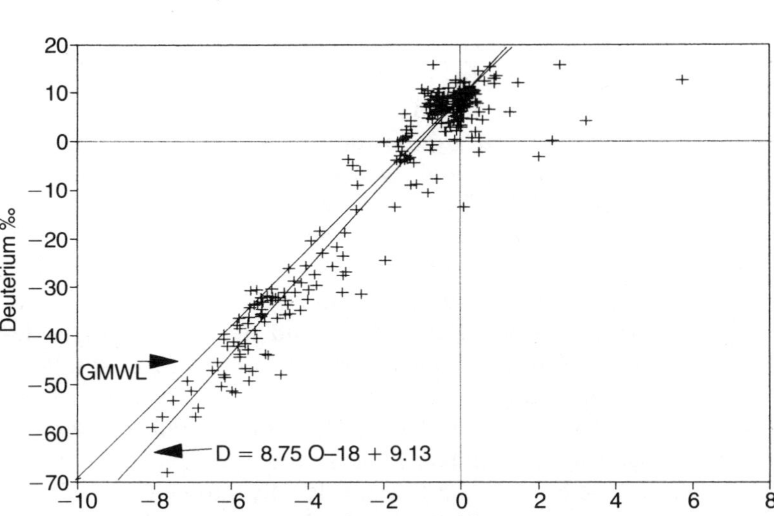

Figure 7.4 Oxygen-18 vs deuterium for precipitation, run-off and groundwater from southern Oman (Al Wusta and Dhofar).

line has been found to be useful in the following discussions in the absence of a more substantial data set.

Run-off samples taken 11 days after the last rain had ceased (Group B), show a strong evaporitic effect reflected in the wide deviation of O-18 and D values from those of the cyclonic water line (Figure 7.3). The extent of deviation is shown in Figure 7.5, in which the evaporative curves from a number of resampled sites are shown.

The time span between resampling after the first rains (lines a and b) was only two days and the deviation is small. Lines c and d represent the effect of evaporation 11 days after the cessation of the rains. In addition two curves show the nature of isotopic enrichment during evaporation over other time periods – one in the month preceding the cyclone (line e), and the other (line f) from Parker (1985). An extended form of line f was used by Parker (1985) to produce a characteristic evaporation line based on water derived from the Wadi Rawnab 1 bore:

$$\delta D = 5.36\ \delta^{18}O - 4.6 \qquad (7.3)$$

Line e is based on groundwater pumped from the Jalmud 1 bore. However it was first sampled some unknown time after evaporation commenced, as is shown by the extent of initial deviation from the cyclonic water line. Despite the differences in time and circumstance, the slopes are fairly uniform with only a minor variation in gradient in line d. The slopes approximate the characteristic evaporation line of Parker (1985), with a similar gradient but variable deuterium excess.

Stable isotope data from Nejd groundwaters had previously indicated rapid recharge. This is supported by direct observations of the disappearance of surface water following Cyclone O6-A. There is no comparable deviation from the meteoric water lines in either Al Wusta or Nejd groundwaters to that observed in Group B surface water only 11 days old (Figures 7.3, 7.5). This observation provides a measure of the rapidity with which groundwater recharge occurs in the limestone

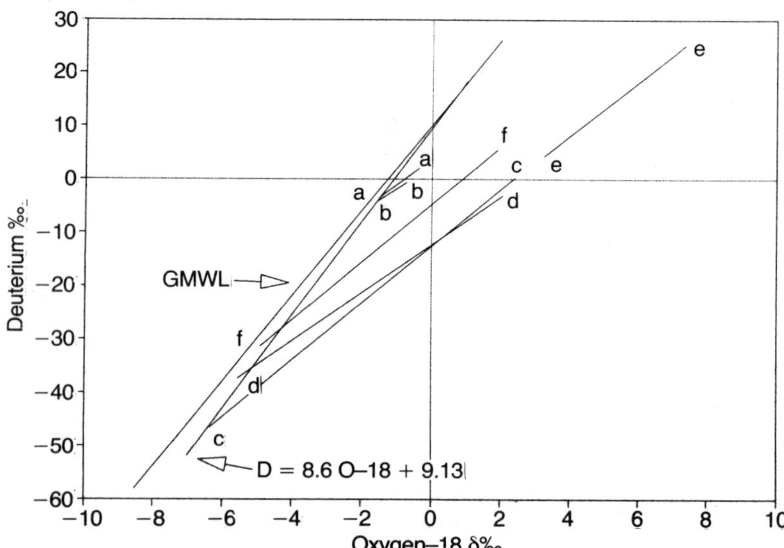

Figure 7.5 Oxygen-18 vs deuterium, showing the nature of isotopic enrichment during evaporation in Al Wusta.

terrains of central and southern Oman. The rapid recharge on the limestone terrains contrasts sharply with the situation observed in the aeolian settings of the Wahiba Sands of eastern Oman (PAWR, 1986, Figure 9.8) where the underlying groundwater shows a strong O–18 and D-deviation, reflecting evaporation from within the near surface regime during the recharge process.

Groundwater

Stable isotope analysis for O–18 and D was carried out on groundwater from 40 bores scattered across Al Wusta, covering both the local and regional flow systems (Table 7.4). These analyses were compared with the cyclonic rainfall and run-off, and with Dhofar (Nejd, Salalah Plain and Dhofar Mountains) groundwaters (Figures 7.4 and 7.6). Al Wusta groundwater, like that from the Nejd, comes from a number of aquifers and spans both modern and ancient **fossil** water, the latter being over 20 000 years old in places. Yet, the isotopic character of the fossil groundwater from the main Nejd aquifer 'C' is essentially the same as that from modern Nejd groundwater – aquifer 'A' (PAWR, 1986, Figure 10.2), suggesting similar recharge conditions for both groundwaters, despite the glacial age of the palaeowater. Both are totally different from Jebel and Salalah groundwater. This means that temporal factors are greatly subordinate to present day rainfall types and their distribution in determining the gross isotopic composition of the groundwaters. Therefore, in this chapter the groundwaters are not subdivided on an age basis, but instead dealt with as areally discrete groups, each with its own range of isotopic and salinity characteristics.

The salinity of Al Wusta groundwater ranges from a low of 390 EC to above 80 000 EC – the lower salinities (< 4000 EC) are from the local flow systems and the water is modern. Nejd regional groundwater is generally less than 4000 EC (Figure 7.6).

The isotopic composition of Al Wusta groundwater clusters along and beneath the cyclonic water line (Figure 7.3). Values lie between those of the two rainfall types observed during the cyclone (Figures 7.3, 7.6). Despite significant salinity differences, there is a strong isotopic overlap between the Al Wusta and Nejd groundwaters (Figure 7.6), and the five varyingly depleted Group A (mixed) run-off waters formed following the later cyclonic storms. This overlap suggests the incorporation in the aquifers of rainfall/recharge of intermediate (mixed) character.

The Nedj groundwater is locally more enriched in O–18 and D than that from Al Wusta, indicating a greater input of less depleted rains, whether monsoonal and/or enriched cyclonic. This is expected, given the closer proximity of the southeastern Nejd to the Dhofar Mountains and the possible impact of the monsoon in these areas.

There is a clear difference between the inland groundwaters and the monsoon derived groundwaters from the Salalah Plain and adjacent Jebel Qara (the fields of which also overlap with the early cyclonic rains from Al Wusta). However, as indicated above, this does not preclude an enriched or monsoonal contribution to the Nejd–Al Wusta groundwater system, but rather suggests that it is a composite water with isotopic characteristics lying outside the pristine monsoonal field.

Chemical evolution of surface waters and groundwaters

The rain which fell during the six-day period was essentially a calcium bicarbonate water (Figure 7.7). Salinities were highest at the beginning of the rain periods, but showed variations throughout. For instance, the initial salinity (measured as EC) during

Table 7.4 Oxygen-18 and deuterium from Al Wusta groundwater

Sample No.	δ‰ Oxygen-18	Deuterium	Locality
90	−5.15	−33.1	Wadi Lisq – sample 2
86	−4.57	−32.7	Anuq 1
97	−5.88	−51.6	Ghariba 2
98	−5.97	−51.3	Mukhaizna 2
66	−7.04	−51.3	Hayma 11
108	−6.26	−50.4	Mukhaizna 1
102	−5.54	−49.2	Qura 3
81	−6.15	−48.6	Reiham 1
99	−4.69	−48.0	Sadad 3
95	−6.18	−48.0	Aseel 1
103	−5.44	−47.3	Sihaam 1
101	−5.63	−46.7	Nafoorah 1
146	−5.76	−44.4	Nimr 22
104	−5.01	−43.9	Hubaira 1
79	−5.09	−43.8	Tannam 1
127	−5.77	−43.8	Sim Sim 1
77	−5.55	−42.8	Saha 1
100	−5.80	−42.2	Sadad 2
84	−6.19	−40.7	Rawnab South 1
82	−5.90	−39.6	Rukh 1
68	−4.79	−36.3	Ma'abar (3 bores)
78	−5.21	−35.6	Faras 2
93	−4.46	−35.4	Jalmud 2
71	−4.18	−34.7	Ma'abar
130	−3.99	−33.5	Nimr 2&3
129	−4.95	−32.7	Maha 1
85	−4.76	−32.2	Salwa 1
90b	−5.05	−31.7	Wadi Lisq
67	−3.96	−31.5	Ma'abar south
110	−2.58	−31.3	Hadh 1
69	−3.08	−31.0	Ma'abar
75	−4.75	−30.5	Al Zaihr
92	−4.92	−30.3	Jazir 2 (MWE)
73	−3.06	−27.6	Ma'abar
106	−2.98	−26.9	Mabrukh 3
91	−4.48	−26.1	Jazir 7 MWE
74	−3.01	−18.8	Al Ajaiz
94	3.24	4.3	Jalmud 1 (trough)[1]
105	7.35	25.1	Jalmud 1 (trough)[1]

[1] Jalmud 1 samples were from groundwater undergoing evaporation in a trough. They do not reflect aquifer conditions.

the more sustained second event ranged from 225 EC at the start to 60 EC some 20 min later and remained so for another hour. The following prolonged overnight rainfall ranged from 110 EC to 140 EC.

Repeated readings were taken over a number of days at a large number of points in order to assess the salinity of run-off and stream flow across the region. Sampling was also carried out at selected sites. Run-off salinities ranged from less than 100 EC to over 3000 EC, however the bulk of the run-off was less than 500 EC.

The isotopic composition of the run-off and groundwater demonstrates only very minor evaporation prior to recharge. Chemical analyses show that the bulk of the salinity increase during the run-off phase is largely due to dissolution of $CaSO_4$, derived from skeletal soils rich in gypsum. An indication of this is given in Table 7.5, which shows two adjacent pools ~10 m apart (samples 137 and 138), with salinities of 210 mg l^{-1} and 840 mg l^{-1} TDS, respectively. Despite the high bicarbonate content of the rains and their being derived from the same rainfall event, both are already sulphate waters. The more saline pool occurred where the surrounding surface had earlier been disturbed.

The isotopic composition of both pools is essentially the same, indicating that no significant evaporation has occurred to explain the higher salinity in sample 138. This implies that the increased salinity is due to mineral dissolution. The relative ionic relationships in Table 7.5 indicate that the salinity difference is largely due to increased $CaSO_4$ and to a much lesser extent, NaCl.

The extent of the dissolution of gypsum by run-off waters is shown more generally in Figure 7.7, in which sulphate is plotted against bicarbonate for surface water samples collected throughout Al Wusta over a nine month period. The rains are essentially bicarbonate waters. As the salinity increases there is little further increase in bicarbonate content, yet the sulphate rises steadily, resulting in a rapid evolution from bicarbonate rain waters to sulphate run-off waters. In contrast, the underlying groundwater is generally much higher in bicarbonate, indicating that significant carbonate dissolution occurs only after the water has entered the limestone aquifer system.

Conclusions

The Al Wusta Region of central Oman has an arid climate with annual rainfall averaging only 10–20 mm. The regional groundwater system is saline and the only potable water comes from thin, locally derived freshwater lenses overlying the saline regional groundwater. Infrequent cyclonic storms, such as those associated with Cyclone O6-A which passed westwards into Oman from the Arabian Sea in early October 1992, cause widespread sheet flooding, run-off and wadi flow which in turn cause rapid recharge of the local groundwater system. Water remains on the surface for only a very short time and is lost mostly by infiltration; by comparison, losses by evaporation are almost negligible. The salinity of the run-off and recharge water is low, mostly <300 mg l^{-1} TDS. This is comparable with fresh groundwater (salinity 250 mg l^{-1} TDS) occurring at 100 m depth beneath the Wadi Rawnab.

The initial oxygen-18 and deuterium content of the rainfall and run-off associated with the cyclone was enriched and that of the later rainfall strongly depleted.

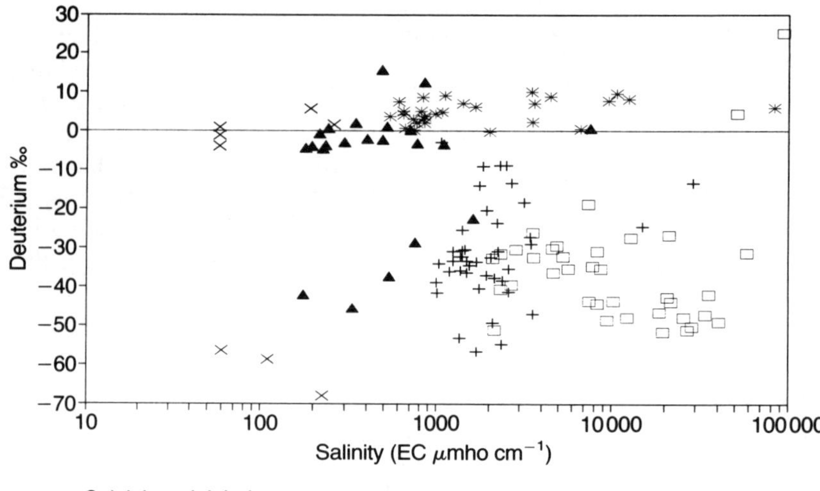

Figure 7.6 Salinity (EC) vs deuterium for precipitation, run-off and groundwater in central and southern Oman.

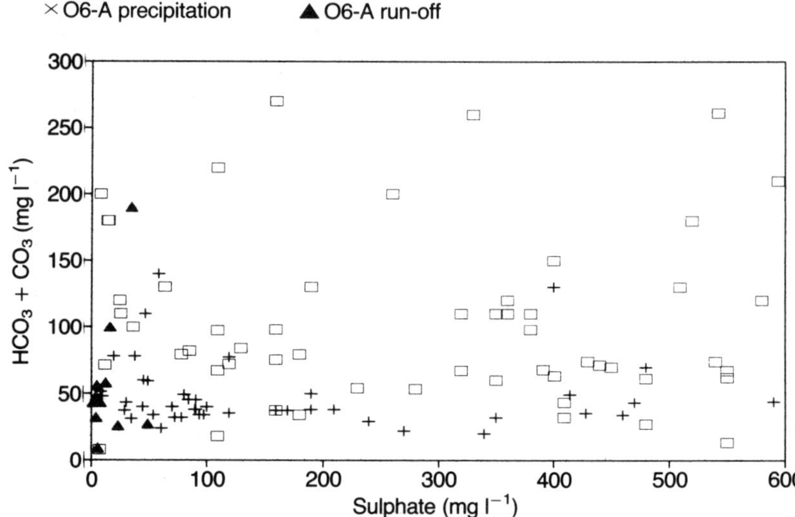

Figure 7.7 Sulphate vs bicarbonate for surface waters in Al Wusta.

Table 7.5 $CaSO_4$ pickup between adjacent pools at Rima

Sample	TDS (mg l^{-1})	(meq l^{-1})	Cl	SO_4	HCO_3	Ca	Mg	Na	$\delta^{18}O$	$\delta^{2}H$
137	210	(meq l^{-1})	0.62	1.64	0.52	1.89	0.02	0.61	−1.41	−3.0
138	840	(meq l^{-1})	2.17	9.79	0.70	9.98	0.02	1.74	−1.30	−3.3
	Difference	(meq l^{-1})	1.55	8.15	0.18	8.09	0.00	1.10		

The isotopic character of groundwater throughout Al Wusta and the Nejd lies between these two extremes, being closest to that observed in mixed run-off following the later (prolonged) rains. It also lies close to a cyclonically derived meteoric water line, and in this respect contrasts sharply with the heavily enriched (evaporated) character of small amounts of remnant surface water present 11 days after the rains ceased. This supports the visual evidence showing very rapid infiltration on the Tertiary limestone surfaces of the region, and very little loss by evaporation.

Both Al Wusta and Nejd groundwaters are quite unlike the O–18 and D enriched monsoonal waters of the Jebel and Salalah Plains. This does not preclude an enriched or monsoonal contribution to the inland groundwater system, but instead suggests a composite water with isotopic characteristics lying outside the monsoonal field.

The initial rainwater is enriched in bicarbonate but the run-off waters quickly dissolve gypsum and evolve to a calcium sulphate water before recharging the aquifer. The high sulphate content in the surface water is not normally accompanied by increased carbonate/bicarbonate content, and higher bicarbonate/carbonate levels generally appear only in the groundwaters, following dissolution of carbonate from the limestone aquifer.

Acknowledgments

For advice and assistance on this project, we wish to thank Harley Young from the Research Section, Len Hutton and the laboratory staff from the D.G. Water Resource Assessment, Oman Ministry of Water Resources, and Andy Herczeg and laboratory staff at CSIRO, Adelaide, Australia. David Parker, Senior Hydrologist at PDO, gave permission to sample the PDO water bores. Permission to publish this chapter was given by H.E. the Minister of Water Resources, Oman.

References

Barrett, E.C., Beaumont, M.J. and Richards, T.S., 1990. *A Satellite Evaluation of the Extreme Rainfall Event of 19th–21 February 1987 in the Sultanate of Oman*. Final Report (part 4) to the Sultanate of Oman Ministry of Environment and Water Resources, 24 pp.

Clark, I.D., Fritz, P., Quinn, O.P., Rippon, P.W., Nash, H. and Sayyid Barghash Bin Ghalib Al Said, 1987. Modern and fossil groundwater in an arid environment: a look at the hydrogeology of southern Oman. In: *Isotope Techniques in Water Resources Development*. I.A.E.A. Vienna, pp. 167–187.

Kirkpatrick, S.W. and Partners, 1978. *Groundwater Supply to Tribal Centre, Hayma*.

PAWR, 1986. *Origin and Age of Groundwater in Oman. A Study of Environmental Isotopes*. Public Authority for Water Resources, Oman, Report 86–7.

Parker, D.H., 1985. *The Hydrogeology of the Cainozoic Aquifers in the PDO Concession Area, Sultanate of Oman*. Petroleum Development Oman LLC.

Power, C.H., Barrett, E.C. and Beaumont, M.J., 1989. *A Satellite Climatology of Arabian Sea Tropical Storms and Cyclones 1966–1987*. Final Report (part 3) to the Sultanate of Oman Ministry of Environment and Water Resources, 85 pp.

Quinn, O., 1986. *Regional Hydrogeological Evaluation of the Nejd, Sultanate of Oman*. Public Authority for Water Resources, Oman, Report OFR-48-86.

Rippon, P.W., 1986. *Results of Pilot Exploratory Drilling and Aquifer Testing in the Nejd, Sultanate of Oman*. Public Authority for Water Resources, Oman, Report 1-86-11.

Taylor, D., Jones P.D., and Wigley, T.M.L., 1990. *Rainfall in Oman: Data Acquisition, Statistics and Climatology*. Final Report to the Sultanate of Oman, Ministry of Environment and Water Resources, 46 pp.

Wushiki, Hisao, 1991. 18-O/16-O and D/H of the meteoric waters in South Arabia. *Mass Spectroscopy*, vol. 39 (5), pp. 239–250.

8 The use of geophysical logging in understanding salinity distribution in alluvial aquifers: a case history from Oman

J.A. Heathcote

Abstract The Batinah Gravels of the north coast of the Sultanate of Oman are very extensive (> 12 500 km^2) and are a major source of potable and irrigation water. In common with many other alluvial sequences, complex permeability distribution with depth leads to the system behaving as a number of local aquifers, with different heads and chemistries. This is particularly pronounced when some of the water is saline.

Geophysical logging was used to investigate a number of wells in the Batinah Gravels near Al Khowdh. Resistivity and SP were successful in defining the location of highly saline zones, and, in some locations, underlying freshwater zones. The same wells were re-logged after development, using temperature, conductivity and a flow meter. A very different picture was obtained because of the complexities of internal flow in the well. A combination of the two sets of logs provides a picture of the natural head distribution and salinity of the aquifer system.

Because of the possible presence of head and chemistry variations in thick alluvial aquifers, it is recommended that exploration wells in such aquifers are drilled with mud and geophysically logged while mud-filled. Observation wells with long open sections should not be constructed because internal flow in the well masks the real head and quality variations with depth in the aquifer.

Introduction

The Batinah Plain extends some 250 km along the north coast of the Sultanate of Oman, and is > 50 km wide in places (Figure 8.1). The coastal edge of the plain is quite heavily populated and forms an important agricultural area. Dates and alfalfa were the traditional crops, but now a much wider range of foodstuffs is grown. Coastal Oman is arid, with an average annual rainfall of 60 mm falling mainly in winter, but several years at a time can pass with no appreciable rainfall. The settlements on the Batinah Plain, including the capital Muscat, situated at the east end of the plain, are thus heavily dependent on groundwater for irrigation water and for human consumption.

The Batinah Plain is underlain by more than 300 m of alluvial gravels (the Batinah Gravels) resulting from the erosion of the Jebel Akhdar mountains which border the plain to the south. Reflecting the geology of the mountains, the principal rock types forming the gravels are dolomite and serpentinite. The gravels appear to have been deposited sub-aerially by wadi systems similar to those of the present day, and are thus poorly sorted with a wide range of grain sizes from large cobbles downwards. In some areas, they also contain heavily cemented horizons apparently arising from diagenetic reaction of serpentinite with dolomite.

As part of the development of the area, increasing use is being made of deep boreholes to supplement the water supplied by traditional dug wells. This permits

Groundwater Quality Edited by H. Nash and G.J.H. McCall. Published in 1994 by Chapman & Hall. ISBN 0 412 58620 7

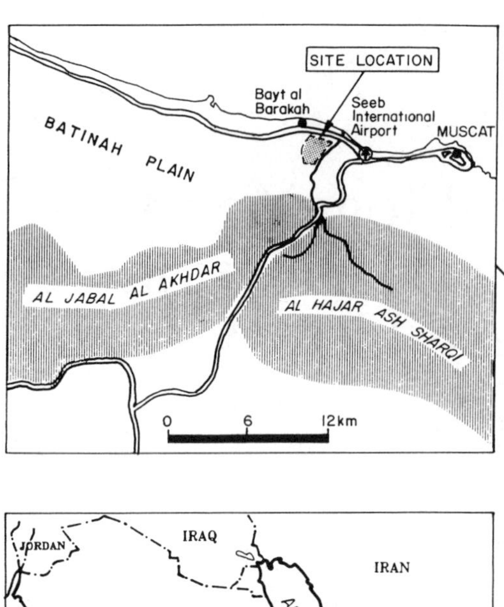

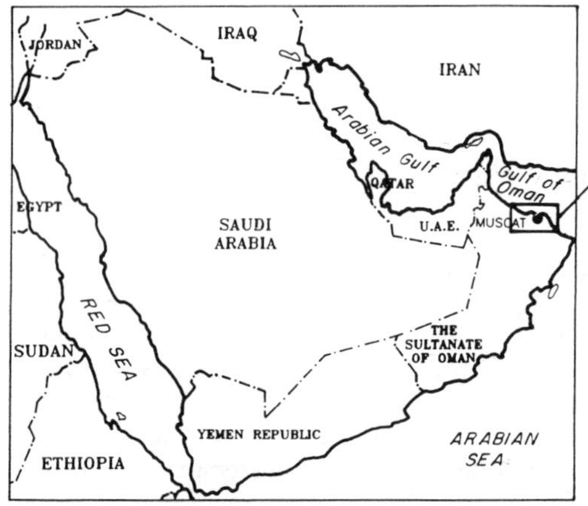

Figure 8.1 Location of study area.

agricultural development along the main highway, inland of the coastal villages, where the water table is too deep for dug wells. An extensive wellfield with an associated recharge dam has been built towards the downstream end of the Wadi al Khowdh, to serve the capital area.

Early drilling of the Batinah Gravels showed that they contained brine at depth, presumably originating from coastal sabkhahs present during the Pleistocene. Increasing development of the gravel aquifer brings the risk of mobilizing this brine, and also of producing modern saline intrusion.

To understand the distribution of saline waters the Public Authority for Water Resources has an ongoing programme of exploratory drilling and geophysical logging. This work was performed in 1984 as part of the programme.

Geophysical logging

Modern instruments are capable of measuring a large number of parameters in a borehole. However, comparatively simple technology, available for the past 60 years (Schlumberger *et al.*, 1933), can yield important information. The most important logs for understanding the Batinah Gravels proved to be the E-log (0.4 m normal resistivity, 1.6 m normal resistivity, and spontaneous potential), conductivity, temperature, and impeller flowmeter. All but the last of these are available as portable equipment.

The electrical resistivity of a rock formation is approximated by (Serra, 1984):

$$R_f = F\,R_w \qquad (8.1)$$

where R_f is the formation resistivity, R_w is the pore fluid resistivity and F is the formation factor, given by:

$$F = k\phi^n \qquad (8.2)$$

where k is a constant in the range 0.65–1.0, n is a constant in the range −2.0 to −2.15 and ϕ is formation porosity (as a fraction).

This derivation assumes that the mineral grains constituting the formation behave as electrical insulators. In practice, there is some electrical conductivity associated with grain surfaces. The effect of this becomes more important as the proportion of fine-grained material (silts and clay) increases. This results in the formation factor becoming smaller than calculated from the equation above. Formations with a relative absence of fine-grained material are thus more resistive for the same formation water conductivity than formations with fine-grained material, and are termed **clean**.

Substituting into equation 8.2, using $k = 0.65$ and $n = -2.15$, for sands with porosity in the range 0.2–0.3 (20–30%), and the value of F is 20.1–8.7 respectively.

Resistivity (R in Ωm) and conductivity (C in μS cm^{-1}) are related by:

$$R = 10\,000/C \qquad (8.3)$$

Thus formation resistivity is a function of both a formation dependent term, and of groundwater conductivity. High groundwater conductivity tends to produce low formation resistivity. Quantitative interpretation of groundwater conductivity from the formation resistivity requires knowledge of the formation factor F. The formation factor can be eliminated by using the different resistivity values from the E-log.

The 0.4 m normal resistivity measures dominantly the resistivity of a sphere of material radius of 0.4 m, centred on the borehole. This includes some of the borehole itself, and a zone of rock that may be quite strongly influenced by migration of drilling fluids into the rock. Correction for the borehole radius is by published charts, e.g. Schlumberger (1977), SPWLA (1979). The 1.6 m normal resistivity measures a correspondingly larger sphere which includes the smaller sphere, but which also includes rock much less likely to be influenced by the invasion of drilling fluid. If it is assumed for simplicity that the 0.4 m normal is totally influenced by drilling fluid, and the 1.6 m normal only by formation fluid, then

$$R_w = R_{mf} R_{1.6}/R_{0.4} \qquad (8.4)$$

where R_{mf} is the resistivity of the drilling fluid (which can be measured readily).

Spontaneous potential arises out of differences in chemistry, particularly salinity, between borehole fluid and formation fluid, where the formation is permeable. The larger the salinity difference, the larger the potential (Serra, 1984).

Conductivity and temperature are self-explanatory. At the high temperatures encountered in Oman, it is important to correct the conductivity logs to a constant temperature (25°C). The Schlumberger correction (Schlumberger, 1977) is a suitable correction for use where temperatures differ considerably from 25°C.

Flow logs are run down and up the hole at a constant speed. By running the log both up and down, the direction as well as the velocity of the flow can be determined, since the velocities add in one direction and subtract in the other. The down trip can be prone to snagging, leading to a variable tool velocity and thus an irregular log.

Borehole KWD2

This hole was drilled to a final depth of 366 m, entirely in the Batinah Gravels. The hole was logged a number of times during its construction: after drilling to 103 m (mud-filled); after drilling to 366 m with casing to 136 m (mud-filled); and several times when the well had been completed with casing to 116 m, and screen from 116 to 365 m (water-filled).

When the hole was cleared of mud by air-lifting at the end of drilling, the water produced had a conductivity of 16 000 μS cm^{-1}. A temperature/conductivity log run a few days after this development is shown in Figure 8.2. The high conductivity water produced during development can be seen clearly in the casing, with the conductivity falling to 1750 μS cm^{-1} at the start of the screened section and remaining fairly steady at this throughout the hole, falling just before total depth (TD). The temperature log shows an increase in temperature of only 2.5°C throughout the screened hole. This is much less than the regional geothermal gradient of approximately 2°C/100 m.

Fluid logging on several subsequent occasions showed the same pattern of temperature and conductivity.

The explanation of this behaviour is derived from the logs run when the hole was mud-filled. These are presented as a composite in Figure 8.3. The caliper log shows a generally uniform hole below the end of the casing at 137 m. The gamma log is rather uniform, reflecting the very low uranium, thorium and potassium content of the ultrabasic source rocks of the gravels.

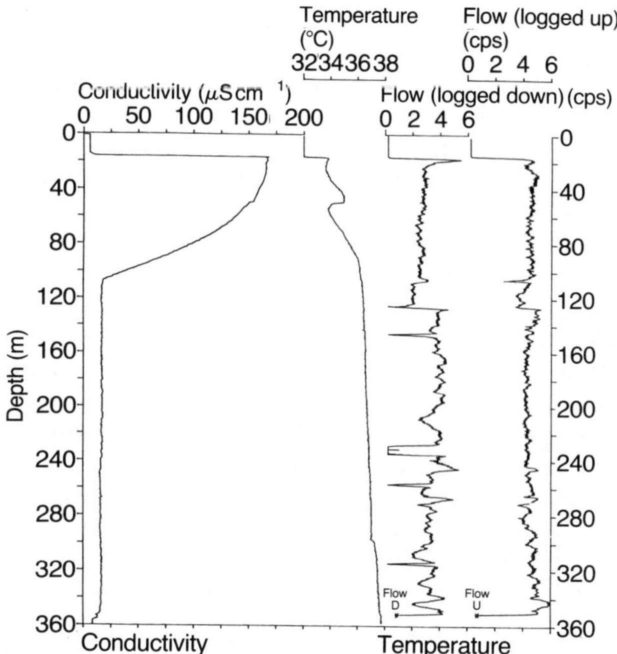

Figure 8.2 Fluid logs in borehole KWD2.

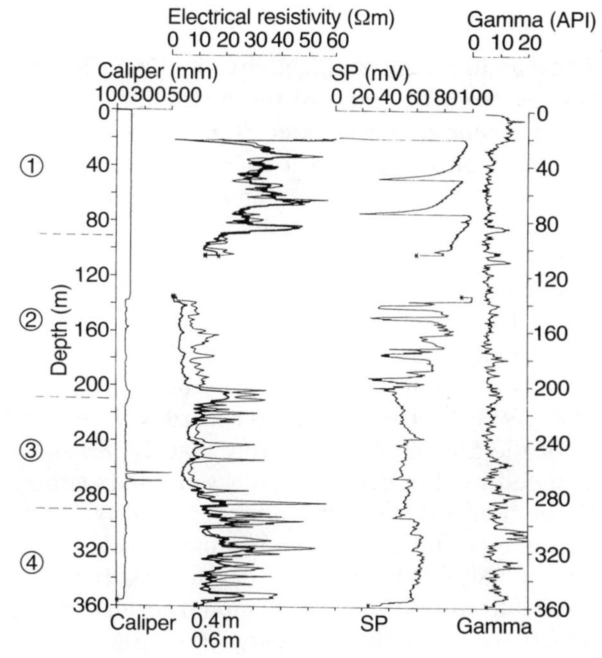

Figure 8.3 Formation logs in borehole KWD2.

The resistivity log has four clear sections: above 90 m both spacings show resistivities continuously above 20 Ωm; from 90 m to 202 m the 0.4 m normal seldom exceeds 10 Ωm, and the 1.6 m normal falls as low as

3 Ωm; from 202 m to 366 m the 0.4 m normal is in the range 10–60 Ωm and the 1.6 m normal is in the range 10–30 Ωm; below 280 m resistivities increase further.

The SP log shows three sections: a relatively straight log above 100 m (the sawteeth are an artifact of the logging process, of undetermined origin); an amplitude of up to 60 mV between 137 m and 202 m; and a relatively straight log between 202 m and 366 m. Mud conductivity during this logging was around 1000 μS cm^{-1} (R_{mf} = 10 Ωm).

The interpretation of the logs, using the techniques outlined above, is that the borehole penetrates a sequence of clean (high resistivity) and less clean (low resistivity) gravels. Formation water conductivity is very variable. Above 90 m, the water quality is good, only slightly more conductive than mud filtrate, but it deteriorates below this depth. Over the range 137–202 m, formation water conductivity can be estimated by substituting in equations 8.4 and 8.3 above.

$$R_w = R_{mf}\, R_{1.6}/R_{0.4} = 10 \times (3/10) = 3 \qquad (8.5)$$

from equation (8.3)

$$C = 10\,000/R = 10\,000/3 = 3333 \qquad (8.6)$$

Thus a minimum conductivity of 3300 μS cm^{-1} is estimated from the ratio of the resistivity logs. However, the value of F estimated from the 0.4 m normal using equation 8.1 is very low (~ 1), suggesting that invasion is far from complete and the 0.6 m normal is seeing a volume of rock containing water more conductive than pure mud filtrate.

Using a more typical value for F of 10 (still low but measured in many wells in the area) and the 1.6 m normal resistivity, the calculated formation water conductivity from equations 8.1 and 8.3 is nearer 30 000 μS cm^{-1}. This is more in accord with the amplitude of the SP signal, considering that the SP signal is attenuated by thin beds and poor sand/shale contrast.

Formation water quality improves sharply at 202 m to less than 4000 μS cm^{-1}, with further steady improvement to 280 m. Below this depth, resistivities are comparable to those above 90 m and formation water conductivity is probably below 1500 μS cm^{-1}. The frequency of clean gravels is less in this section than above 90 m, but greater than the interval above 280 m.

The flow log shown in Figure 8.2 is somewhat spoiled by snagging. By comparison of the up and down runs, and making due allowance for the different sensitivity of the instrument when running up and down, there is an indication of an upward water flow with a velocity of up to 2 m min^{-1}, from around 290 m to around 130 m. In 100 mm screen, this is equivalent to a flow rate of approximately 1 l s^{-1}.

The upward flow is an explanation of the very low temperature gradient observed in this well, described above. Warm water from depth has little time to cool as it rises through the well. Temperature decreases rapidly in the static water within the casing, where water temperatures are expected to be very similar to the temperature of the surrounding formation. It reaches a typical value for shallow groundwater (32°C) at the top of the hole.

Commentary

The fluid logs, run under natural conditions, give clear evidence of natural flow within the borehole. Even without the impeller flowmeter logs, the high and constant temperature is clear evidence of upward flow of deeper groundwater.

The resistivity logs show the presence of a freshwater aquifer below 280 m, overlain by a formation containing saline water, and a freshwater aquifer near the surface. The fluid logs show that the deep freshwater aquifer has the highest head and flows up the hole, the water dissipating into the overlying saline aquifer. This movement of fresh water from the bottom of the hole to the top results in the conductivity log recording a low value throughout the screened part of the borehole, despite the presence of saline water in part of the formation.

Under the drawdown caused by the airlift development, water is produced from all three zones and therefore the produced water is a mixture and is brackish. On the basis of this result, this area would have been abandoned as containing no fresh water below the near surface aquifer. The logging has shown that there is a freshwater aquifer at greater depth, which could be exploited.

Borehole RGS1

This hole is situated a couple of km from KWD2 along a line parallel to the coast. The hole was drilled to 180 m entirely through gravels and was logged while mud-filled. Mud filtrate conductivity varied slightly during drilling, over the range 1500–2000 μS cm^{-1}. The hole was completed with blank casing to 21 m and screen to TD. Fluid logs run after the completed hole was cleaned out are given in Figure 8.4.

The resistivity logs show that the formation water is fresher than the mud filtrate to a depth of around 45 m since above this point the resistivity of the uninvaded formation exceeds that of the invaded formation. In the interval 45–75 m, the two resistivities are similar suggesting that the formation water is somewhat brackish. Below 75 m, the formation water becomes more saline. A rapid fall in resistivity below 90 m, together with an SP signal amplitude of more than 60 mV, indicates that the formation water has become highly saline at this

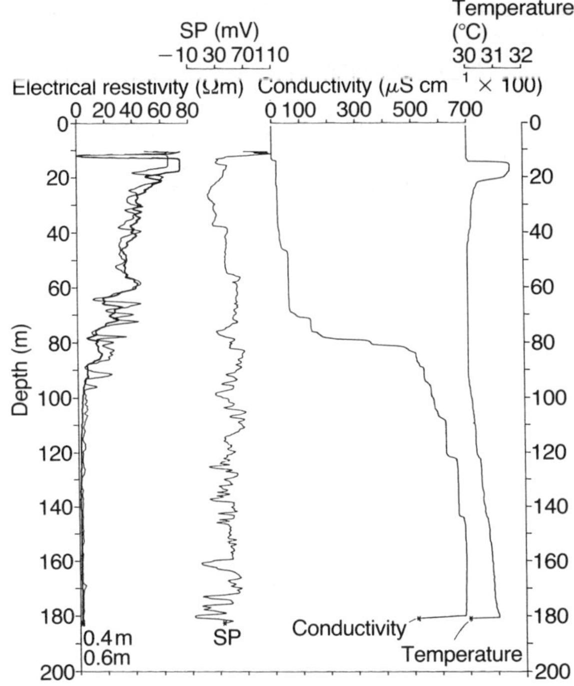

Figure 8.4 Geophysical logs in borehole RGS1.

depth. Using techniques discussed previously, the salinity certainly exceeds 6000 μS cm^{-1} and is more likely to be in the range 50 000–100 000 μS cm^{-1}.

The conductivity log in the clean hole shows an increase in conductivity at 45 m from 2000 μS cm^{-1} to 6000 μS cm^{-1}, corresponding to the information from the resistivity logs, and then a sharp saline interface at 80 m, with the conductivity rising to over 60 000 μS cm^{-1}. The temperature log shows a negative gradient in the upper part and is difficult to interpret in terms of flow directions. Vertical flow was below the limit of detection with an impeller flowmeter.

The most significant fact is that the saline interface in the borehole is 10 m higher than the position of the change in salinity in the formation, indicating that the head of the saline water is higher than that of the fresh water. Flow along the borehole could therefore result in saline water contaminating the freshwater aquifer. The log run in the mud-free hole does not indicate the correct position of the saline transition.

Conclusions

The logs from these two holes show that the views of a borehole deduced from pumped behaviour, logging under natural (flow) conditions, and logging while mud-filled, are very different. Only by logging mud-filled, which prevents natural flow in the borehole in most circumstances, can the true distribution of formation water salinity be seen. Once the mud in the hole is displaced by water and the borehole developed, the observed conductivity distribution in the well is dominated by the zone of highest head, and controlled by complex flow patterns within the hole. The behaviour on pumping cannot be predicted from these data.

The difference between the position of the saline interface in an open well, and its position in the aquifer, as demonstrated by RGS1, casts doubt on monitoring possible saline intrusion using such open wells.

The simple measurements and interpretation techniques described above can reliably detect saline water, the accuracy with which its composition (conductivity) can be estimated is low, with errors greater than one order of magnitude.

Irregular head and conductivity profiles show that the Batinah Gravels formation in the area studied, behaves as a number of different aquifer units, despite the absence of regional confining layers. This has obvious implications for estimating the water resource available and optimizing its exploitation.

Recommendations

It is clear that when investigating a thick alluvial aquifer in an area where saline water may be present, the exploration wells should be drilled with mud, and geophysically logged while mud-filled, so that the true salinity distribution can be understood.

Monitoring wells should not be open over long distances, but should be constructed essentially as piezometers. Wells with long open sections will inevitably flow internally with two consequences: firstly, the measurements made in them may be impossible to understand in terms of aquifer salinity, and secondly, flow along the borehole may produce aquifer contamination.

Production wells, which will have longer screened sections and which might not be drilled with mud, can be designed only in the context of a proper understanding of the head and salinity.

Acknowledgment

The borehole logging described herein was performed under contract to the Public Authority for Water Resources, Sultanate of Oman, whose permission to use the data and to publish this chapter is gratefully acknowledged. The conclusions and recommendations made are entirely the responsibility of the author.

References

Schlumberger, C., Schlumberger, M., Leonardon, E.G., 1933. A new contribution to sub-surface studies by means of electrical measurements in drill-holes. *American Institute of Mining and Metallurgical Engineers, Technical Publication 503.*

Schlumberger Ltd, 1977. *Log Interpretation Charts, 1977 Edition*. 83 pp.

Serra, O., 1984. *Fundamentals of Well-log Interpretation.* Elsevier, 423 pp.

Society of Professional Well Log Analysts (SPWLA), 1979. *The Art of Ancient Log Analysis*. Houston.

9 Salinity and shallow wells: a view from the field

M. Woodhouse

Abstract The relevance of salinity monitoring in the context of community-based shallow wells is considered with reference to practical experience in Kibwezi, Kenya. Non-potability and variations in salinity are shown to cause a significant reduction in the usage of new water points. Observations of salinity monitoring are summarized with reference to the concerns of consumers that salt-free sites can be identified and that trends in rising salinity can be rectified. Land management is suggested as a practical salinity control mechanism.

Introduction

Many community-based shallow well programmes are run on tight budgets because organizations out of necessity try to keep their construction and operational costs down. Funding is rarely available for water quality monitoring or research, nor to cover siting of wells, using appropriate techniques to predict salinity. Monitoring of the salinity of well water can, however, easily be introduced as a routine project activity, as was found in Kibwezi, Kenya (Figure 9.1), but the interpretation of the acquired data can be difficult and may fall short of providing a basis for acting on communities' concerns. Observations have shown that water is not used when it is too saline to be palatable: in the case of Kibwezi, the water in 14% of the wells was found to be above the potability level (determined on the basis of consumers' comments) for at least part of the year. This means that newly constructed water points are not fully used and that finances therefore are not being utilized to maximum benefit. This picture is complicated by the fact that the salinity of well water changes dramatically at some wells within a single year, and in some wells is found to be too saline ever to be potable.

Figure 9.1 Location of the Kibwezi Division in Kenya.

Communities which are prepared to invest their own efforts in well construction usually have two main concerns regarding salinity; firstly, how can they be sure that a well will not tap groundwater which has a present or future risk of unacceptable salinity; secondly, if the water is saline, can the salts be removed? In relation to the latter concern, existing desalination technologies are generally not appropriate to the rural situation: reverse osmosis techniques require power supplies and capital, solar distillation is more promising but expensive for the volumes concerned, whilst chemical balancing of the water is too sophisticated for most rural situations – though it is interesting to note that some communities do blend two

Groundwater Quality Edited by H. Nash and G.J.H. McCall. Published in 1994 by Chapman & Hall. ISBN 0 412 58620 7

waters to produce acceptable water quality. In any case, an alternative salt-free site may not always be present, and such sources may be deep and thus expensive to exploit. For this reason shallow wells continue to be the most acceptable solution.

Kibwezi water project

The African Medical and Research Foundation (AMREF) began to assist communities in the construction of shallow wells in Kibwezi in 1984. A three-person technical team was stationed in the area and to date has assisted 150 communities to construct supplies. The aim of the project, to transfer the technology for construction and maintenance to the local people, has been achieved through practical and formal training. The local management body of the project has subsequently extended the aims of the project to encompass developing its financial sustainability and to assist schools in the area to develop water supplies and sanitation facilities. All this work has been funded through charitable contributions and local self-help. Funds are currently available for a further two years' work.

The Kibwezi division covers an area of 8000 km^2 and is inhabited by some 200 000 people. It is a semi-arid area of marginal agricultural potential. The rainfall in the area is erratic and unreliable. The weathered bedrock aquifer, which is exploited by the shallow wells, is of metamorphic origin: it comprises a complex mosaic of interconnected and penned sections and little is at present known about the flow within the mosaic. Recharge of the aquifer is from local rainfall and is estimated to equate to about 5% of the annual average rainfall. The average well depth is 10 m with the top part being lined with bricks made locally by community groups.

Well siting

In the past, wells have been sited with little or no consideration for local hydrogeological conditions. Particularly, when a well is to be hand-dug by the community, there is a need to have a high degree of assurance that not only will there be enough water but also that it will be potable. An efficient community group requires two months of regular work to dig and line a well, and this amounts to a considerable commitment of their resources. For this reason, the Kibwezi project is keen to adapt new geophysical techniques to assist in well-site selection. Better results are being obtained using electromagnetic (EM) profiling in the case of these shallow wells, to differentiate between saline and fresh water. The cost of such equipment is substantial, however, more than £8000. The equipment is portable and it is possible to survey several hundred line metres per day with it.

Low cost alternatives to a geophysical approach for well siting depend mainly on experience. In Kibwezi, the native vegetation has been found to provide quite accurate indicators of shallow salt-free groundwater. One particular tree, *Acacia robusta*, which indicates shallow groundwater, has never yet been seen to grow in the vicinity of a well with salinity (EC) which is 1500 $\mu S\ cm^{-1}$ (Woodhouse, 1991).

In the case where a number of groundwater sources are already being exploited, it is possible to map salinity in the area and thus to avoid digging wells in particularly saline sites. The further application of geophysics to assist in such mapping exercises would clearly be of great benefit. However, in most cases all that is done is to measure the salinity once the well has been dug and a pumping test has been carried out. It is common for well siting and digging to continue without any comprehensive interpretation of available salinity data. Thus, communities often find that they have acquired a new well, an apparent improvement on their former situation, but which is of limited value because it is salty. Also, the salinity may increase as the dry season approaches and with the passage of time.

Measuring well salinity

A robust self-contained conductivity meter can be obtained for an outlay of about £200. Metal or plastic probes are more suitable than glass for field use: a meter with a built in temperature probe is most convenient. Conductivity probes are very simple to operate and do not require any special operator skills. The meters do require routine calibration against a standard solution, however. This is important to ensure consistency of data. It is common practice that the person visiting a well to sample the water will take a temperature and conductivity reading on site. To obtain a real understanding of salinity distribution and variation with time, monitoring needs to be planned in such a way that geographical distribution can be inferred as well as annual trends, including the effects of periods of intense rainfall, long-term trends and similarities between groups of wells. Local people could easily monitor salinity, provided funds are available to purchase a number of conductivity meters: due to the distance between the wells, the costs and logistics otherwise make routine, regular monitoring visits prohibitively expensive.

Table 9.1 Average well salinities for two well programmes

Salinity ($\mu S\ cm^{-1}$)	Percentage of wells in range	
	Project 1	Project 2
<500	2.5	10
501–1 000	30	17
1 001–1 500	30	21
1 501–2 000	12.5	21
2 001–2 500	17.5	7
2 501–3 000	2.5	7
>3 000	5.0	17
Mean EC	1 480	1 957

Project 1: Lake Kenyatta, Kenya, February 1988, wet season, 40 wells in sample, coastal area.
Project 2: Kibwezi, Kenya, June 1989, dry season, 29 wells in sample.

Table 9.2 Consumer objections to salinity

Well salinity ($\mu S\ cm^{-1}$)	Percentage of consumers objecting to taste
483	17
734	0
781	60
1 349	0
1 458	18
1 880	63
1 990	0
2 800	29
6 980	91

Acceptable levels of salinity

Although the European Community recommends a guideline limit for potable water of 1000 $\mu S\ cm^{-1}$, this is a hard target for many shallow well programmes in developing countries to attain. Some representative values from Kenya are given in Table 9.1.

To determine the acceptability of water supplies in Kibwezi, water consumers were asked for their opinions on the taste of their well water. This is a notoriously difficult question for which to obtain a satisfactory answer because perceptions vary so much between individuals (WHO, 1984). Not surprisingly, the results of this enquiry, shown in Table 9.2, are inconclusive: however, for the wells and consumers in Kibwezi, it would seem that about 40% of the people refuse to drink water that is above 3000 $\mu S\ cm^{-1}$.

During one survey, carried out in July 1989, it was found that 14% of the well waters had salinities in excess of 3000 $\mu S\ cm^{-1}$. Some wells experience a marked increase in salinity towards the end of the dry season. Table 9.3 illustrates the average monthly rainfall and salinity for Thasyo well, Kibwezi.

Ad hoc salinity data have been collected from the Kibwezi wells during the past five years and the following four conditions are recognized:

(1) in some wells salinity rises and falls inversely with rainfall (as is shown in the case of the Thasyo well), with a time lag of one to three months;
(2) some wells have a permanently high salinity (e.g. Kalulu well, 4500 $\mu S\ cm^{-1}$);
(3) some wells have an initial low salinity, but levels increase subsequently (e.g. Likoni well, which, when measured each year, showed a value of 570 $\mu S\ cm^{-1}$ in March 1988, rose to 1300 $\mu S\ cm^{-1}$ in March 1990 and reached 2550 $\mu S\ cm^{-1}$ in March 1991); and
(4) some wells have a low salinity when dug and remain so (e.g. the Kwa Mailo well, which has been consistently below 500 $\mu S\ cm^{-1}$ for 5 years).

Mechanisms of salinization

The concept of salinity management for these shallow aquifers stems from the fact that each well appears to tap only a very localized groundwater catchment. Abstraction is balanced by local rainfall falling in the vicinity of the well. If 5% of the annual rainfall infiltrates and recharges the aquifer, and assuming an average annual rainfall of 650 mm, then rain over a circular area of 100 m radius around a well would be sufficient to provide an average well yield of 3000 $l\ d^{-1}$.

The above calculation has so far been partially substantiated for some wells in Kibwezi. Rainfall is rapidly translated into recharge and well water levels are seen to rise a few days after a major storm. On average, 45% of the rain in Kibwezi falls in only three storms (Fenner, 1982). Whilst recharge appears to be a very quick and localized event, the mechanism for rapid recharge remains unknown.

Since the groundwater in most of the wells shows rising salinity in the dry season and decreases in the rains, it is currently assumed that recharge decreases well-water salinity. However, it is to be expected that percolating recharge water leaches in fresh salts, yet these are not noticed in any quantity until well levels drop in the dry season.

The first consequence of a water point being established in Kibwezi is the change of land use. Often, indigenous bush is cleared and maize and beans planted. Soil erosion is commonly unchecked and this results in increased run-off, changing the seasonal stream morphology and producing less permeable stream-bed deposits and hence reducing groundwater recharge. This suggests that increasing recharge, in a relatively small area around the well, through simple land-use management may result in a lowered salinity.

Table 9.3 Average monthly rainfall and well-water salinity at Thasyo well, Kibwezi 1988–1991

Month	J	F	M	A	M	J	J	A	S	O	N	D
Average monthly rainfall (mm)	42	30	77	113	29	2	1	1	2	28	172	115
Average monthly well water salinity (μS cm^{-1})	N/a	N/a	535	N/a	N/a	533	N/a	588	2 280	1 914	1 900	N/a

Conclusion

The situation in Kibwezi is not unusual. Local recharge feeding shallow aquifers must represent the situation dealt with in many existing rural drinking water programmes in developing countries and is likely to be the subject of future such investigations. Since, in general, there is inadequate monitoring of well-water salinity, the percentage of wells unused for drinking for part of or all the year due to high salinity is unknown (though the estimate for Kibwezi is 14% in the dry season). Recovering this 14% by attempting salinity management through increasing recharge by the use of water conservation techniques represents a promising approach, at the present time, because desalination techniques are beyond the resources of most rural communities. One possible further benefit of managing salinity through increasing recharge may be increased yield. New well siting techniques should also be adopted for identifying salt-free sites.

The effectiveness of water supply projects in developing countries is constrained by our lack of field knowledge of water quality. For a relatively small financial outlay, field testing equipment could be purchased to cover salinity (and also bacteriological quality) and the resulting water quality information would then be available to improve the effectiveness of groundwater supplies. Our definition of the 'sustainability of a water supply' is at present focused on yield, but considerable benefit would be obtained by adding water quality to the definition, because quality is no longer just an aesthetic consideration but is influencing the long-term usefulness of the water.

References

Fenner, M., 1982. Features of the rainfall at Kibwezi, Kenya. *Journal of East African Agriculture and Forestry Research Organisation*, vol. 45 (1) pp. 83–91.

WHO, 1984. *Guidelines for Drinking Water Quality*, Volume 2. Geneva.

Woodhouse, M., 1991. Natural indicators of groundwater in Kibwezi Division, Kenya. *J. East Africa Natural History Society and National Museum*, vol. 81, No. 197.

Part Three

Impact of Human Activity on Groundwater Quality in Rural Areas

C.F. Ward

Whilst the definitions of 'rural' and the associated population densities vary around the world, a characterisitic common to many rural households and communities is that they are not served by regional reticulated water supplies or sewerage systems; rather they rely on locally developed water sources and on-site sanitation systems such as pit latrines and septic tanks. In rural areas water is required principally for domestic use and to support farming activities, including livestock watering and irrigation, and some small industries, such as brewing and food preparation.

Throughout the developing world, groundwater has been identified as the key to the provision of safe drinking water for people in rural areas and in arid and semi-arid regions it is often the only reliable source of water. Small demands and scattered settlements mean that adequate water supplies may be obtained from low-yielding aquifers such as the shallow weathered basement aquifers referred to by **Herbert** and by **Connelly and Taussig**. Such aquifers are easily mismanaged and overpumped, leading to deterioration in water quality and, in coastal areas, to saline intrusion.

Shallow, unconfined aquifers are always vulnerable to pollution from activities at the surface. Furthermore, since water can be accessed by shallow hand dug wells there may be no control over the siting, design and construction of wells which, lacking sanitary protection, may in turn become conduits of pollution. **Sutton** reports an investigation into the contamination of wells and springs by water users in Zambia.

The single biggest use of water is for irrigation, accounting for 73% of all water use; 15% of the world's cultivated land is irrigated, and by definition this activity occurs in rural areas. The environmental impact of irrigated agriculture is enormous; natural ecosystems are disrupted, micro-climates are modified, and hydrological processes such as evapotranspiration and percolation are generally intensified. Most irrigation utilizes surface water and, in the absence of adequate drainage systems, groundwater levels beneath irrigated areas always rise; indeed excess irrigation water is sometimes the main source of groundwater recharge, as reported by **Attia** from the Nile Valley in Egypt. Furthermore this groundwater is always prone to degradation by salts leached from the soil; the rate of salinization depends on the dilution effect of groundwater flow.

The impact of human activity on groundwater quality falls into three broad categories:

(1) contamination from point sources, e.g. on-site sanitation systems (see **Connelly and Taussig**), waste tips, manure heaps, fertilizer stores (**Daly**), and direct contamination of wells and springs (**Sutton**). The areal extent of pollution from these point sources may be limited, as found by **Daly** in Ireland.

Groundwater Quality Edited by H. Nash and G.J.H. McCall. Published in 1994 by Chapman & Hall. ISBN 0 412 58620 7

(2) changes in groundwater quality resulting from changes in land use and agricultural practices, e.g. vegetation clearance, new cropping patterns, tillage and irrigation practices, waste disposal including irrigation with waste waters, intensification of livestock grazing, application of pesticides, herbicides etc. (see for example the papers by **Chilton, Lawrence and Stuart, Connelly and Taussig**). Such pollution which is characterized by its dispersed or diffuse sources may be inadvertent, slow to show and therefore remain undetected although it may be widespread.

(3) water quality changes resulting from pumping. Commonly, under natural conditions, water of different quality occurs in distinct layers within aquifers and pumping causes groundwater from different depths to mix. In particular it is common for more saline water to occur at depth with fresher water above; pumping causes upconing of this saline water and in coastal aquifers may cause saline intrusion. **Attia** describes this scenario as it applies to the groundwater resources of the Nile Valley in Egypt.

In each case human activities degrade the chemical or microbiological quality of the groundwater. Microbiological contamination of groundwater tends to be associated with developing countries and indeed, with more than half the population of the world (principally in developing countries) exposed to water sources containing pathogens, this is a major concern. In developed countries, concerns are rather centred around the issue of chemical quality. However, in the world's rural areas this dichotomy is less clear-cut due to the intensive use of chemicals in agriculture and the spread of infectious diseases via animals as well as humans. In particular the high levels of fertilizer application which have accompanied the intensification of agriculture throughout the world are inevitably leading to higher nitrate levels in groundwaters and thus adding to the problem of provision of safe water to rural populations. **Chilton *et al.*** present an example from Sri Lanka where nitrate levels now exceed the World Health Organization's recommended maximum concentration.

The chapters in this part draw attention to the main (but not the only) pollutants of concern in rural areas. These are:

- total dissolved salts (TDS) and chlorides – particularly beneath irrigated land;
- nitrates – from fertilizers, human and livestock excreta;
- pesticide and herbicide residues; and
- faecal bacteria and viruses – including pathogens from human and animal wastes.

The actual extent of groundwater contamination in rural areas is unknown. The following chapters provide rare exceptions to a general paucity of information on this topic. The situation is aggravated by our incomplete knowledge of the hydrogeology of many rural regions, the variability in background concentrations of pollution indicators and the presence of several possible pollution sources in close proximity. A theme common to four of the chapters in this part is the uncertainty regarding the origin of nitrate contamination where agriculture and unsewered sanitation co-exist. Also, of course, groundwater is no respecter of settlement boundaries and urban-derived pollutants may pervade an aquifer, rendering a hazard to water users down gradient. So remedial measures which tackle only one pollution source may not be effective.

Water quality considerations rarely figure in the planning stage of development interventions (except to ascertain that quality is initially adequate for the planned use). Long-term effects of land and water (mis-)management are not considered. By the time water users begin to reject water as unpalatable or unsuitable for irrigation it is too late. Remediation is not possible in the short term and the resource is effectively lost. **Herbert's** *Rural Water Supplies Organiser* is a useful reminder of the holistic view which must be adopted at the planning stage of any water supply project for anticipated health benefits to be realized.

Groundwater always suffers from being out of sight hence out of mind and is still a matter of widespread ignorance. Routine monitoring of either groundwater levels or quality is rare and non-existent in many countries. Even where groundwater may be developed as a key resource, e.g. as a source of irrigation water, the effects of abstraction may be considered carefully but little or no attention be given to the threat which irrigation poses to the quality of the resource. Similarly, in relation to waste disposal practices, including waste-water irrigation, attention may be given to monitoring the quality of surface waters, soil and plants whilst the long-term effects on groundwater quality are neglected. Such practices reflect assumptions that pollutants will be attenuated in the unsaturated zone and/or that groundwater flow will dilute and disperse contaminants sufficiently that they will not represent an environmental or health hazard. This depends to a great extent on the hydrogeological conditions of the specific locality. Flow through fissures and macropores reduces the effectiveness of the unsaturated zone as a natural treatment system so that bacteria and viruses do reach the saturated zone. Chlorides and nitrates are very mobile and so are always liable to be present in groundwater recharge even where water tables are at considerable depths.

This apparent lack of concern for water quality

perhaps reflects the lack of feasible alternative supplies or remedial action – sewerage and piped, treated water supplies are prohibitively expensive for widespread adoption. Also, in developing countries, priority is given to increasing the number of water supply points, and not to maintenance or upgrading of existing sources. This appears to be a consequence of the initial political response to the United Nations International Drinking Water Supply and Sanitation Decade.

The papers which follow indicate that quality degradation of shallow groundwaters is an inevitable consequence of human interference with pre-existing ecosystems at the surface. Further discussion of the impact of agricultural activities on groundwater is to be found in a book of that title edited by Vrba and Romijn (1985). Where all land is utilized, as in much of Asia, statements such as 'locate the water source up-gradient of the cultivated land and contamination sources' are clearly unhelpful. The alternative of water treatment is neither affordable nor practicable for many rural communities. Given the pressures on the land and the immediate economic returns from agricultural production, it seems unrealistic to expect a return to extensive farming practices. The most feasible approach would seem to be to minimize groundwater pollution by more careful water management and by protecting sources from direct contamination. The World Health Organisation's *Guidelines for Drinking Water Quality Control in Small Communities* and the UNEP/WHO project referred to by **Sutton** and described in more detail by Lloyd and Helmer (1991) are therefore to be welcomed.

Whilst we really do not know what impact our activities are having on the quality of our groundwater resources, it is clear that their general degradation is a legacy of our generation.

References

Lloyd, B. and Helmer, R., 1991. *Surveillance of Drinking Water Quality in Rural Areas*, Longman Scientific & Technical, Harlow, England

Vrba, J. and Romijn, E., 1986. *Impact of Agricultural Activities on Ground Water*, International Contributions to Hydrogeology vol. 5, International Association of Hydrogeologists, Hanover, Germany.

10 The impact of tropical agriculture on groundwater quality

P.J. Chilton, A.R. Lawrence and M.E. Stuart

Abstract Intensification of agriculture, often by means of irrigation, can produce groundwater quality problems, particularly with respect to nitrate, salinity and pesticides. To study this, field programmes have been carried out by the British Geological Survey in four countries, covering a wide range of climates, farming practices and hydrogeological conditions. Agricultural surveys have established cropping regimes and fertilizer use and identified pesticides for study. The results presented suggest that, as for temperate agriculture, the leaching of nitrate is dependent on soil and aquifer types and cropping regime. Residues of the selected pesticides have been detected at the water table. The risk posed by these residues to drinking water supplies will be dependent on their persistence in aquifers and their rate of movement in groundwater from beneath the cultivated land to drinking water sources. The assessment of risk to potable water supplies from agrochemical use is discussed.

Introduction

In developing countries groundwater is often the best and sometimes the only source of cheap, potable water. It is attractive as a supply option because it is often conveniently available close to where the water is required, it has excellent natural quality, which is generally adequate for potable supply with little or no treatment, and the capital costs of development are relatively low. In addition, development in stages, to keep pace with rising demand is usually more easily achieved for groundwater than for surface water. As a consequence, hundreds of millions of people in both urban and rural areas depend on groundwater for domestic supplies, often drawn from shallow aquifers which can be highly vulnerable to pollution.

Concern has been expressed about the impact on groundwater quality from intensification of agriculture, often by means of irrigation, and the associated increases in fertilizer and pesticide use. This is observed particularly in the Latin American and South Asian regions. In these areas, rates of application of nitrogen fertilizer are high, and up to three crops a year can be raised by intensive, irrigated cultivation. It is clearly important that the risks to groundwater quality posed by intensive agriculture should be assessed, so that any necessary control measures can be introduced.

Developing countries have shown the highest rates of increase in nitrogen fertilizer usage during recent years (Figure 10.1), which has tripled since 1975. This increase has had enormous benefits in food production, with, e.g., a quarter of the growth in rice production in Asia attributed to increased fertilizer use. In developing countries the highest applications are to plantation crops, sugar cane, coffee, cocoa, pineapple and oil palm, although usage on vegetables is becoming more important.

Total consumption of pesticides also continues to grow; annual growth rates reached 12% in the 1960s but now are around 3–4% (Conway and Pretty, 1991). Increases in pesticide usage are now more rapid in many developing countries than in the developed world. Pesticide use has also produced major benefits for agriculture. By reducing pest and disease attacks and weed competition they have contributed significantly to improved crop yields and to the reduction in

Groundwater Quality Edited by H. Nash and G.J.H. McCall. Published in 1994 by Chapman & Hall. ISBN 0 412 58620 7

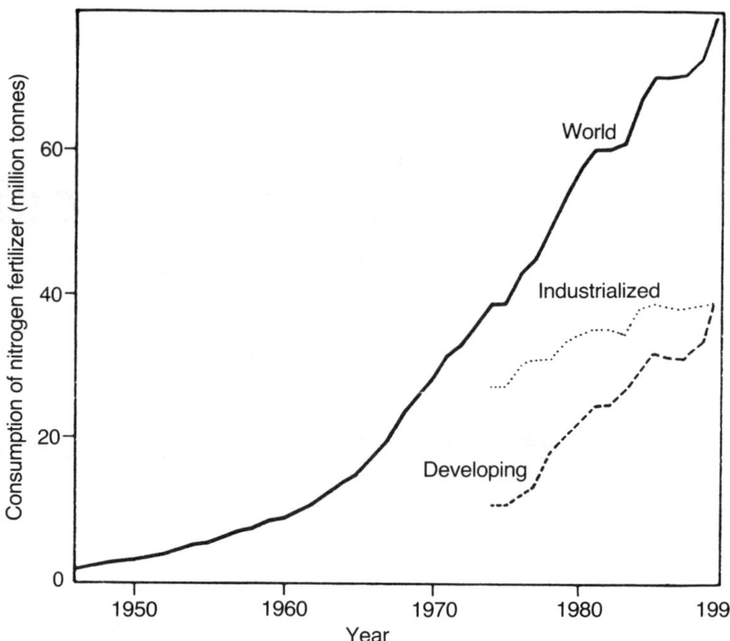

Figure 10.1 Consumption of nitrogen fertilizer, 1946–1989.

variable costs of production, including labour. Losses to pests in many tropical crops are put in the range 10–40% (Walker, 1987), although it is difficult to separate losses in the field from losses during storage and distribution post-harvest. As for fertilizer use, it is difficult to isolate the benefits of pesticide use from other agronomic improvements, and estimates of direct benefits, often used to justify increased agrochemical use, must be treated with caution.

Environmental problems associated with the intensification of agriculture have been apparent in the northern hemisphere for some years. The link between a combination of expanding cultivated areas and increased fertilizer use and nitrate concentrations in groundwater has been extensively researched in both Europe and North America. Expensive measures to control nitrate pollution from agricultural sources are being implemented. More recently, pesticides originating from intensive cultivation have been identified in groundwater used for potable supply in Europe and North America. While much work in temperate zones relates to cultivation without irrigation, nevertheless the same principles can be expected to apply to irrigated agriculture and many of the same approaches can be used.

Thus, where intensive, tropical agriculture is being carried out in regions overlying shallow aquifers, significant pollution of groundwater can be expected where soils are thin, well-drained and permeable. The results of very limited monitoring of groundwater nitrate concentrations in such areas suggests that considerable leaching losses from the applied fertilizer can occur, which can be a significant economic loss to farmers. In most countries, however, there is inadequate monitoring of groundwater quality to establish the extent of nitrate pollution from agriculture, and it is in any case often difficult to distinguish between pollution from this source and pollution from unsewered sanitation. Analysis of nitrate in water is not difficult or costly; relatively simple methods can be performed on small sample volumes and, as national monitoring programmes become developed, further evidence of nitrate pollution of groundwater from intensive cultivation can be anticipated.

Awareness of the leaching of nitrate has led to concern about the possible leaching of pesticides. Evidence of significant pesticide occurrences in groundwater has only been available for a relatively short time from Europe and North America; hardly any routine monitoring of pesticides in groundwater is currently undertaken in developing countries. The study of pesticides in groundwater systems presents substantial problems because (Foster *et al.*, 1991):

- a wide range of compounds is in common agricultural use, many of which break down into toxic derivatives. Analytical scanning of water samples for all or many of these would be prohibitively expensive. Monitoring organizations require knowledge of local pesticide usage to select compounds for analysis.
- very sophisticated analytical procedures and relatively large volumes of sample are required as some compounds are highly toxic at very low concentrations which are close to detection limits.
- considerable care in sampling is required to avoid sample modification, contamination or volatile loss.

In addition to these important technical difficulties, significant scientific uncertainty remains in two key areas. Firstly, much of the available information on the physicochemical properties of pesticides originates from manufacturer's trials. These are invariably performed on 'standard, fertile, organic clay soils' from temperate regions. There is a lack of information generally on the properties of pesticides in groundwater and aquifers as opposed to soils, and an even greater scarcity of data pertaining to tropical environments. Secondly, there are fundamental questions about the movement of pesticides through aquifers. The present project has attempted to examine these issues.

Nitrate

The cultivation and management of farmland has a major influence on groundwater quality and groundwater recharge rates. Some land uses are capable of causing serious groundwater pollution. Nitrogen fertilizers applied to the soil are subject to various complex processes. These include volatile losses, incorporation into the soil organic pool, uptake by crops, run-off and leaching below the soil (Madison and Brunett, 1985). Agricultural soils contain large quantities of nitrogen in an organic form, as much as 5000 kg N ha^{-1} in some temperate regions. This nitrogen can be mineralized to soluble nitrate which can then be leached below the root zone. Thus, while the nitrate leached in a particular year does not necessarily come directly from the applied nitrogen, nevertheless the overall rate of mineralization and leaching is related to fertilizer application rates.

The British Geological Survey project attempted to address the following questions:

- What is the current situation with regard to groundwater quality in developing countries?
- What are the prospects for future groundwater quality, given current trends in agricultural development?
- What are the implications for health?
- What options are there for improved management of agriculture to reduce its detrimental impact on groundwater and to ensure the quality of potable supplies is adequately protected?

Whilst high nitrate concentrations in groundwater have been widely reported and the leaching of fertilizer nitrogen has in many cases been suggested as the probable cause, it is important to recognize that other sources of nitrate exist. These include geological sources, principally evaporite deposits, palaeo-recharge in semi-arid areas, irrigation with waste water, manure dung heaps, unsewered sanitation and atmospheric deposition. For example, in semi-arid areas, nitrate derived from pre-existing vegetation types can produce elevated nitrate concentrations in groundwater receiving little or no modern recharge.

Measurement of nitrate concentrations in groundwater and assessment of the potential for leaching of nitrate from intensive agriculture was a component of the study in four field areas. This involved regular sampling from existing wells and boreholes and the construction of new investigation boreholes. Liaison with local agricultural organizations was required to establish cropping regimes and fertilizer use in the study areas. The results are summarized in Table 10.1 and described briefly below.

Land use and groundwater nitrate concentrations

In areas where intensive agriculture and unsewered sanitation occur together, determining the relative contributions of each of these to the eventual nitrate concentrations in groundwater is not easy. Thus, by itself, routine monitoring of nitrate concentrations cannot be used to assess the impact of agriculture on groundwater quality. Any attempt to evaluate the impact, needs to separate and quantify these two sources, and this was an important aspect of the study and was achieved by different approaches appropriate to the conditions in each field area, as described below.

In Sri Lanka, intensive, irrigated horticulture is being carried out on permeable, well-drained sandy soils over a shallow sand aquifer (Figure 10.2). Triple cropping with heavy applications of nitrogen fertilizer results in significant losses of nitrogen, and high nitrate concentrations in groundwater. A very good correlation was observed between land use and nitrate concentrations in the underlying groundwater. Nitrate concentrations in groundwater from wells in the intensively cultivated areas were typically in the range 10–50 mg NO_3–N l^{-1} equivalent to 45–220 mg nitrate l^{-1} (Figure 10.3), in comparison with nitrate concentrations of less than 2 mg NO_3–N l^{-1} for groundwater drawn from wells located in non-cultivated areas or within coconut plantations which receive minimal fertilizer applications. The correlation of nitrate concentrations with land use is maintained because the intensive abstraction from shallow wells for irrigation restricts groundwater flow to localized **cells**, and prevents mixing and dilution with groundwater from non-cultivated areas. Monitoring of lysimeters and piezometers beneath a research plot indicated that nitrogen-leaching losses are typically in excess of the equivalent of 25% of the applied rate. This produces an average annual increase in the groundwater nitrate concentration of 2–3 NO_3–N mg l^{-1}.

Barbados has a long history of sugar cane cultivation on large plantations, although there is currently a rapid

Table 10.1 Summary of results of nitrate investigations

	India	Sri Lanka	Mexico	Barbados
Aquifer type	Thin alluvium	Coastal dune sands	Thick alluvium	Coral limestone
Depth to water (m)		1–3		20–80
Principal crops	Rice	Onions, chillies	Wheat	Sugar cane
Crops/year	2	2–3	1	1
Length of period of intensive cultivation (yr)	20	10–20	<30	<30
Fertilizer type	Urea	Urea, triple super-phosphate	Urea, anhydrous ammonia	24–0–18
Application rates (kg N ha^{-1} a^{-1})	200	400–500	120–220	125
Current groundwater nitrate concentrations (mg NO_3–N l^{-1})	2–5	10–30	2–5	5–9
Leaching losses (kg N $ha^{-1}a^{-1}$)	Small	60–120	?Small	35–70

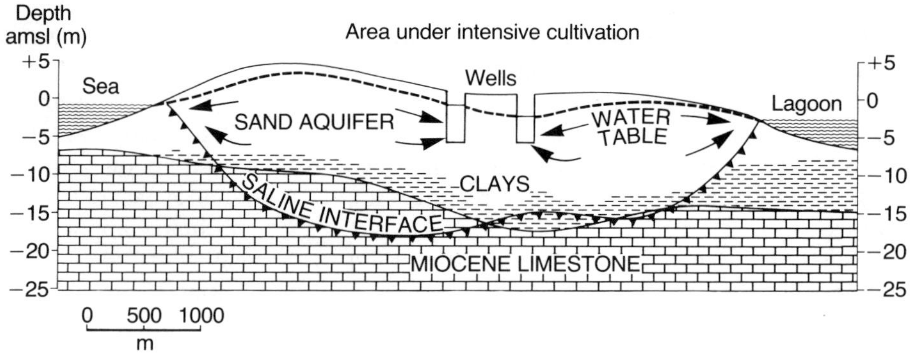

Figure 10.2 Hydrogeological cross-section through the Kalpitiya Peninsula, Sri Lanka.

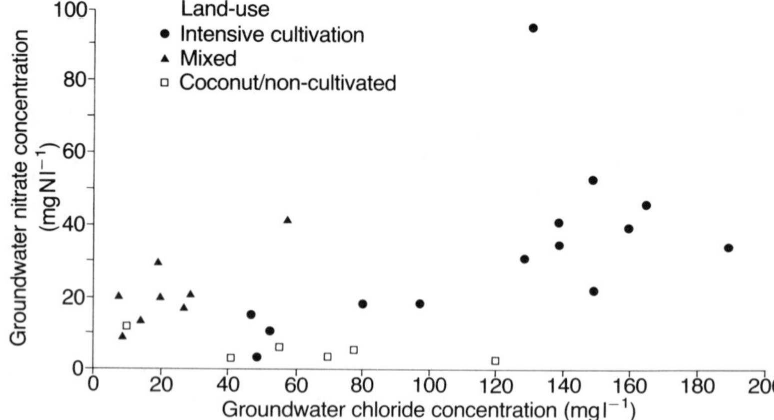

Figure 10.3 Plot of nitrate vs chloride for shallow wells, Kalpitiya Peninsula, Sri Lanka.

increase in horticultural production. The two catchments studied remained under sugar cane in 80% of the cultivated area. Sugar cane receives about 130 kg N ha^{-1} a^{-1}. Some of this may be subject to direct leaching, but sugar cane is a relatively efficient user of nutrients, because of continuous crop cover with strong root development. Current nitrate concentrations in most wells average 6–8 mg NO_3–N l^{-1},

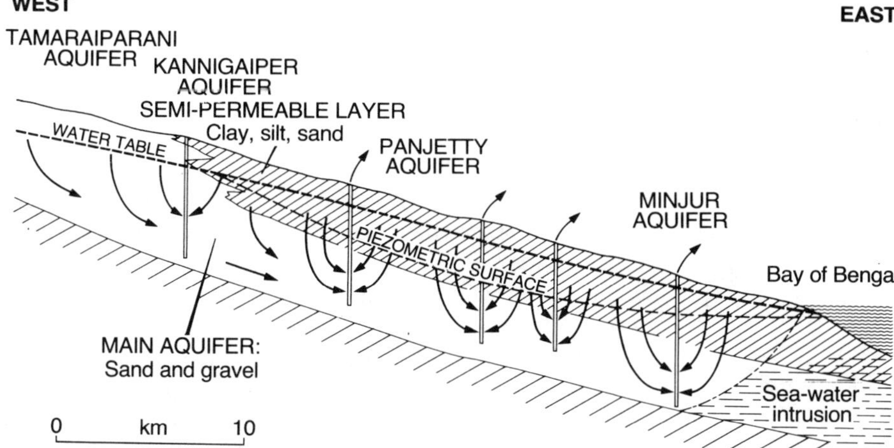

Figure 10.4 Hydrogeological section, Madras study area, India.

showing the impact of land use activities. In the dominantly agricultural areas, leaching losses are calculated to be 30–60 kg N ha^{-1} a^{-1}. In urban areas, where population density rises from 4 to 31 persons ha^{-1}, nitrate concentrations above 10 mg N l^{-1} reflect this additional loading. The overall quality situation is one of universal but modest impacts from cultivation upon which are superimposed locally higher nitrate concentrations arising from unsewered sanitation. The latter source of pollution also results in elevated faecal coliform counts in some urban catchments.

A similar picture emerges from the study in Mexico. Since 1960, inorganic nitrogen fertilizers have progressively replaced organic animal wastes and crop residues. Application rates have probably remained constant over the last decade at about 120–220 kg N ha^{-1} a^{-1} according to crop. The two main forms of fertilizer are urea and ammonia. No information on groundwater nitrate concentrations before 1989 is available and eight locations distributed over an irrigation district were selected for regular sampling. In addition, nitrate profiles were obtained from deep cored boreholes. Average nitrate concentrations ranged from 1–9 mg NO_3–N l^{-1}, with the highest concentrations coinciding with towns using unsewered sanitation.

In Madras, the alluvial aquifer studied, consisted of a two-layered system; a shallow, less permeable layer, some 10–15 m thick, overlying a highly permeable gravel aquifer. This deeper aquifer provides part of the water supply for Madras city. The layering within the aquifer produces a groundwater system in which vertical flow dominates in the upper, less permeable part and horizontal (regional) flow in the main aquifer (Figure 10.4). Typical agriculture in the area consists of two paddy crops and a groundnut crop, each receiving at least 60 kg N ha^{-1} a^{-1}. Monitoring of groundwater quality from piezometers constructed in the upper aquifer immediately beneath ricefields enabled the quality of the recharge from cultivated soils to be assessed. Nitrate concentrations were low to moderate (2–5 mg NO_3–N l^{-1}). One possible explanation for these low nitrate concentrations is that denitrification is active in the anaerobic conditions of the flooded, saturated soils. This has been reported elsewhere in the region (Venkateswarlu and Sethunathan, 1978). In this situation, the main losses of nitrogen from the soil appear to be volatilization, denitrification within the soil and crop uptake. Leaching of nitrate to groundwater is a relatively small component of the nitrogen mass balance.

Implications for future groundwater quality

These four case studies have shown the wide variation in nitrate leaching losses that can occur from agricultural soils. This variability results from differences in soil and crop types, fertilizer application rates and irrigation practices. The four areas described provide reasonable coverage of the range of these variables.

Groundwater is most vulnerable to nitrate leaching where:

- the soil and unsaturated zone are thin and permeable;
- several crops a year are grown;
- fertilizer inputs are high; and
- excess irrigation leads to rapid leaching of nutrients beyond the root zone of the crop.

Where these conditions exist, a rapid build-up of nitrate concentrations in groundwater can be expected. In Sri Lanka, intensive cultivation began some 10–20 yr ago, and already groundwater nitrate concentrations exceed 20 mg NO_3–N l^{-1} in many of the irrigation wells. The increase in groundwater nitrate concentration will also depend on the volume of water stored in the aquifer. The thin sand aquifer of the Kalpitiya Peninsula, Sri Lanka, has a low overall volume of storage compared to the thicker alluvial aquifers in Mexico and India. The scope for dilution is limited, and a rapid build-up of nitrate is to be expected. The average increase in nitrate concentration with time is thought to be 2 mg NO_3–N l^{-1} a^{-1} which is difficult to discern over short periods of monitoring, especially as current groundwater nitrate concentrations average 10–50 mg NO_3–N l^{-1}.

The situation in Barbados is somewhat analogous to that in Sri Lanka in that there is rapid flow to the water table through the permeable soil and unsaturated zone and aquifer storage is low. However, nitrate concentrations are lower than in Sri Lanka (6–10 mg NO_3–N l^{-1}) due to:

- the moderate fertilizer applications (125 kg N ha^{-1} a^{-1} compared with up to 500 kg N ha^{-1} a^{-1} in Sri Lanka, Table 10.2);
- the deeper-rooted, more extensive crop cover; and
- the very limited irrigation of sugar cane.

Further, in contrast to Sri Lanka, the groundwater system is characterized by rapid, regional flow and mixing, largely controlled by the major abstraction wells. Consequently, the quality of water from these wells averages inputs from a large area, including urban as well as agricultural land use. Comparison of data from the two catchments suggests that nitrate inputs from urban areas may be greater than from agriculture. However, horticulture has started to replace sugar cane and a significant increase in nitrate leaching to groundwater can be anticipated. Prior to implementation of large-scale changes in cropping patterns, their potential impact on groundwater quality should be considered.

Groundwater in the study area in Mexico has generally low nitrate concentrations (<5 mg NO_3–N l^{-1}), lower than in the study areas in Sri Lanka and Barbados. This reflects both the generally moderate nitrate application rates and also the different hydrogeological setting. The aquifer in Mexico is deep and possesses considerable storage; most irrigation and water supply boreholes are screened at depths of 50–120 m. This has two consequences. Firstly, there is a considerable time-lag, probably of the order of 25–40 yr, between solutes leaching from the soil zone and arriving at the screened section of abstraction boreholes. Secondly, dilution within the aquifer will reduce nitrate concentrations significantly. There are, however, some indications of higher nitrate concentration in the upper part of the aquifer. These are probably derived by leaching from cultivated land in the last ten years, when fertilizer application have been higher. Nitrate concentrations in potable supplies could, therefore, exceed 10 mg NO^3–N l^{-1} as the 'front' of more recent recharge moves down through the aquifer.

The leaching of nitrate from paddy cultivation as practised at the Madras site does not appear to pose a general risk to groundwater quality. This may not be the case where paddy is cultivated locally on more permeable soils, or where significantly greater applications of nitrogen are made.

Pesticides

All pesticide compounds pose a significant environmental health risk, since they are designed to be persistent and toxic. Evidence of significant occurrences in groundwater has only been available for a relatively short time in Europe and North America. Here the pesticides most frequently observed in groundwater are herbicides (i.e. triazines, carboxy acids and phenyl ureas), carbamate insecticides and soil sterilants such as the chloropropanes. Most occurrences have been in shallow aquifers. Hardly any routine monitoring of pesticides is currently undertaken in developing countries.

Before the risk posed to groundwater by pesticides can be assessed it is essential to consider the following questions:

- Which pesticide compounds are most likely to be leached to groundwater?
- What are the most probable pathways?
- Are these pesticide residues currently detected in water supply boreholes likely to be approaching equilibrium concentrations?

Pesticide leaching from the soil

The natural processes which govern the fate and transport of pesticides can be grouped into the following broad categories: leaching, volatilization, degradation, sorption and plant uptake (Figure 10.5). Plant uptake is usually a small component. The mode of application and action of the pesticide are important factors in relation to soil leaching, since those targeted at plant roots and soil insects are more likely to be leached than those applied to the leaves. Volatile losses occur from the soil particles, from the plants and from soil moisture.

Pesticide compounds may degrade in the soil by microbial or chemical processes to produce metabolites and ultimately simple compounds such as ammonia and carbon dioxide. Soil half-lives for compounds in widespread use range from 10 d to years, but for the most mobile pesticides are normally less than 100 d. Insecticides used for soil treatment may persist in the soil for sufficient time to allow significant leaching to groundwater to occur. Moreover, some derivatives from partial oxidation or hydrolysis may be as toxic and mobile as the parent compound.

Most pesticide compounds have water solubilities in excess of 10 mg l^{-1} (10 000 μg l^{-1}), and this is not a limiting factor in leaching from soils. The mobility of pesticides in solution in soil water will normally depend upon their affinity for organic matter and/or clay minerals. Pesticides that are strongly sorbed on to

Table 10.2 Susceptibility to leaching of pesticides used in the Hampton Catchment, Barbados

Active ingredient	Use	Type	Acute oral toxicity[1]	Mobility class[2]	Solubility[3]	Soil half-life[3]	Total application (kg a^{-1})	Average application[4] (kg ha^{-1} a^{-1})	Mean input to catchment[5] (kg ha^{-1} a^{-1})	Average concentration in groundwater (for 5% leaching) ($\mu g\ l^{-1}$)
Asulam	H	TC	IV (4 000+)	?	3	?	7 293	2.83	1.1	18.3
Ametryn	H	T	III (3 080)	4	1	4	6 570	3.59	0.99	16.5
Atrazine	H	T	IV (2 000)	?	5	4	6 334	4.25	0.95	15.8
Methylarsonic acid	H	–	IV (1 800)	?	1	?	6 127	3.39	0.92	15.3
2–4D Amine	H	Ph	II (700)	?	4	2	5 996	3.38	0.90	15.0
Glyphosate salt	H	–	IV (4 000+)	2	1	3	4 753	2.86	0.71	11.8
Paraquat	H	Py	II (150)	9	1	2	4 542	3.74	0.68	11.3
Ioxynil octanoate	H	–	II (110)	?	6	1/3	1 651	0.75	0.25	4.2
2–D Isooctyl		Ph	II (700)	3	6	?	1 651	0.75	0.25	4.2
Diazinon	I	OP	II (300)	5	5	4	6 240	15.7	0.94	15.6
Acephate	I	OP	II (900)	?	1	?	3 724	19.7	0.56	9.3
Benomyl	F	C	V (10 000+)	?	6	?	2 247	8.11	0.34	5.7
Chlorothalomil	F	–	V (10 000+)	?	6	?	1 638	8.71	0.25	4.2

1 WHO classification 1988–89 on LD50 (mg kg^{-1}) rates but adjusted to include dermal toxicity and other factors.
2 Based on K_{oc} and K_{ow} partition coefficients (1 = most readily leached – 9 = unlikely to be leached (strongly sorbed)).
3 From *Pesticide Manual*, 8th edition, 1987.
4 Average application rate for the plantations using the compound.
5 Total application divided by area of catchment.

H = herbicide, I = insecticide, F = fungicide, T = triazine, Ph = phenoxyl acid, OP = organo-phosphorus, TC = thiocarbamate, C = carbamate

Solubility classes
1=>100 g l^{-1}
2=10–100 g l^{-1}
3=1–10 g l^{-1}
4=0.1–1 g l^{-1}
5=0.01–0.1 g l^{-1}
6=<0.01

Soil half-life classes
1=<10 d
2=10–30 d
3=30–100 d
4=100–300 d
5=>300 d

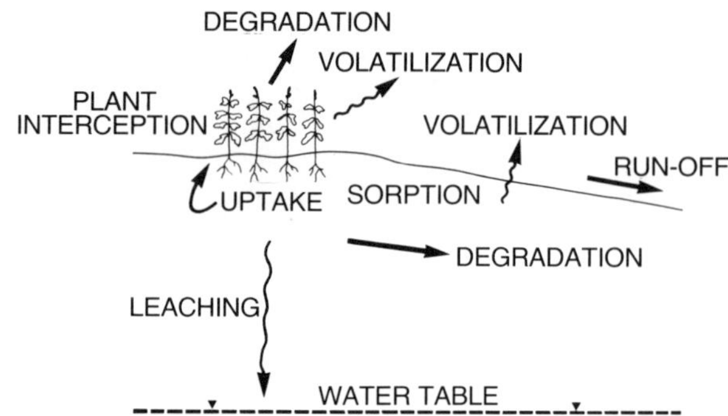

Figure 10.5 Processes determining the fate of pesticides.

organic matter or clay particles are likely to be retained in the soil rather than leached to groundwater.

Transport in the unsaturated zone

Preliminary estimates can be made of the possible transport of pesticides from soils into groundwater systems, based on the physicochemical properties of the pesticides themselves and on knowledge of groundwater flow and aquifer properties gained from previous investigations. Pesticide compounds leached from permeable soils into the unsaturated zone enter an environment which contains less clay minerals and organic matter and has a greatly reduced indigenous microbial population. The attenuation processes which affect pesticides are likely, therefore, to be much less active beneath the soil zone. Thus the mobility and persistence of all pesticide compounds should be many times greater in the unsaturated zone than in a typical agricultural soil (Lawrence and Foster, 1987; Bouwer, 1987).

The development of preferential flow in the unsaturated zone is of major importance in the consideration of pesticide transport into aquifers. This term is taken to include all forms of rapid downward movement in macropores and fissures, effectively by-passing the matrix. Preferential flow can be caused by numerous factors (Thomas and Phillips, 1979; Bevan and Germann, 1983; Bowman and Rice, 1986) and is often associated with instability of downward flow in situations where a permeable formation is overlain by a somewhat less permeable soil horizon (Samani, Cheragi and Willardson, 1989).

Preferential flow could permit low- to moderately-persistent components to reach the water table even where there are relatively thick unsaturated zones.

Summary of field investigations

Given the complexity of pesticide transport in the soil and subsurface environment outlined above, and the fact that the behaviour of a particular compound may vary significantly depending upon conditions in the soil and aquifer, for example, moisture content, pH, dissolved oxygen (Chilton and West, 1992), predicting pesticide residue concentrations in groundwater presents many difficulties and uncertainties.

The fieldwork has included regular sampling from existing wells and broreholes, and the construction of new boreholes. Agricultural surveys have established cropping regimes and identified the most important pesticides in regular use in the study areas. In Barbados, atrazine and ametryn were selected because they are widely applied at relatively high rates (Table 10.2). They are believed to be mobile and persistent and, in the case of atrazine, have been detected in groundwater in the USA.

In Sri Lanka, monitoring of the leachate beneath an experimental field plot indicated that 10% of first application of the insecticide carbofuran to an onion crop was leached to groundwater. However, leaching of the second application, one month later, was considerably reduced (*c.* 1% of the applied rate) and this was attributed to enhanced microbial degradation in the soil. The following season the application was increased 10-fold and the percentage leached to the water table increased to 50% of the applied rate. Clearly the rate of leaching to the water table, even at the same site, can vary considerably and depends on the quantity applied and whether previous applications have been made recently (Figure 10.6).

It is important to establish the degradation pathway when assessing the risk to groundwater supplies, firstly because the metabolites produced could be of very different toxicities to the parent compound and to each other, and secondly because the persistence of the different metabolites can vary. For example the carbofuran phenol metabolite appears more stable than carbofuran in groundwater.

Once in the saturated zone, dilution together with degradation are probably the most important processes attenuating pesticide concentrations. The attenuation will depend on the travel time to the water supply well, the storage within the aquifer, the decay constant (or half-life) of the pesticide and the percentage area of the catchment to the water supply well that the pesticide is applied.

An indication of the range in capacity for aquifers to dilute contaminants can be best illustrated by two examples; a porous sand aquifer and a fissured limestone aquifer. The shallow aquifer beneath the Kalpitiya Peninsula, Sri Lanka, consisted of a highly porous sand, which despite being only about 15 m thick, probably stored 5–8 times as much as the average annual recharge. The potential for dilution in this aquifer is clearly considerable. Even greater dilutions are possible for deep boreholes in thick granular aquifer systems. By contrast, the storage available within the fissured limestone aquifer of Barbados

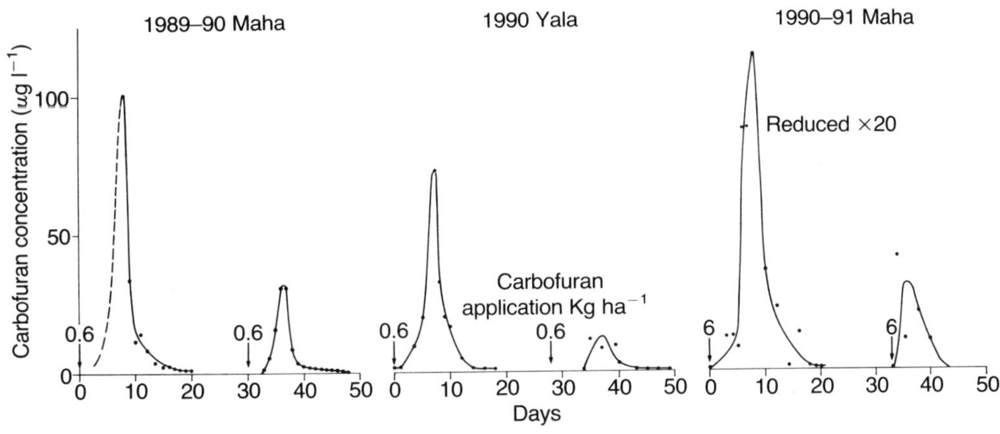

Figure 10.6 Carbofuran in lysimeter drainage over three cropping seasons in Sri Lanka.

Table 10.3 Estimated possible pesticide concentrations in groundwater due to dilution within the aquifer

Pesticide application rate (ai) ($kg\ ha^{-1}\ a^{-1}$)	Percentage leaching loss	Concentration at water table* dilution capacity in aquifer		
		Thin, fissured aquifer[1] ($\mu g\ l^{-1}$)	Fissured or thin porous aquifer[2] ($\mu g\ l^{-1}$)	Porous thick aquifer[3] ($\mu g\ l^{-1}$)
2.0	1–10	10–100	1.0–10	0.1

* Assumes uniform mixing with groundwater through full thickness of aquifer.
[1] Total storage within aquifer equivalent to 200 mm.
[2] Total storage within aquifer equivalent to 2000 mm.
[3] Total storage within aquifer equivalent to 20 m.

probably represented only once or twice the average annual recharge. It is clear that dilution of pesticide residues by mixing with waters stored within the aquifer can significantly reduce pesticide concentrations but that this dilution can vary by 1–2 orders of magnitude depending upon the porosity of the aquifer and the depth of the groundwater flow paths (Table 10.3).

Degradation and dilution can be very effective in reducing pesticide residues to acceptable concentrations. Field observation in Sri Lanka demonstrated that these processes reduced residue concentrations at the water table, immediately beneath the cultivated plot, from 60 $\mu g\ l^{-1}$ to $< 5\ \mu g\ l^{-1}$ within four days.

The persistence of relatively similar compounds in groundwater can vary considerably and this is thought to explain why atrazine is detected widely in the groundwater of Barbados but not ametryn.

Both of these compounds are used widely, and applied at about the same rate (4 kg (ai) $ha^{-1}\ a^{-1}$) to sugar cane. Further, both compounds are regarded as relatively mobile and likely to be leached to groundwater. However, whilst atrazine is detected at concentrations in the range 0.5–3 $\mu g\ l^{-1}$, throughout the year, in all public supply boreholes sampled, ametryn was rarely detected. The most likely explanation is that degradation within the aquifer reduces ametryn to below detectable concentrations during migration through the vadose and saturated aquifer zones. The concentrations of atrazine in the pumped groundwater of Barbados represent the equivalent of 1% of the applied pesticide being leached to groundwater although the actual leaching loss from the soil may be significantly higher.

Whilst this chapter has focused on the uncertainties and difficulties in predicting pesticide behaviour in the subsurface, it is perhaps worthwhile attempting to suggest likely maximum concentrations of pesticide residues in groundwater. First leaching losses are likely to be in the range 1–10% of the applied rate for mobile compounds on relatively permeable soils. Most pesticides are applied at the rates of 0.5–2 kg(ai) ha^{-1} and so the concentrations in groundwater immediately beneath cultivated land are likely to be less than 100 $\mu g\ l^{-1}$ even in aquifers with a limited dilution capacity. In Sri Lanka, where carbofuran was applied at 12 kg ha^{-1} the concentration at the water table beneath the cultivated land was only 60 $\mu g\ l^{-1}$ (Table 10.4).

Further reductions in residue concentrations are likely to occur in pumped groundwater supplies as a result of dilution with recharge derived from non-cultivated sources and degradation of residues. Those compounds which are most likely to pose a significant risk to groundwater supplies are those which are both persistent in groundwater (i.e. have a half-life of > 100 d) and have a low drinking water guideline value.

Table 10.4 Summary of pesticide investigations

Country	Pesticide compound	Breakdown products	Rate of application (kg(ai) ha^{-1})	Maximum concentration detected at water table ($\mu g\ l^{-1}$)	Range of concentration in abstraction ($\mu g\ l^{-1}$)
Sri Lanka	Carbofuran	–	12/crop	60	Not detected
		3-OH-7-phenol.	–	10	Not detected
India	Carbofuran	–	1.25/crop	–	Not detected
		7-phenol	–	Occasionally at detection limit	Not detected
Barbados	Atrazine	–	3.6/a	Not determined	0.5–3
		N-Desethylatrazine	–	Not determined	0.2–2
	Ametryn	–	4.3/a	Not determined	Occasional low positive samples

Conclusions

This research has demonstrated that intensive agriculture can produce a serious deterioration in groundwater quality as a result, principally, of the leaching of nitrogen fertilizers. The degree of deterioration will depend, amongst other factors, upon the crop type and rates of fertilizer application, the permeability of the soil and unsaturated zone and the potential for dilution within the aquifer.

In the case of the shallow sand aquifer in Sri Lanka, which represents a highly vulnerable aquifer type, groundwater nitrate concentrations are greatly in excess of WHO guidelines. The lower nitrate concentrations in Barbados and Mexico result from lower fertilizer application rates and, in the latter, from greater dilution. Groundwater nitrate concentrations beneath paddy appear to be low as a result of denitrification.

The greatest risk to groundwater supplies from pesticides will be where:

- aquifers are overlain by permeable soils and travel times to the water table are short;
- the aquifer is of low porosity and consequently dilution is small; and
- the pesticide is relatively stable.

This research suggests that degradation rates within aquifers can vary significantly between pesticide compounds, even where they belong to the same group, as is demonstrated by the triazines in Barbados.

References

Bevan, K. and Germann, P. 1982. Macropores and water flow in soils. *Water Resources Research*, vol. 18, pp. 1311–1325.

Bouwer, H., 1987. Effect of irrigated agriculture on groundwater quality. *Journal of Irrigation and Drainage Engineering*, vol. 113, pp. 4–15.

Bowman, R.S. and Rice, R.C., 1986. Transport of conservative tracers in the field under intermittent flood irrigation. *Water Resources Research*, vol. 11, pp. 1531–1536.

Chilton, P.J. and West, J.M., 1992. Aquifers as environments for microbial activity. *Proceedings of the International Symposium on Environmental Aspects of Pesticide Microbiology, Sigtuna, Sweden, August 1992*, pp. 293–304.

Conway, G. and Pretty, J., 1991. *Unwelcome Harvest: Agriculture and Pollution*. Earthscan Publications.

Foster, S.S.D., Chilton, P.J., and Stuart, M.E., 1991. Mechanisms of groundwater pollution by pesticides. *Journal of the Institution of Water and Environmental Management*, vol. 5, pp. 186–193.

Lawrence, A.R. and Foster, S.S.D., 1987. The pollution threat from agricultural pesticides and industrial solvents. *BGS Hydrogeology Research Report*, 87/2.

Madison, R.J. and Brunett, J.O., 1985. Overview of the occurrence of nitrate in groundwater of the United States. In: *National Water Summary 1984 – Water Quality Issues, USGS Water Supply Paper*.

Samani, Z., Cheragai, A., and Willardson, L., 1989. Water movement in horizontally-layered soil. *Journal of Irrigation and Drainage Engineering*, vol. 115, pp. 449–456.

Thomas, G.W. and Phillips, R.E. 1979. Consequences of water movement in macropores. *Journal of Environmental Quality*, vol. 81, pp. 149–152.

Venkateswarlu, K. and Sethunathan, N., 1978. Degradation of carbofuran in rice soils as influenced by repeated applications and exposure to aerobic conditions following anaerobiosis. *Journal of Agricultural and Food Chemistry*, vol. 26, pp. 1151.

Walker, P.T., 1987. Losses in yield due to pests in tropical crops and the value of decision-making. *Insect Science Applications*, vol. 8, pp. 665–671.

11 Groundwater quality in the Nile aquifer system and desert fringes in Egypt

Fatma Abdel Rahman Attia

Abstract Egypt covers an area of slightly over 1 000 000 km^2 of the arid belt in North Africa. It has a population of 56 million, of whom 99% are concentrated along the coastal zone and in the Nile Valley and Delta, where the population density is about 1300 people per km^2. Accordingly this chapter deals with the Nile system and its fringes, where 90% of Egypt's population is concentrated. Rainfall seldom exceeds 200 mm yr^{-1} on the Mediterranean coast, and decreases rapidly inland to be almost nil at Cairo. The main source of water in the country is the River Nile. The construction of the Aswan High Dam ensured the seasonal and long-term regulation of water releases. The total live storage capacity of the lake is 130 × 10^9 m^3.

Groundwater in the alluvium of the Nile aquifer system, although the almost only renewable groundwater in the Nile Valley and Delta, cannot be considered a resource in itself; it is replenished from the Nile water as a result of irrigation activities. Accordingly the Nile system is generally considered a regulation reservoir analogous to the Aswan High Dam. Groundwater in the major part of Nile aquifer system and its periphery is generally of good quality. On the other hand, groundwater quality in the other aquifers is generally marginal, with the exception of the Nubian Sandstone aquifer.

The hydrogeological conditions of the Nile aquifer system are summarized to identify the sources of recharge and discharge affecting groundwater quality. Groundwater quality trends are presented together with the most important water types, which indicate the close relation between groundwater and Nile water. Sources and the extent of pollution are categorized according to their impact on groundwater suitability. Finally, quality constraints on groundwater development in the Nile aquifer system and desert fringes are summarized, and recommendations made concerning the long-term sustainability of groundwater development.

Introduction

Egypt covers an area of ~ 1 000 000 km^2 (Figure 11.1). It is divided into four regions:

(1) the Nile Valley and Delta, including El Fayum depression and Lake Nasser;
(2) the Western Desert, including the Mediterranean littoral zone and the New Valley;
(3) the Eastern Desert, including the Red Sea littoral zone, the islands and the high mountains; and
(4) the Sinai Peninsula, including the littoral zones of the Mediterranean, the Gulf of Suez and the Gulf of Aqaba.

The country lies for the most part within the temperate zone, and the climate varies from arid to extremely arid. The air temperature frequently rises to over 40°C in daytime during summer, and seldom falls to zero in winter. The average rainfall over Egypt as a whole is only 10 mm yr^{-1}. Along the Mediterranean littoral zone, where most of the winter rain occurs, the annual average rainfall is less than 200 mm yr^{-1} decreasing rapidly inland. The evaporation rates are high, being in excess of 3000 mm yr^{-1}.

Groundwater Quality Edited by H. Nash and G.J.H. McCall. Published in 1994 by Chapman & Hall. ISBN 0 412 58620 7

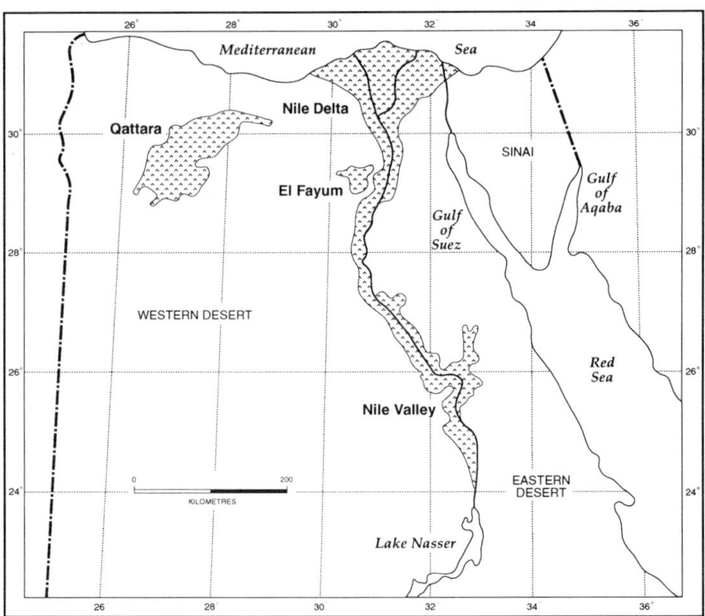

Figure 11.1 Location map of Egypt.

Hydrogeological environment

The hydrography of Egypt comprises two systems:

(1) a system related to the Nile; and
(2) a system related to the rainfall in the past geological times, particularly in the Late Tertiary and Quaternary.

The Nile system comprises the Valley and Delta regions. These are morphological depressions filled with Pliocene and Quaternary sediments. The Nile enters Egypt at Wadi Halfa, south of Aswan. This area is at present occupied by the High Aswan Lake. From Aswan the Nile meanders until it reaches Cairo. About 20 km north of Cairo, the river divides into two branches, each of which meanders separately through the Delta to the sea. Within the Nile flood plain there are extensive man-made drainage systems, especially in the traditionally cultivated land. Some of these extend to the areas reclaimed for agriculture from the desert fringes of the flood plain. Such drainage systems discharge to the Nile itself or to the sea.

The other hydrographic system in Egypt is the complex network of dry streams (wadis), the development of which dates back to wet periods in the Tertiary and Quaternary. This system covers more than 90% of the surface area of Egypt, including the Western Desert, the Eastern Desert, and Sinai. The main catchment areas drain towards the Nile Valley and Delta, to the coastal zones, and to inland depressions.

The landscape in Egypt can be broadly divided into the elevated structural plateaux and the low plains, which include the fluviatile and coastal plains. These geomorphological units play a significant role in determining the hydrogeological framework. The structural plateaux constitute the active and semi-active watershed areas. The low plains can contain productive aquifers and are also, locally, areas of groundwater discharge.

The Nile flood plain is mainly the site of agricultural activities. On its fringes, large areas of land have been reclaimed from the desert for agriculture, relying on river water or groundwater for irrigation.

The hydrogeological framework of Egypt comprises six aquifer systems (RIGW, 1993), as shown in Figure 11.2.

(1) The Nile aquifer system, assigned to the Quaternary and Late Tertiary, occupies the Nile flood plain region and the desert fringes where 90% of Egypt's population lives.
(2) The Nubian Sandstone aquifer system, assigned to the Paleozoic–Mesozoic, occupies the Western Desert for the most part.
(3) The Moghra aquifer system, assigned to the Lower Miocene, largely occupies the western edge of the Delta.
(4) The Coastal aquifer systems, assigned to the Quaternary and Late Tertiary occupy the northern and western coasts.
(5) The 'karstified carbonate' aquifer system, assigned to the Eocene and to the Upper Cretaceous, predominates in the northern part of the Western Desert.
(6) The 'fissured and weathered hard-rock' aquifer system, assigned to the Precambrian. This predominates in the Eastern Desert and Sinai.

With respect to groundwater quality, the most explored aquifer system is that of the Nile and its desert fringes. The potential of the second source of groundwater,

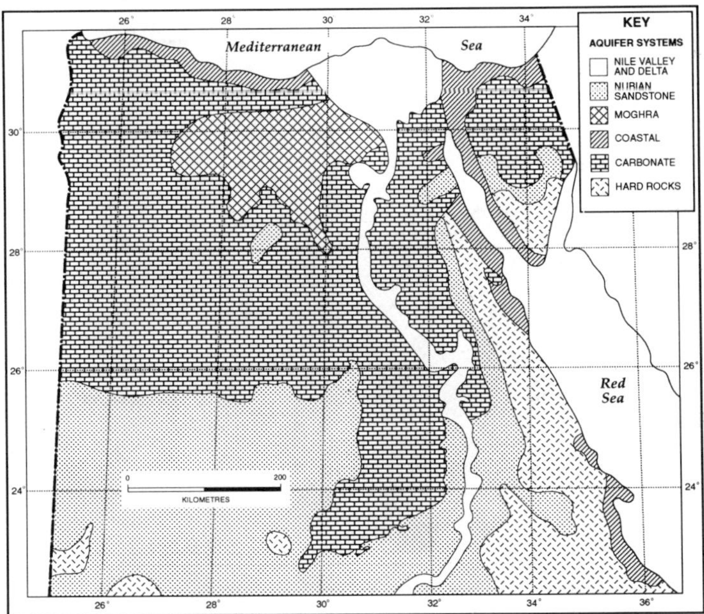

Figure 11.2 Distribution of the main aquifer systems in Egypt.

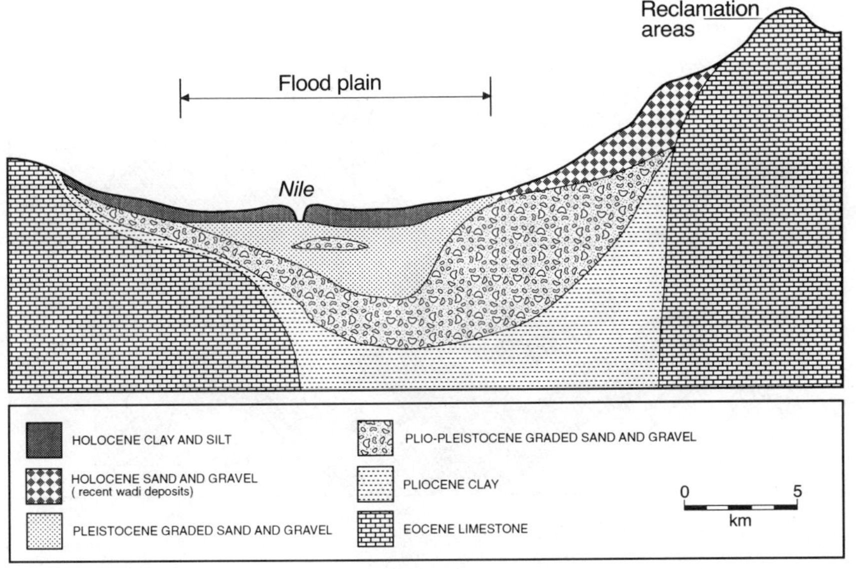

Figure 11.3 A typical geological section across the Nile valley.

in other aquifer systems outside the Nile basin is still under investigation – such aquifers dominate the Eastern and Western Deserts and Sinai. This paper will therefore only deal with the Nile system and fringes.

Hydrogeology of the Nile aquifer system and desert fringes

The Nile Valley aquifer system occurs within Quaternary sediments deposited in a structurally controlled graben cutting into the Eocene–Upper Cretaceous limestone plateau (Said, 1990). Figure 11.3 shows a typical geological cross section in the Nile Valley (vertical scale exaggerated).

The Nile aquifer system consists of Pleistocene graded sand and gravel, while its extension in the desert fringes includes Pliocene–Pleistocene sediments. In the centre of the flood plain, the aquifer is semi-confined by a 10 m average thickness of silt-clay, becoming phreatic on the edges of the flood plain and on the desert fringe. The aquifer is underlain by Pliocene marine clays. The maximum saturated thickness of the aquifer ranges from 300 m in the Valley to 800 m in the Delta. The aquifer transmissivity ranges from 20 000 $m^2 d^{-1}$ in the centre of the flood plain, to less than 500 $m^2 d^{-1}$ at the edges.

Average groundwater heads decrease gradually from 65 m (amsl) at Aswan to 15 m (amsl) at Cairo, being <1 m (amsl) in the north of the Delta. The general flow direction is south–north but deviations are found in the vicinity of the Nile levee (Figure 11.4).

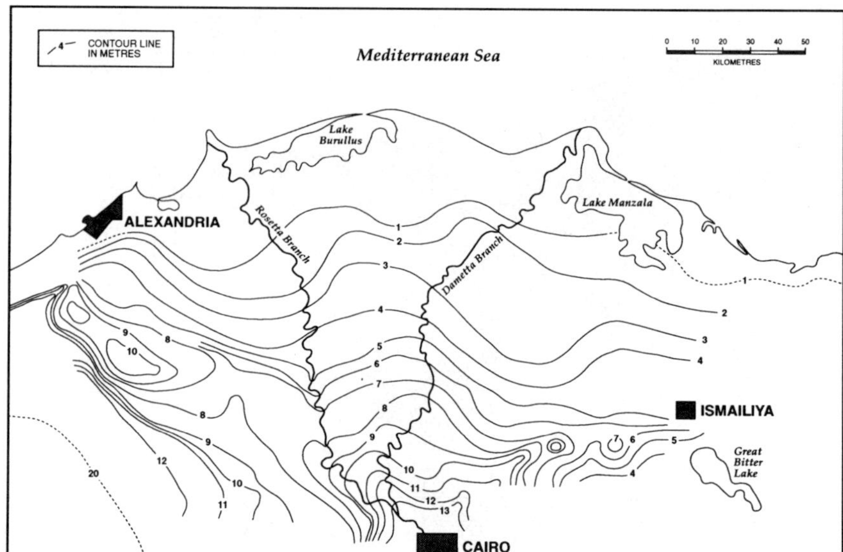

Figure 11.4 Average groundwater contour map for the Nile Delta and its fringes in 1992 (m above mean sea level).

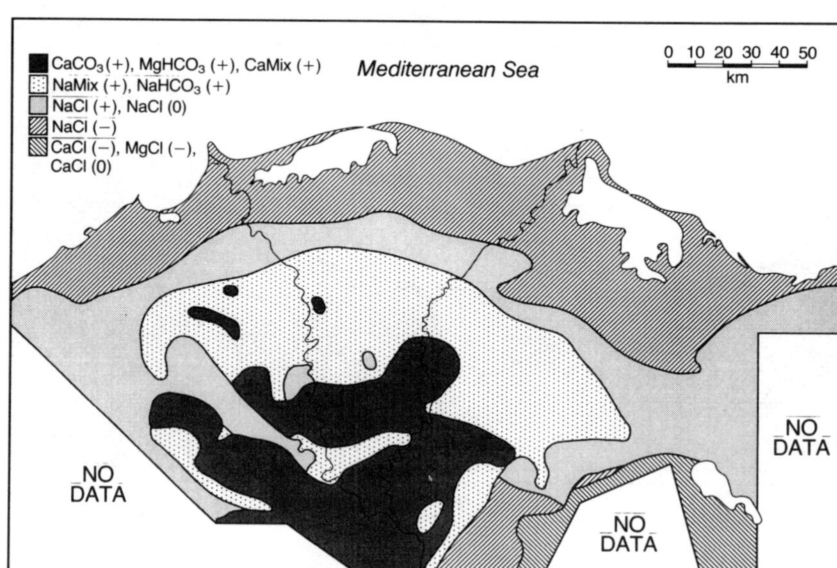

Figure 11.5 Prevailing groundwater types in the Nile Delta and its fringes.

The main source of recharge is percolation from irrigation water. Recharge varies according to the type of soil, source of irrigation water, irrigation method, and the presence of artificial drainage networks. In sandy areas with basin irrigation diverted from the river and no artificial drainage, percolation losses vary between 1 and 2.5 mm d^{-1}. In silty areas with artificial drainage, percolation losses may be less than 0.5 mm d^{-1} (Farid and Tuinhof, 1991; Warner *et al.*, 1991). Discharge from the aquifer is through seepage to the river (more than 3×10^9 m^3 yr^{-1} (Attia, Allam and Amer, 1986), or groundwater extraction by means of wells. The total annual extraction from the Nile aquifer system and desert fringes in 1991 was about 4.4×10^9 m^3 (RIGW, 1993). Discharge from the aquifer also takes place by upward groundwater leakage in the north of the Delta (Farid and Amer, 1986).

Groundwater quality

Groundwater quality in the system depends to a large extent on the quality of the recharge source. In the flood plain and desert fringes the main source of groundwater recharge is irrigation water percolating through agriculture soils. Locally, seepage from carbonate and sandstone rocks may also occur. From surface water to groundwater the total dissolved salts (TDS) in water increases from an average of 250 mg l^{-1} to an average of about 600 mg l^{-1} in the Valley and southern part of the Delta. In the northern part of the Delta, groundwater becomes brackish to saline due to sea-water intrusion (Figure 11.6). About half of the Delta is occupied by brackish to saline groundwater.

The chemical types of groundwater in the Nile Delta and fringes are shown in Figure 11.5 (RIGW/IWACO,

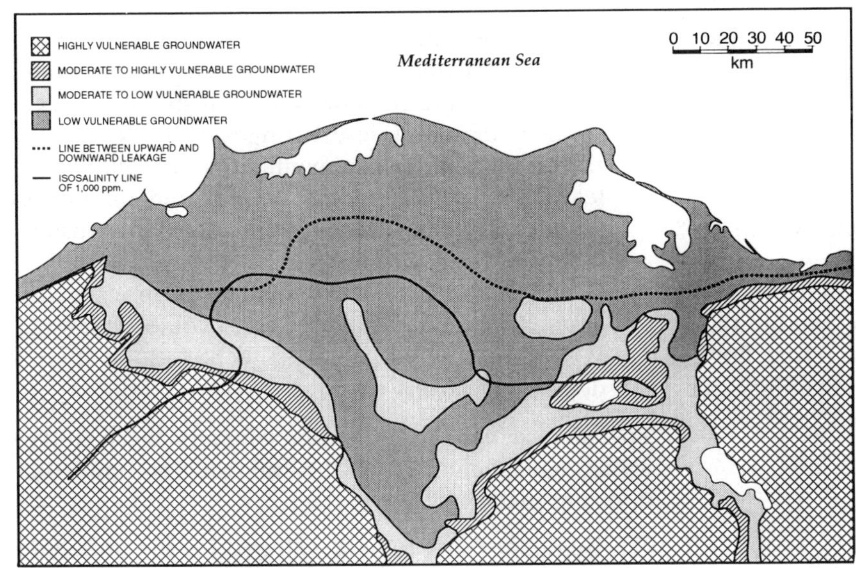

Figure 11.6 The vulnerability of groundwater to pollution and average salinity in the Nile Delta region.

1992a). The classification is made according to Stuyfzand (1986), based on the prevailing anions and cations and the [Na + K + Mg] content (corrected for sea-water contribution). The most important families are [Ca + Mg] and [Na + K + NH_4] for the cations, and [HCO_3 + CO_3], [SO_4 + NO_3 + NO_2] and Cl for the anions. The anion family may be labelled 'mix' if none of them takes up more than 50% of the sum of the anions. Groundwater is attributed to three classes, indicated by +, 0 or −, indicating a surplus, equilibrium or deficit in [Na + K + Mg]. A deficit implies sea-water intrusion, whereas a surplus indicates recharge from Nile fresh water (irrigation).

In the south, the $CaHCO_3$(+) and $MgHCO_3$(+) groundwater types indicate continuous recharge from Nile water to the aquifer. Northwards, the $NaHCO_3$(+) and NaMix(+) water types indicate a decrease in recharge. This zone is followed to the north, west and east by a zone with NaCl(+) and NaCl(0) groundwater types, indicating a continued decrease in recharge. Near the coast, the groundwater type is NaCl(−), indicating that the groundwater is being replaced by sea-water. The NaCl(−) type of groundwater is also found east of Cairo due to recharge from deeper Tertiary aquifers.

In the Nile Valley, similar types of groundwater are encountered with the exception of sea-water.

Groundwater pollution

The main sources of groundwater pollution in the Nile aquifer system and desert fringes are from agriculture (fertilizer and pesticide application) and domestic installations (sewerage and septic tanks) on the one hand, and from return flow (salinity), on the other hand.

The vulnerability of groundwater to pollution is largely determined by the lithology and thickness of the top soil layers, depth to groundwater, and direction of natural vertical groundwater flow. In the Delta, four areas are distinguished (Figure 11.6). These comprise the following:

(1) the reclaimed desert areas with moderate to high vulnerable groundwater, due to the presence of sandy formations with high infiltration and low adsorption capacities, although groundwater is relatively deep;
(2) the traditionally cultivated area with moderate to low vulnerable groundwater due to the presence of a clay cap;
(3) the transition zone between the old land (traditionally cultivated) and the reclaimed areas with highly vulnerable groundwater due to the presence of sandy soils and a shallow groundwater table; and
(4) the northern part with very low vulnerable groundwater due to the presence of a top clay cap and upward groundwater flow.

Groundwater vulnerability in the Valley can be classified in a similar way, but including the three first categories only, as there is no upward flow. The main pollutants encountered in groundwater are as follows:

(1) faecal coliforms (> 100 f.c./100 ml), found at shallow depths in regions with high vulnerable groundwater;
(2) nitrate (70–100 mg l^{-1} and expected to increase with time due to the increase in fertilizer application): in regions with high vulnerable groundwater, concentrations decrease with depth;
(3) phosphates, generally less pronounced, being retained by the upper soil layers;
(4) ammonium, limited to very localized areas;

(5) chlorinated pesticides, in highly vulnerable groundwater;
(6) organo-phosphorus and carbamate (up to 5 g l^{-1}), at shallow depths only;
(7) iron and manganese (believed to originate from clay and silt sediments), found at low concentrations in groundwater within the Nile system at depth; and
(8) groundwater of high salinity (>1500 ppm), in the north Delta due to sea-water intrusion, and in the fringes of the Valley and Delta due to return flow from irrigation and the decreased availablility of groundwater for dilution (thin aquifers).

Groundwater development

Groundwater in the Nile aquifer system is mainly replenished from irrigation water losses. Accordingly, it cannot be considered a resource in itself, but a regulation reservoir analogous to the Aswan High Dam. The Nile aquifer system can, nevertheless, have a strategic role in the overall water management of the country. The contribution of groundwater depends on the hydraulic and hydrochemical properties of the aquifer system. These properties determine the economic development of the groundwater, represented by the pumping head and the suitability of groundwater for various uses. Even if groundwater is initially suitable, pumping may result in the deterioration of quality through various means, e.g.:

(1) seepage from low quality sources (septic tanks, adjacent formations, drainage water);
(2) upconing of saline water; and
(3) seawater intrusion.

Conclusions

It can be concluded that, with respect to groundwater quality, the main factors constraining groundwater development are the following.

(1) In the northern part of the Delta, which represents about half of the Delta area, groundwater is either brackish or saline, being unsuitable for development.
(2) In the south of the Delta, the thickness of the fresh groundwater increases, allowing more opportunity for groundwater development, especially because groundwater withdrawals are replenished from fresh water (Nile and irrigation canals). Nevertheless, the location of well screens and the rate of pumping should be carefully designed to control upconing of deep saline groundwater.
(3) In the Nile Valley flood plain, generally, no qualitative restrictions are encountered. Groundwater withdrawals are replenished by fresh Nile water.
(4) On the edges and desert fringes of the Valley and Delta, the aquifer is unconfined and shallow. This leads to pollution of the aquifer and dilution of pollutants, thus affecting the long-term sustainability of these aquifers.
(5) In the urban areas, the shallow portions of the aquifer are heavily polluted from domestic sewage. Well screens should accordingly be located at great depths.

To enhance the role of groundwater in the country's water management plan, the following recommendations are made.

(1) Guidelines should be developed for the protection of groundwater from deterioration. These should include monitoring, assessment of environmental impacts of external activities on groundwater, and assessment of environmental impacts of groundwater development.
(2) Investigations should be mounted into the possibilities of recharging the aquifers located on the fringes of the Nile system to maintain their long-term sustainability.
(3) Methods should be developed for the conservation of run-off water in the coastal areas; and for the proper use of such waters. This will probably be of great importance in the near future for the development of the coastal areas where groundwater is saline and surface water is difficult to obtain. Various methods are known to be successful under such conditions; among these are skimming and the use of 'scavenger' wells. Testing of such methods on pilot scales is strongly recommended.

References

Attia, Fatma Abdul Rahman, Allam, M.N. and Amer, A.M., 1986. A hydrologic budget analysis for the Nile Valley in Egypt. *Journal of Groundwater*, vol. 24, 4, pp. 453–459.

Farid, M.S. and Amer, A., 1986. An approach to handle seawater intrusion in the Nile Delta Aquifer. *Proceedings of the 9th Water Intrusion Meeting, Delft, The Netherlands.*

Farid, M.S. and Tuinhof, A., 1991. Groundwater development planning in the desert fringes of the Nile Delta. *Proceedings of the 'Round Table meeting' on Planning for Groundwater Development in Arid and Semi-arid Regions, Cairo, Egypt*, pp. 125–134.

RIGW/IWACO, 1992a. *Hydrogeological Map 1:500,000 of the Nile Delta, Cairo.*

RIGW/IWACO, 1992b. Guidelines with respect to groundwater quality. *TN 77.* 01300–92–04, Cairo.

RIGW, 1993. *Groundwater Potential for Water Security*. Technical report, Cairo.

Said, R. (ed.), 1990. *The Geology of Egypt*. Balkema, Rotterdam.

Stuyfzand, P.J., 1986. A new hydrochemical classification of water types: principles and application to the coastal dunes aquifer system of the Netherlands. *Proceedings of the 9th SWIM, Delft*, pp. 641–655.

Warner, J.W., Gates, T., Fatma Abdul Rahman Attia and Mankarious, W.F., 1991. Determination of vertical leakage for Nile Valley. *Journal of Irrigation and Drainage Eng.*, vol. 117, 4, pp. 515–533.

12 Groundwater quality in the Nore River Basin in the southeast of Ireland

E.P. Daly and L. Woods

Abstract Groundwater is a major source of water supply in many parts of rural Ireland. Well surveys carried out in a number of areas have detected extensive contamination of low yielding wells. Local point sources are considered to be the principal sources of this contamination. However, analyses of most of the major springs in the two principal aquifers in the country show that groundwater is generally of good quality.

A detailed water quality investigation in a large river basin in the southeast of Ireland is used to examine this apparent contradiction. The results of this study show that the waste produced by local point sources has a far greater impact on the capture zones of low yielding wells than it does on that of high yielding boreholes or springs. Hence the latter give a much better reflection of the overall groundwater quality in the aquifers of the southeast of Ireland.

Introduction

Groundwater from low yielding boreholes and dug wells is used extensively as a source of domestic and farm water supplies in many parts of the Republic of Ireland which are not serviced by regional systems. A number of surveys of groundwater sources carried out in different parts of Ireland have indicated a significant amount of groundwater pollution. However, the experience of hydrogeologists (Daly and Daly, 1982; Daly, 1985a; Wright, 1987; Daly, Geraghty and Aldwell, 1989) working in the country suggests that groundwater is generally of good quality. This apparent contradiction and the impact of local pollution on overall groundwater quality is considered in this chapter, using the results of a detailed study in the Nore River Basin, a large catchment in the southeast of Ireland.

Nore River Basin

The Nore River Basin (Figure 12.1) covers an area of 2530 km^2 and is typical of the geology/hydrogeology, land use practices and general level of development in the southeastern part of Ireland. A comprehensive groundwater investigation (Daly, 1983, 1994) was carried out in this area between 1975 and 1981 with additional hydrochemical sampling thereafter. Subsequently, a detailed study of groundwater quality was carried out in 1990 (Woods, 1990).

There are three upland areas (150–500 m) separated by two lowlands (50–150 m) in the catchment. The basin is drained by the River Nore and its major tributaries, the Erkina, Dinin and Kings Rivers. The upland areas are generally characterized by rolling hills, with impermeable soils or blanket peat. There are substantial areas of high, medium and low permeability soils in each of the lowlands.

Most of the land is devoted to agriculture and the principal enterprises – livestock production and dairy farming – are dependent on intensive grassland cultivation. Tillage and forestry are also important in certain areas. There are a number of urban centres (population of 1000–18 000) which are essentially market towns with some light industry and food processing. The population of the Nore River Basin is ~ 75 000. Apart from the urban centres and associated ribbon development, the population is widely dispersed in small

Groundwater Quality Edited by H. Nash and G.J.H. McCall. Published in 1994 by Chapman & Hall. ISBN 0 412 58620 7

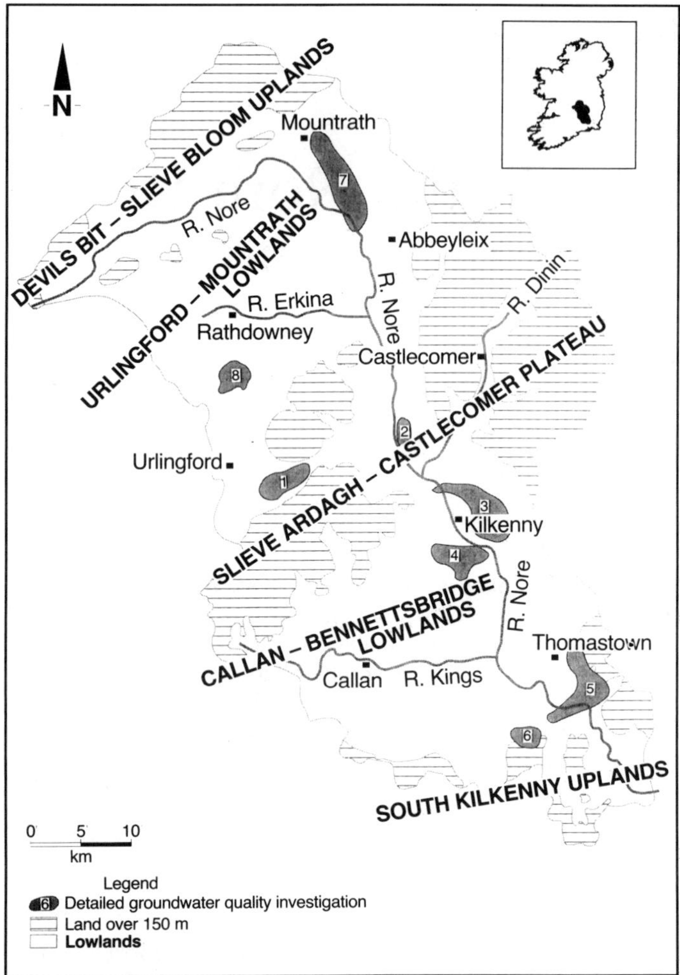

Figure 12.1 The Nore River Basin location map.

towns, villages and farms (mostly between 20 and 60 ha).

Groundwater is a significant source of industrial and regional water supplies and is also the principal source of domestic and farm water supplies in many rural areas. There are over 2500 mainly low-yielding boreholes, dug wells and springs in use in the catchment. These bored wells are generally 30–60 m deep, with yields ranging from 2 to 20 m^3 d^{-1} and they are simply constructed.

Background

A number of studies of groundwater sources in Ireland (Daly and Daly, 1982; Aldwell, Thorn and Daly, 1988; Deacon, 1990; Arcon Mines Ltd, 1992; Bennett, 1993) have shown significant contamination of substantial numbers of wells especially those using small amounts of groundwater for domestic and/or farm use. The distribution of contaminated wells generally appears to be random and they have frequently been found to be in close proximity to uncontaminated wells (Daly, 1981). Some concentration of polluted waters has been found in urban centres (Daly, 1981; O'Callaghan, 1984) and in particularly vulnerable areas such as the karst region of mid-Galway (Daly, 1985b). It is generally concluded that the main sources of this pollution are farmyards, septic tanks and urban areas (Daly and Daly, 1982; Wright, 1987; Aldwell, Thorn and Daly, 1988) as indicated by the presence of nitrates, potassium, chloride, ammonia and *E. coli*. Although inorganic fertilizers have probably raised the background levels of nitrate in groundwater, they are not thought to be causing significant contamination (Daly and Daly, 1984).

A number of factors are considered to be responsible for the situation described above.

- Until quite recently the regulations governing the location, design and management of farm and domestic waste storage and disposal systems were generally inadequate (Daly, 1985, Doyle, Henry and Thorn, 1986). They largely ignored the vulnerability of groundwater, the variability of hydrogeological conditions and the potential of these storage and disposal systems to cause groundwater pollution.
- The regulations for the location, design and management of private water supply wells are minimal and the majority are constructed with no professional advice.
- Enforcement of the regulations for private wells and waste disposal systems is minimal and normally only takes place at the planning stage and not over an extended period.

A survey (Daly, Geraghty and Aldwell, 1989) of most of the large springs in the Republic of Ireland in 1985/86 found that groundwaters in the unconfined parts of the principal limestone and sand and gravel aquifers were of good quality. Such groundwaters are considered to provide a reasonably accurate reflection of the overall quality in the upper and most vulnerable parts of these aquifers. These results confirm the views of Daly and Daly (1982); Daly (1988) and Wright (1987).

Irish agriculture, although starting from a relatively low base, developed dramatically on Ireland's accession to the Common Market in 1973. Agriculture became more intensive as illustrated by the 2.5-fold increase in the overall application of nitrogen fertilizer between 1974 and 1985. With the imposition of production quotas in the late 1980s and more recently the 'setaside' of some land, levels of inputs (nitrogen) and outputs have stabilized, so it is this current level of agricultural activity that is reflected in the groundwater quality that we see today.

Geology

The geological succession in this area is similar to that elsewhere in the southeast of Ireland and is summarized in Table 12.1. It consists of Lower Palaeozoic basement rocks uncomformably overlain by a thick sedimentary sequence of Upper Devonian to Upper Carboniferous strata. These strata have been folded into a large syncline with minor secondary anticlines and synclines. They are regularly faulted throughout the area and most are well jointed. Bedrock is normally overlain by Quaternary deposits which consist mainly of tills, sands and gravels and peat bogs of variable lithology and thickness.

Hydrogeology

The strata in the Nore River Basin have been subdivided into four major aquifers and four minor aquifers with the remainder being classified as aquitards (Table 12.1). The aquifers crop out over 40% of the area of the basin. A number of features of the hydrogeologic regime in this area are particularly important in relation to groundwater quality.

- Recharge is relatively high (200–600 mm) and the bulk of it occurs between October and March.
- Groundwater is frequently vulnerable to pollution because the Quaternary deposits and soil are either quite thin and/or permeable.
- The water table is generally <10 m below the surface, with an annual fluctuation of <5 m.
- In the bedrock aquifers only fissure flow exists and storage is relatively low. Permeability and storage tend to be greatest in the near surface layer (up to 20 m below the minimum water level) which frequently includes saturated Quaternary deposits and weathered bedrock.
- The bulk of the groundwater throughput occurs in the unconfined parts of the major aquifers (Table 12.1). Most of it moves relatively rapidly, at shallow depths, along short flow paths and discharges, frequently via large springs, into the many effluent streams which cross these aquifers.

Hydrochemistry

Samples were collected for chemical analysis at over 500 locations in order to establish the background hydrochemistry of the groundwaters in the different strata in the catchment. These sources are fairly well distributed on a geological and geographical basis throughout the basin. The samples were analysed for major ions, hardness, iron and manganese. Information on water quality and potability was obtained as a by-product of the hydrochemical study.

More than 50% of the area is underlain by limestone or limestone-derived Quaternary sediments. Hence, carbonate dissolution is the main hydrochemical process and calcium and magnesium bicarbonate are found in all the major rock types. Sodium bicarbonate waters (Misstear *et al.*, 1980) and very soft waters (Total Hardness <100 mg l^{-1} $CaCO_3$) are also found in restricted parts of the basin.

Water quality investigations

A number of studies by the Geological Survey of Ireland (Daly *et al.*, 1980; Daly, 1981; Daly, Geraghty and Aldwell, 1989) have shown that background concentrations of nitrate, potassium and chloride in Irish groundwaters, are less than 2 mg l^{-1} (as N), 1.3 mg l^{-1} and 12 mg l^{-1}, respectively. Experience has show that waters, in which the levels of any of these parameters are 3–4 times the background concentrations (Daly, 1981), also have abnormal levels of some other parameters such as total hardness, total dissolved solids, manganese, and frequently contain *E. coli.* Hence, concentrations of nitrate, potassium and chloride of 6–8 mg l^{-1} (as N), 4–5 mg l^{-1}, and 30–40 mg l^{-1}, respectively, have been found to be a useful guide for distinguishing between uncontaminated waters and those showing significant contamination. While these levels of the three parameters are lower than the maximum admissible concentrations of the EC directive (CEC, 1980), they do indicate very significant human influence on water quality in the immediate area of a source.

The concentrations of nitrate, potassium and chloride were used as the basis to separate contaminated from the uncontaminated waters in the Nore River Basin. Overall, almost 24% of the waters were found to be contaminated at the time of sampling. This proportion is considered to underestimate the problem as in most cases it was the intention, in sampling, to avoid contaminated waters. A slightly more detailed breakdown of the results (Table 12.2) shows that 31% of the sources in the lowlands were contaminated at the time of sampling as opposed to 15% in the uplands. The data suggest that sources in the non-carbonate rocks are less vulnerable than those in the carbonate rocks. This is most likely to be due to differences in the hydraulic regime and permeability. Furthermore, considerably more waters in the more permeable and generally unconfined aquifers, the limestones, dolomites and sands and gravels were found to be contaminated.

In the bar chart (Figure 12.2) the sampled sources

Table 12.1 A summary of the geological succession in the Nore River Basin

		Distribution	Principal formations present	Lithologies	Approximate thickness (m)	Hydrogeological significance
Quaternary		Widespread		Till, peat and sand and gravel	Normally <10 m but can be up to 30 m	Sand and gravel aquifers
Carboniferous	Westphalian	Castlecomer Plateau/Slieve Ardagh	Clay Gall Sandstone and Moyadd Coal/Glengoole Sandstone	Sandstones, shales and coal seams	250–320/up to 350	Minor sandstone aquifers
	Namurian	Castlecomer Plateau/Slieve Ardagh	Bregaun Flagstone, Killeshin Siltstone and Luggacurren Shale	Shales and sandstones	390–460	Aquitard
	Dinantian	Urlingford-Mountrath and Callan–Bennettsbridge Lowlands	Cullahill Limestone	Thick, clean coarse limestones, thin shales and clay wayboards	225–400	Major limestone aquifer
			Aghmacart Limestone	Thin, fine limestones with occasional shales	125–160	Aquitard
			Waulsortian Mudbank Complex	Cherty limestones, over-lying 'reef' limestones that are extensively dolomitized	50–220	Major dolomite aquifer
			Sub-reef Limestones	Thin-bedded limestones, argillaceous limestones, oolites and shales. The succession is dolomitized in places	300–500	Aquitard except for minor oolite aquifer
		The foothills of the Devils Bit–Slieve Bloom/South Kilkenny Uplands	Porter's Gate	Mainly mudstones/siltstones underlain by sandstones/limestones	30–80	Aquitard
Old Red Sandstone			Clonaslee Flagstone/Kiltorcan Sandstone	Sandstones with mudstones and conglomerates	40–230	Major sandstone aquifer
		Devils Bit–Slieve Bloom/South Kilkenny Uplands	Slieve Bloom/Carrigmaclea	Mudstones, sandstones and conglomerates	30–270	Aquitard
Lower Palaeozoic Strata			Capard/South Lodge, Ahenny and Carricktriss	Greywackes, siltstones and mudstones/greywackes and aureole rocks.	Over 1 000	Aquitard
Leinster Granite		South Kilkenny Uplands		Mainly granodiorite		Aquitard

Table 12.2 Numbers of contaminated and uncontaminated sources sampled in the Nore River Basin

		Lowland			Upland		
		Uncontaminated	Contaminated		Uncontaminated	Contaminated	
		Sample numbers		%	Sample number		%
Bedrock	Aquifer	87	43	33	83	10	11
	Aquitard	61	24	28*	73	11	13
Quaternary		41	19	32	40	13	25
Total		189	86	31	196	34	15

* Over most of these lowlands contamination is 21% except for two small areas where 75% (9) of sources are contaminated.

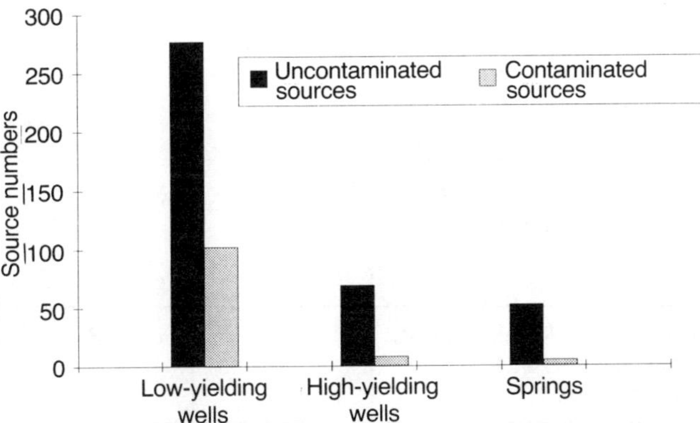

Figure 12.2 Water quality of the different types of source in the Nore River Basin.

are subdivided into three groups, low- and high-yielding wells, and springs. A high proportion of the low-yielding wells are contaminated, but only a very small number of the larger yielding wells and springs, which tap a significant volume of groundwater, showed any evidence of contamination. The one major source that does signify contamination over a sizeable area results from the spreading of waste by a large industry.

The contaminated groundwaters generally appear to be randomly distributed throughout the region. However, in a few areas there does appear to be a concentration of waters showing evidence of significant contamination. Hence, in the summer of 1990, a more detailed groundwater quality investigation was carried out in six areas (Numbers 1 to 5 and 7, Figure 12.1) characterized by varied hydrogeological conditions, land use and level of contamination (Table 12.3). The results of a similar investigation carried out in another area (Number 6, Figure 12.1) in 1988 were also included in this study.

The overall vulnerability of each area was assessed and a number of water sources were sampled for chemical and bacteriological analysis. At each sampling location the potential sources of contamination, well head completion and protection and land use practices were recorded. A description of each area and a summary of the results are given in Table 12.3. More than 40% of the groundwaters contained *E. coli* and a further 34% showed chemical evidence of significant contamination. In the case of the latter, the cover of unsaturated Quaternary deposits was considered to be effective in attenuating the bacteria (Woods, 1990). All five dug wells which tap very shallow groundwater were found to be contaminated. The three high-yielding wells and the one large spring sampled were found to have good groundwater quality.

The effluent from point sources such as septic tanks and general farmyard waste was considered to be the main source of contamination (Woods, 1990), with urban areas, old refuse dumps and possibly inorganic fertilizers being important in a few specific areas. Inadequate well construction and the proximity of wells to the waste sources are clearly contributory factors in the contamination of low-yielding wells. The results of this study confirmed the conclusions of the previous groundwater surveys mentioned above.

Discussion

The impact of these point sources of pollution on the overall quality of groundwater in the Nore River Basin depends on the general hydrogeological conditions. Groundwater in substantial parts of most strata in the Nore River Basin is vulnerable to pollution, owing to the nature of the Quaternary cover and shallow water tables. The large volumes of recharge to most of the aquifers move relatively rapidly, at shallow depths and

Table 12.3 Description of the areas investigated in 1990 and a summary of the results

Area	Geology		Land use	Sources sampled	
	Bedrock	Quaternary		Total	Contaminated
(1) Tubbrid	Cullahill Limestone extensively karstified	Sands and gravels normally <6 m thick	Rural area, intensive animal grazing and silage making	6	2
(2) Clintstown	Cullahill Limestone fissured close to the surface	Sands and gravels with fine sands and clays in places, 3–9 m thick	Rural ribbon development and intensive animal grazing	10	9
(3) Kilkenny city, north and east	Namurian Shales, Cullahill (karstified at top) and Aghmacart Limestones	Thick (8–26 m) sand and gravel or till	Some urban area, intensive animal grazing and silage making	7	4
(4) Kilkenny City, south and west	Cullahill Limestone significantly karstified	Sand and gravel (10–12 m) or till (3–11 m)	Some urban area, intensive animal grazing and tillage	8	8
(5) Thomastown	Kiltorcan and Old Red Sandstones and Lower Paleozoic strata	Mainly thin (<5 m) till or clay	Rural area, animal grazing at low stocking rates	5	4
(6) Kiltorcan Hill	Kiltorcan Sandstone	Thin (<5 m) till	Rural ribbon development, animal grazing at low stocking rates	5	3
(7) Kilbricken area	Dolomitized Waulsortian, Waulsortian and Sub-Reef Limestones	Up to 10 m of sand and gravel or 6–10 m of till	Rural area, animal grazing at variable stocking rates and silage making	6	6

along short flow paths. Contaminants deposited on or into the geological environment will generally receive considerable dilution, be confined mainly to limited parts of the upper layer and be flushed relatively quickly through the system owing to the generally restricted nature of the circulation in these aquifers. Therefore, it is unlikely that contaminants will move to the deeper (>30 m) parts of the aquifers unless induced by pumping.

Most low-yielding wells being used for farm and domestic water supply in this area and in the southeast of Ireland in general are accompanied by at least one potential point source of pollution within 50 m and another within 100 m. Hence these sources will be centrally located within zones of capture of most of these wells.

In the Nore River Basin a borehole or spring producing 1000–4000 $m^3 d^{-1}$ under normal hydrogeological conditions will be recharged within an area of the order of 1–10 km^2 (100–1000 ha). Similarly, a well providing a supply of 2–20 $m^3 d^{-1}$ will be recharged mainly from an area of 0.2–5 ha. Hence, in general, the waste produced by a farm and/or domestic house will have a far greater impact on the groundwater quality in the zone of capture of a nearby low-yielding well, where the potential for dilution and attenuation is limited, than several waste sources will have in the case of a high-yielding well or spring. This explains the existence of numerous contaminated low-yielding wells from aquifers which provide good quality groundwater to high-yielding wells and springs. The latter give a much better reflection of the overall groundwater quality in the aquifers of the Nore River Basin and in the southeast of Ireland in general.

This hypothesis was confirmed by a six-week pumping test in the summer of 1992 in the dolomite aquifer in the western part of the basin (Number 8, Figure 12.1). A cone of depression in excess of 2 km^2 was created by the combined discharge rate of 4400 m^3 d^{-1} from two high-yielding wells. Within this area 4 of the 8 low-yielding wells had earlier shown evidence of significant contamination. However, the two high-yielding wells pumped good quality groundwater (Arcon Mines Ltd., 1992).

Conclusions

Many of the low-yielding wells in the Nore River Basin are contaminated by the effluent from point sources of pollution. This effluent is generated close to these wells and its impact is limited to quite a small area and confined mainly to the upper parts of the geological environment. The investigation reported here has shown that the overall quality of groundwaters is good and of potable quality as indicated by the analyses of waters from large springs and high-yielding wells. A similar situation arises in the adjoining River Barrow catchment (Daly and Daly, 1982) and is likely to occur elsewhere in the southeast of Ireland.

Acknowledgments

The authors acknowledge the permission of the Director of the Geological Survey of Ireland to publish this chapter. Most of the samples collected during this investigation were analysed by the State Laboratory at Abbotstown.

References

Aldwell, C.R., Thorn, R.H. and Daly, D., 1988. Point source pollution in karst areas. In *Karst Hydrogeology and Karst Environment Protection. Proceedings of IAH 21st congress, Gulin, China*, Part 2, pp. 1046–1051.

Arcon Mines Ltd., 1992. *Environmental Impact Statement, Galmoy Mine Project*. 5 volumes and 10 technical reports.

Bennett, P., 1993. Groundwater monitoring in Northern Ireland. *Proceedings of 13th Annual IAH Groundwater Seminar, Portlaoise*.

CEC (Council of the European Communities), 1980. Council directive of 15 July, 1980 relating to the quality of water intended for human consumption (80/778/EEC). *Official Journal of the European Communities*, No. 1229/25.

Daly, D., 1985a. Groundwater quality and pollution in Ireland with particular reference to the effects of agricultural activities. In *Impact of Agriculture on Groundwater in Ireland*. Proceedings of a meeting held in Ireland. Contribution to the UNESCO international hydrological programme. IHP, Dublin, 269 pp.

Daly, D., 1985b. Groundwater in County Galway with particular reference to its protection from pollution. Unpublished report, Geological Survey of Ireland, 97 pp.

Daly, D., 1988. Groundwater pollution in Ireland: a review. *Proceedings of 11th Environmental Health Conference, Kilkenny*, p. 36–46.

Daly, D. and Daly, E.P., 1984. A review of nitrate in groundwater and the situation in Ireland. *Irish Journal of Environmental Science*, vol. 3, No. 1.

Daly, D., Lloyd, J.W., Misstear, B.D.R. and Daly, E.P., 1980. Fault control of groundwater flow in the aquifer system of the Castlecomer Plateau, Ireland. *Q.J. Eng. Geol. London*, vol. 13, pp. 167–175.

Daly, E.P., 1981. Nitrate levels in the aquifers of the Barrow River Valley. Unpublished report, Geological Survey of Ireland.

Daly, E.P., 1983. Water in the landscape: groundwater resources in Laois. In J. Feehan (ed.), *Laois, an Environmental History*. Ballykilcavan Press, Stradbally, Co. Laois, pp. 59–70.

Daly, E.P., 1994. The groundwater resources of the Nore River Basin. *Geological Survey of Ireland*, Report Series, RS94/1.

Daly, E.P. and Daly, D., 1982. A study of the nitrate levels in the aquifers of the Barrow River Valley, Ireland. *Proceedings of the IAH International Symposium, Impact of Agricultural Activities on Groundwater, Prague, Czechoslovakia*, vol. XVI, Part 1, pp. 183–194.

Daly, E.P., Geraghty, M.F. and Aldwell C.R., 1989. Water quality of the major springs in the Republic of Ireland in 1985/86. Unpublished report, Geological Survey of Ireland.

Deacon, A. 1990. Water quality of private water supplies in County Wexford. *The GSI Groundwater Newsletter*, No. 17.

Doyle, M., Henry, H. and Thorn, R.H., 1986. Septic tanks: a survey. *The GSI Groundwater Newsletter*, No. 2.

Misstear, B.D.R., Daly, E.P., Daly, D. and Lloyd, J.W., 1980. The groundwater resources of the Castlecomer Plateau. *Geological Survey of Ireland,* Report Series, RS 80/3, 31 pp.

O'Callaghan, J., 1984. A study of groundwater quality in two aquifers in Kilkenny. Unpublished M.Sc. dissertation, Trinity College, Dublin.

Woods, L., 1990. Groundwater quality in the Nore River Basin, Republic of Ireland. Unpublished M.Sc. dissertation, University College, London.

Wright, G.R., 1987. A review of groundwater pollution occurrences in Ireland. *Report to the Water Pollution Advisory Council*, 18 pp.

13 Microbiological quality of groundwater supplies in rural Zambia

S.E. Sutton

Abstract Over a period of four years, drinking water quality has been monitored, both at source and at point of consumption in a rural part of Zambia. The analysis of microbiological quality was undertaken, partly to see whether long-term monitoring was necessary to ensure safe supplies, and partly to assess the effects of low cost improvements to traditional sources and of disinfection of wells through use of bottled laundry bleach.

The results indicated that levels of contamination were generally low, perhaps because of the low population density and consequent low mobility of coliforms, and also the relatively good practices of users in relation to water collection and storage. The risks to health, however, varied significantly according to source type, and could be reduced greatly with low cost interventions, including health education.

Introduction

Water quality, particularly in relation to faecal coliforms, has been measured on a regular basis in a rural water supply project in Western Zambia, funded by the Norwegian Government. In the early stages, the water quality studies were connected to the WHO/UNEP Microprojects to test the applicability of existing *WHO Guidelines for Drinking Water Quality Control in Small Communities* (WHO, 1984–85).

The project area covers over 120 000 km^2 of land which includes the Zambezi flood plain, and surrounding higher forest lands which contain a series of dambos (small enclosed ablation hollows) in which groundwater is relatively near the surface (see Figure 13.1). In the forest areas, water normally lies more than 20 m below ground level. The aquifer over most of the region is the wind-blown Kalahari sands, which are very fine and uniform in grain-size distribution and exceed 300 m in depth. Most sources tap this aquifer, although in dambos there may also be a perched aquifer.

The rural population is approximately 500 000, living mainly in scattered settlements of <1000 inhabitants.

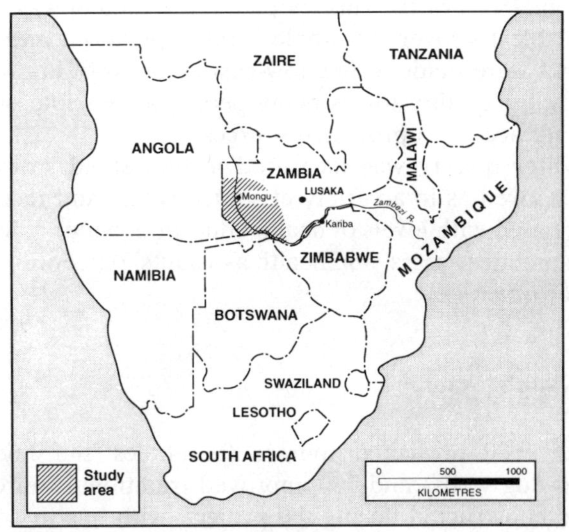

Figure 13.1 Map showing area of study in southern Zambia.

The largest communities lie along the edge of the flood plain, above the level to which water reaches after the rains, but within easy reach of the cultivatable land along the lower valley edge.

Groundwater Quality Edited by H. Nash and G.J.H. McCall. Published in 1994 by Chapman & Hall. ISBN 0 412 58620 7

Population densities are usually less than 5 per square kilometre, distances between communities are large and access through soft sand is often difficult. Boats and sledges are common modes of transport. Bilharzia, malaria, cholera and dysentery are endemic to the country, although cholera is rare in this particular area. Health services are provided by both traditional healers, and conventional health services. The latter consist of some 60 rural health centres, each staffed by a clinical assistant and a nurse who provide mainly curative medicine and immunization at the centre, and a health assistant who is an extension worker and whose job it is to visit communities and improve their hygiene and health practices. This includes water utilization, sanitation, as well as many other aspects of preventive health care.

Interventions in sanitation have been few, and pit latrines are uncommon, except at institutions such as schools and rural health centres. Whilst the lack of conventional sanitation is generally regarded as a health risk, with such low population densities it may be that existing practices of shallow burial of faeces, or decomposition by sunlight provides less risk than the concentration of faeces in (often poorly maintained) latrines. Sanitation projects have to date tended to concentrate on peri-urban rather than rural areas.

For the project in Western Zambia, funding was set at a level which would allow construction of sufficient boreholes and lined wells to provide protected supplies to some 50% of the rural population. Funding was not available to cover the whole rural population, and so efforts were made to find low-cost methods of improving water quality for as many people as possible who had no access to protected sources.

Water quality was measured in household storage jars as well as in a variety of source types, and results correlated with levels of health education, water collection practice, users' and health assistants' perception of water quality etc.

Figure 13.2 Water in this scoophole (in sand) is 10 cm deep, necessitating regular deepening. Because of this, the sides are continually caving in, making the hole wider. Thus, access to the well, without standing in the water, becomes difficult. This example is from Mwandi, Liyoyelo.

Source types

Over 1000 protected sources (boreholes and lined, hand-dug wells) and 100 improved traditional sources were constructed during the project, with health education and the establishment of community based maintenance systems forming integral parts of the work. Boreholes vary in depth, with rotary drilled wells reaching up to 110 m and percussion drilled or jetted wells seldom exceeding 15 –20 m. All have handpumps installed, a drainage channel, usually with some crop grown on the outflow, and often with a fence to keep animals away. Hand-dug wells are lined with concrete rings and seldom exceed 10 m in depth. They are mounted with a windlass and bucket on a chain. Most have a lid for the access hole in the top slab, and have a protective surround at ground level. Few are fenced, but all have drainage-channels.

Traditional sources tend to be wide scoopholes in sand (Figure 13.2), but may have narrower openings and steeper sides where there is some overlying peat, which is more stable. They can seldom exceed a depth of 2 m without the sides slumping in, and so as water levels fall, nearer sources often have to be abandoned and new ones dug further from the village where water levels are still sufficiently shallow. Few villages have previously done anything to improve the sources, apart from sometimes putting a piece of wood across the opening, to allow people to collect water without stepping in it. Usually, users each bring their own jug, pan or old cooking oil container with which to bail water into a bucket, gourd, bowl or other carrying vessel. This is in contrast to the lined wells where only one vessel is put directly into the water, and this well bucket is stored on the top slab of the well, away from the dirt and mud which tends to accumulate around any source.

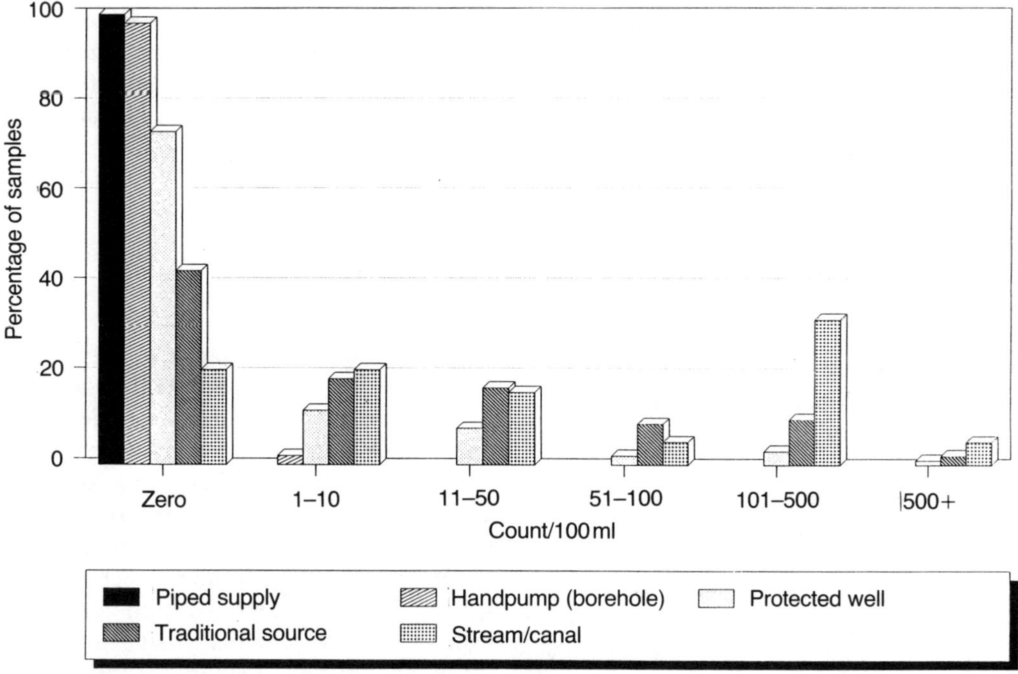

Figure 13.3 Water quality from different sources (faecal coliform).

To obtain a picture of the proportion of the population using a specific source type in different physical environments and with various levels of access to protected supplies, surveys were carried out among 15 000 school children. It was assumed that this method presents a good composite picture of:

- the proportion of the population with no access to protected sources;
- the proportion with access but choosing not to use them; and
- those who use them for some, if not all purposes.

The similarity of results from different schools in the same physical environment (i.e. riverside, plain edge, forest and dambo) and the contrast between those in different environments suggests that results were fairly representative.

The survey showed that surface water was mainly used for drinking only where it was both perennial in large quantities, and crocodiles did not present a risk to health. It was the main source of drinking water for <10% of the population. Generally, people preferred to use groundwater for drinking, and if there was access to a handpump or shallow lined well with bucket and windlass they would use this water for drinking and cooking, but not necessarily for washing, it being more convenient to go to a stream or river to bathe. Traditional sources (Figure 13.2) were used by >50% of the population for drinking water, and by many more for washing clothes and soaking cassava. Both these activities require large amounts of water, and using traditional sources usually cuts down the amount of water carrying needed.

Comparative quality of different source types

The source of water, in most cases, is surface or shallow groundwater, but the type of source construction has a major effect on water quality (Figure 13.3). This is because almost all contamination is through the source itself (particularly from contact of the bailing vessel with contaminants at the surface), not as contamination of the aquifer from other points such as latrines.

These results show that those sources most vulnerable to contamination through the method of abstracting water, and through return of surface water to the source, present most health risk to the user. Piped supplies and handpumps on well-sealed boreholes eliminate the risk of bacteria of faecal origin entering the water supply. Handpumps on jetted wells presented a very low risk, whilst one in four shallow wells with windlasses showed low levels of contamination, and only one in ten was significantly polluted (counts of more than 10 faecal coliform colonies per 100 ml). This compares with traditional scoop-holes where one in three was significantly polluted and less than half of the samples were free of faecal coliforms. There was no significant difference in quality between the wet or dry season (Utkilen and Sutton, 1989).

Levels of contamination were generally lower than for a similar study in Malawi (Lewis and Chilton, 1984), where around one in five boreholes indicated the presence of faecal coliforms, and where lined wells showed faecal coliform levels more comparable to the unprotected traditional sources of the Zambian rural supplies.

Sanitary inspecting or laboratory testing?

The cost of sample collection and of running a laboratory with microbiological facilities is high, and the results are not always reliable. An experiment was therefore undertaken in comparing independent sanitary inspections with faecal coliform counts, for 60 shallow lined wells.

Routine monitoring of sources involved the collection of samples from >100 sources, twice a year. Samples were collected in autoclaved sterile bottles, stored for transport in heavy wooden boxes with icepacks to keep them cool (but not frozen). They were transported to the provincial laboratory within six hours of collection, and samples were then filtered and incubated using standard media and methods (WHO, 1984–85). Health assistants who were not involved with the sample collection were then asked to inspect the sources in their area.

The health assistants identified 26 sources at risk, out of the 60 they visited. Of these, 25 were found to contain faecal coliform. None of the wells identified as of low or no risk contained any faecal coliform, and it was therefore concluded that, in most cases, sanitary inspection by qualified people at rural health centre level, provided a sufficient level of information for monitoring purposes. Higher levels (involving laboratory back-up) might be required on a temporary basis, firstly to design and calibrate inspection forms relevant to a particular environment, secondly to train field officers in sanitary inspection and thirdly to provide additional information at times of cholera outbreak.

Contamination by consumers

Users may contaminate the supply by introducing dirt into the source directly, through using dirty buckets or by standing in the water, or indirectly, by allowing the accumulation of dirty water at the surface which then percolates back into the well if this is inadequately sealed. Health education, plus the improvement of well designs and construction can do much to reduce this risk. However, such efforts are largely wasted if users contaminate water further during transport and storage.

Studies of household water quality (Sutton and Mubiana, 1989), suggest that in this area, people have relatively good water collection, transport, and storage practices. This means that any improvement to a source will generally be carried through as a benefit to the user. In a study of water quality in 200 household storage vessels, 80% had no faecal coliform, and 95% had less than 10/100 ml.

Improving water supplies with minimum intervention

Traditional water supplies

Communities had to approach health assistants with a request for help, and then decide on what improvements they felt would provide greatest benefit at least cost. The health assistant provided some ideas for technical solutions, but most came from the community, which also provided all labour and all materials, except cement (which the project provided in the early stages, so that demonstration sites could be established). As the benefits of the improvements became more widely acknowledged, and more communities became involved, requests for assistance increased, the subsidy on cement was reduced to 50%, and assistance was provided to help the community to set up its own management and decision-making frameworks, and to undertake health education which the management committee would promote both by example and further teaching.

Interventions which were most commonly adopted were:

- raising the ground around the source to stop run-off into it,
- casting a small top slab with an opening too small to allow a bucket to be used; slabs were generally of a size which could be picked up by three or four women and moved to a new site, if water levels dropped too low to allow convenient abstraction; the slabs were supported on wooden cross-beams, and the small opening was proposed by users to reduce the risk of dirt being brought in, and as a measure to provide cooler water;
- supplying a mug, a saucepan, or a can on a pole (Figure 13.4) for use as bailers. These were made communal property, so that only one vessel was used, and this was kept clean and placed by the source;
- closing the access hole at times when not in use. Some villages wove a small mat, others cast an old saucepan into the slab, and then used the saucepan lid to close off the source;
- providing a fence and a roof over the source (a few villages only); and
- casting concrete rings using rush mat or oil drum shuttering, or installing an oil drum to allow greater

Figure 13.4 Concrete slab, placed over well, with insect saucepan, lid and communal scoop at Mwandi, Ikwichi.

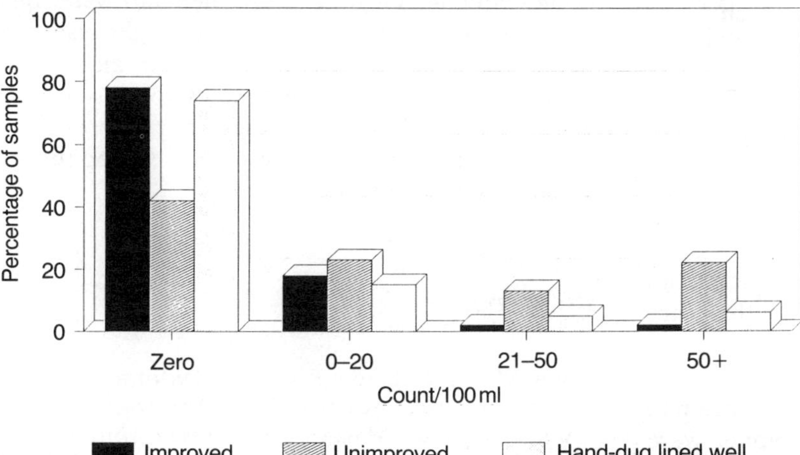

Figure 13.5 Traditional source improvement (faecal coliform counts).

depth and stability, and so a more reliable supply. However, the oil drum on its own provided only a short-term solution because of the aggressiveness of the water, which meant the drum corroded within one or two years.

At the same time, education on water use, water related diseases, problem solving, collection and use of communal funds etc., was undertaken, partly using participatory and partly formal education methods.

The result was that there was a significant improvement in water quality, at least for the first year after construction, which was as long as monitoring was possible. In this time, the proportion of sources with faecal coliform was halved, and the overall distribution of water quality became similar to, if not slightly better than that for lined wells with windlasses (Figure 13.5).

The improvement in quality was not purely a theoretical benefit. Users perceived the improvement, with 90% mentioning better taste and smell, and over half feeling that reliability and quantity had also improved.

The main unprompted reasons given for liking the new supply were that it gave cooler water, and was better protected from animals and run-off. The perception of benefits by users meant that nearby communities soon decided to undertake similar works, sometimes on their own, sometimes with the health assistants.

Improving water sources with minimum intervention

Shallow well chlorination

One in ten shallow, lined wells, with bucket and windlass assembly for lifting water, had shown medium to high faecal coliform counts. Efforts were made to reduce this level of contamination, since such wells are

Table 13.1 Shallow well chlorination

Step	Procedure
1	Explain to the community what is going to be done and why. The next time they will carry out the procedure without outside help.
2	Measure the depth of water in the well, using the bucket and chain. Bail out until only an arm's length of water is left in the well.
3	Leave one or two full buckets of water by the well.
4	Pour one 800 ml bottle of locally available laundry bleach into the well, followed by the two buckets of water to ensure good mixing.
5	Leave the water for a minimum of one hour (preferably two), plus someone at the well to warn people not to use it.
6	Bail water out again, using bailings to wash around the well. Continue till level of chlorine is aceptable to users.
7	Encourage the community to do this if water becomes smelly, tastes bad or goes cloudy.

expensive to construct, and therefore benefits should be significant and long-lasting.

Where contamination could not be traced back to an outside source, such as a pit latrine (of which there are very few), the well was disinfected. The procedure was as indicated in Table 13.1.

This procedure was carried out on 15 wells which had been persistently badly contaminated. None were found with faecal coliform during the six months after treatment. Chlorine levels during treatment reached over 75 mg l^{-1} with new bottles of bleach and exceeded 25 mg l^{-1} even with the oldest bottles (despite chlorine content decreasing with age). The method proved acceptable to consumers and was also used on wells with no conventional contamination, but which users regarded as polluted because of such things as lizards falling in the water, or worms found in the bottom. Without the disinfection procedure, people would not go back to using such wells, even though the animal debris had been removed. It appeared that the additional disinfection procedure was sufficient for users to feel that the source was then clean and usable again.

Conclusions

Rural water quality, in terms of bacteriology, was found to vary markedly according to the source type. Compared with many countries, the overall levels of contamination were generally not high, and the number of users per source was low (usually fewer than 200 per handpump and 100 per traditional source). In this situation it was found that low cost interventions could bring about significant improvements to water quality, and some of the approaches used might also provide benefits in areas where user numbers and contamination levels are higher. Because of good user practices in water collection and storage, any improvement to quality at the source was likely to be carried through as a benefit to the consumer.

References

Lewis, W.J. and Chilton, P.J., 1984. Performance of sanitary completion measures of wells and boreholes used for rural water supplies in Malawi. *Challenges in African Hydrology and Water Resources, Harare Symposium*. IAHS Publ. no. 144.

Sutton, S.E. and Mubiana, D., 1989. Household water quality in rural Zambia. *Waterlines*, vol. 8, 1, pp. 20–24.

Utkilen, H. and Sutton, S.E., 1989. Experiences and results from a water quality project in Zambia. *Waterlines*, vol. 7, 3, pp. 6–9

WHO, 1985–86. *Guidelines for Drinking Water Quality*. Vol. 3, *Drinking Water Quality Control in Small Community Supplies*.

14 Nitrate contamination of groundwater in the Kutama and Sinthumule districts of Venda, South Africa

R.J. Connelly and D. Taussig

Abstract High nitrates in groundwater are a common occurrence in parts of Southern Africa. The main sources are nitrogenous fertilizer or sewage effluent. During investigation for improved groundwater supply to the Kutama and Sinthumule area in South Africa, high nitrates were identified in areas where the source was not clear. No artificial fertilizers are used and no pit latrines are nearby. Following drilling and testing, it was found that the cause of high nitrate was dryland cropping, which, under the climatic and physical conditions, favour nitrification and subsequent leaching of nitrate into groundwater.

Introduction

The Venda Department of Water Affairs (VDWA) required an investigation to determine the most appropriate source of water supply and means of distribution for supplying the present and future water requirements of the Kutama and Sinthumule districts of Venda, in South Africa (Figure 14.1). During the investigation of the groundwater resources, several boreholes were found to contain high levels of nitrate. High nitrate levels in drinking water can cause methaemoglobinaemia in infants and has been linked to stomach cancer in adults. This chapter discusses the activities and results of an investigation into the reasons for high nitrates which have been recognized as a major problem in water supplies in rural and developed areas all over the world.

A nitrate concentration of 45 mg l^{-1} as NO_3 is regarded as the maximum permissible level, in drinking water supplies, by the World Health Organization (WHO). However, Terblanche (1991) records a number of researchers who state that only when the concentration of nitrate in drinking water reaches 90 mg l^{-1} will it become the main component of the total nitrate intake of adults. This may indicate that the WHO limit is over-conservative but that it should not be raised to more than 90 mg l^{-1} for adults.

Objectives of the investigation

The investigation aimed to meet the following objectives:

- to determine the source of nitrate;
- to investigate the migration of nitrates; and
- to determine the long term effect on the water supply and recommend suitable remedial works.

The overall approach to the investigation was to define the extent of high nitrate, then to evaluate the possible sources of nitrate and the mode by which it could be transported through the soil profile into the groundwater. The work included a survey of existing boreholes and water quality, exploratory borehole siting and test pumping, soil sampling and textural classification, chemical and bacteriological analysis.

Geographical setting

The Kutama and Sinthumule area occupies some 19 900 ha of flat-lying country south of the Soutpansberg. The area is roughly rectangular in shape with a length of

Groundwater Quality Edited by H. Nash and G.J.H. McCall. Published in 1994 by Chapman & Hall. ISBN 0 412 58620 7

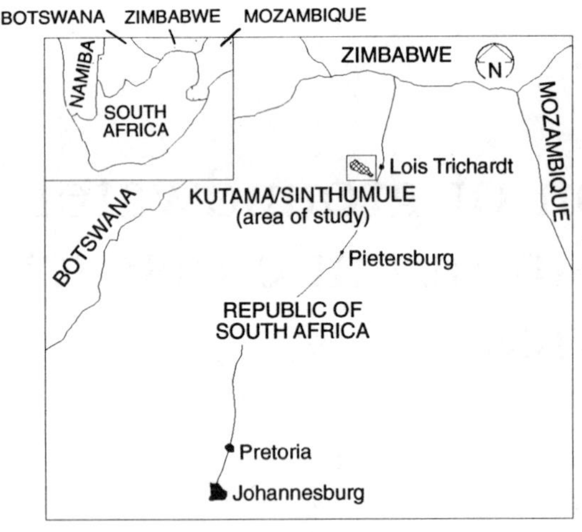

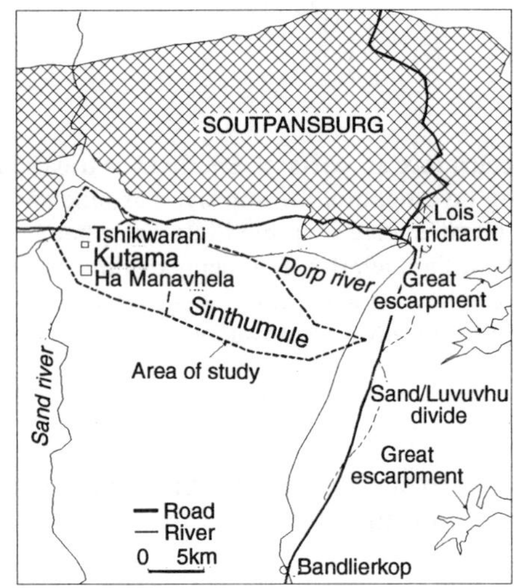

Figure 14.1 Kutama Sinthumule locality plan.

29 km aligned west–northwest and a breadth of 8 km. The position of the study area is shown in Figure 14.1.

The Kutama and Sinthumule areas can be defined as arid to sub-arid climatic regions with an average annual rainfall of 450 mm. The monthly potential evaporation exceeds the mean monthly rainfall, resulting in a water deficit throughout the year. However, the maximum rainfall on a daily basis can be in excess of evaporation. This factor is significant when determining rainfall infiltration and aquifer recharge. The seasons are characterized by mild winters and very hot summers.

The area lies within the southern margin of the Limpopo Metamorphic Province and is underlain by the Baviaanskloof Gneiss of Archean age. Xenolithic rafts of older rocks known as the Banderlierkop Metamorphic Suite occur within the Gneiss and are shown as pyroxenite outcrops. In the northwest of the study area, these highly metamorphosed rocks are overlain unconformably by the Sibasa Formation which comprises volcanic rocks of Proterozoic (Mokalian) age.

Alluvium composed of sand and silt occurs in relativelynarrow zones in close proximity to the Dorp and Sand Rivers.

Much of the study area is overlain by transported sands of recent origin and calcrete has formed in places. The depth of weathering varies from 10 to 30 m in the Gneiss and Soutpansberg volcanics. The general geology is shown in Figure 14.2.

In the Kutama and Sinthumule districts, the soils on the Pietersburg Plateau have a moderate infiltration potential, a low storage potential and a low erosion potential. The soils occurring in the stream beds have a moderate infiltration potential, a high storage potential and a moderate erosion potential.

The natural vegetation of the area is of secondary open shrubland type and consists mainly of *Acacia* spp. on the plain and rocky slopes. The grass cover consists of sweet grassland types of *Aristida* spp. and *Eragrostis* spp. The vegetation is in a highly stressed condition due to heavy overgrazing by cattle and goats. Large arable fields adjoin most of the villages and these are fenced off from the remainder of the area which is used mainly for communal grazing. Crop production is usually maize with no artificial fertilizers or irrigation.

Results

A survey was carried out to locate and sample boreholes outside the limits of the main study area to assess the extent of the nitrate problem on a more regional basis. Seventeen boreholes were located within approximately 10 km of the area which showed high nitrate values but not generally as high as in the study area. The results indicate a mean concentration of 34.4 mg l^{-1} NO_3 with a maximum and minimum of 155 mg l^{-1} and 0.44 mg l^{-1} respectively. It would appear from these results that high nitrate levels are not restricted to the Kutama and Sinthumule districts, but that the occurrence of high levels of nitrate in groundwater is a common feature in the surrounding area. Water-level measurements within the study area ranged from 8 to 30 m below surface with a general hydraulic gradient from southeast to northwest (Figure 14.3). Three main aquifer types have been identified in the study area from the results of production drilling: A **regional aquifer** extends over most of the study area and is encountered in the subsurface at the interface between weathered and fresh bedrock. This aquifer is only 0.3–3 m thick at between 20 and 30 m below ground surface level and is partly confined by the overlying weathered material. **Structural aquifers** comprise fault and fractures zones, dyke contacts and lithologic contacts. These aquifers usually have high transmissivities but low storage but can be in hydraulic contact with the

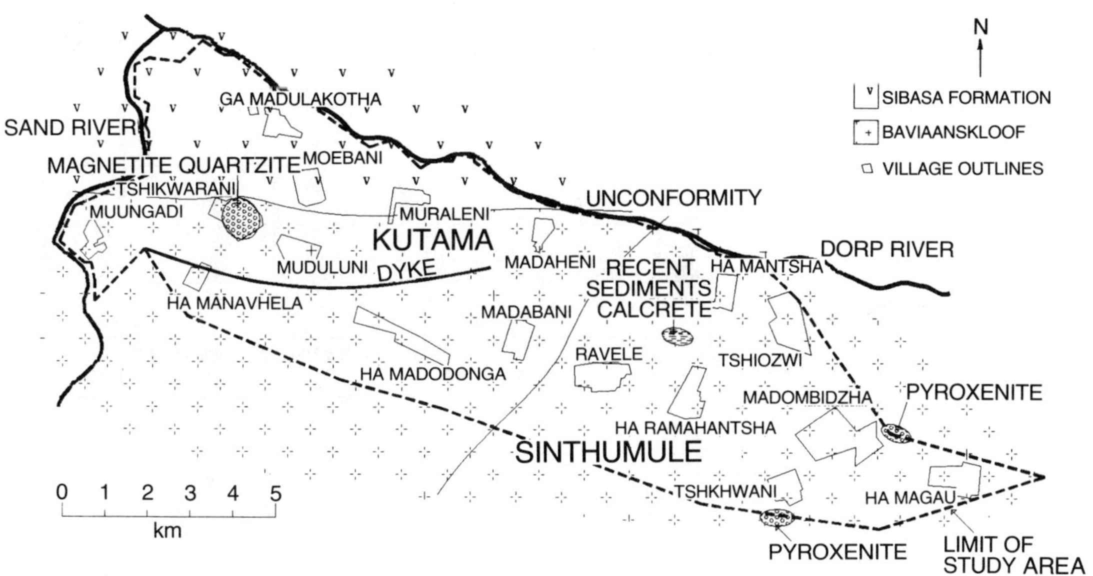

Figure 14.2 Main geological features at Kutama and Sinthumule.

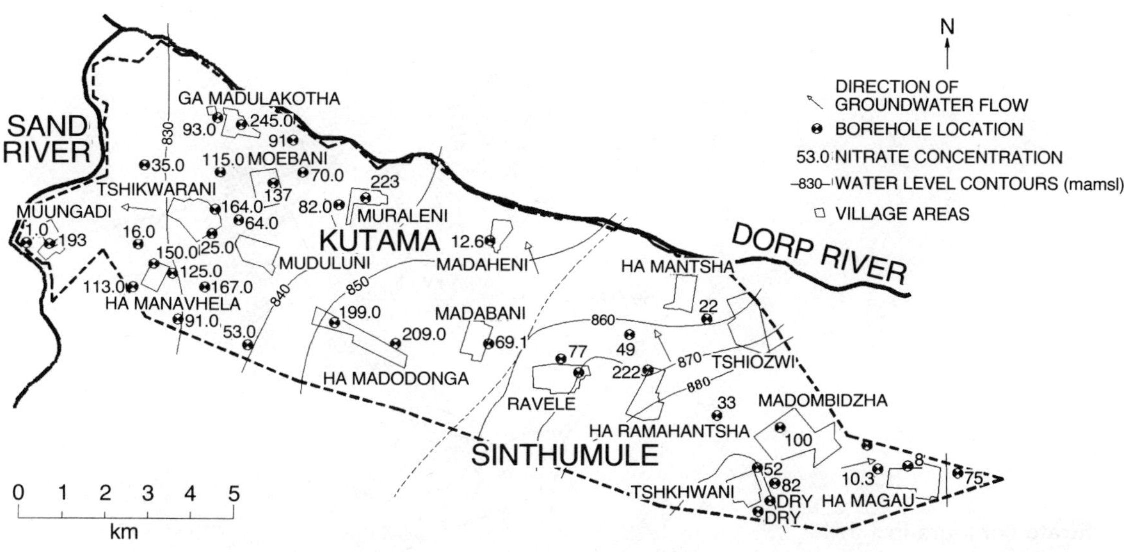

Figure 14.3 Plan showing the static water level for the Kutama and Sinthumule districts.

regional aquifer. **Alluvial aquifers** are found along the larger river courses. However, they are thin and not laterally extensive and are not therefore regarded as important aquifers in the study area.

It is estimated that average annual recharge will vary between 2 and 5% of the mean annual precipitation which is typical of the region based on previous studies.

Seven holes were drilled in the vicinity of Tshikhwane and Ha Manavhela for water supply. Two of the boreholes were dry, and the remaining five had yields of between 3.5 and 12 $m^3 h^{-1}$. During test pumping the nitrate concentrations were found to be high in all of the boreholes located in the extensive cultivated areas, ranging from 50 to 167 mg l^{-1}. In all cases the boreholes were located up hydraulic gradient of the villages so pollution from pit latrines is unlikely. The borehole located down hydraulic gradient of Ha Manavhela, in an area of natural vegetation had the lowest nitrate concentration of 16 mg l^{-1}. There was no significant variation in nitrate concentrations in any of the boreholes during sustained pumping, although an observed decrease of up to 10 mg l^{-1} could indicate dilution of nitrate levels as deeper groundwater is drawn into the borehole during pumping.

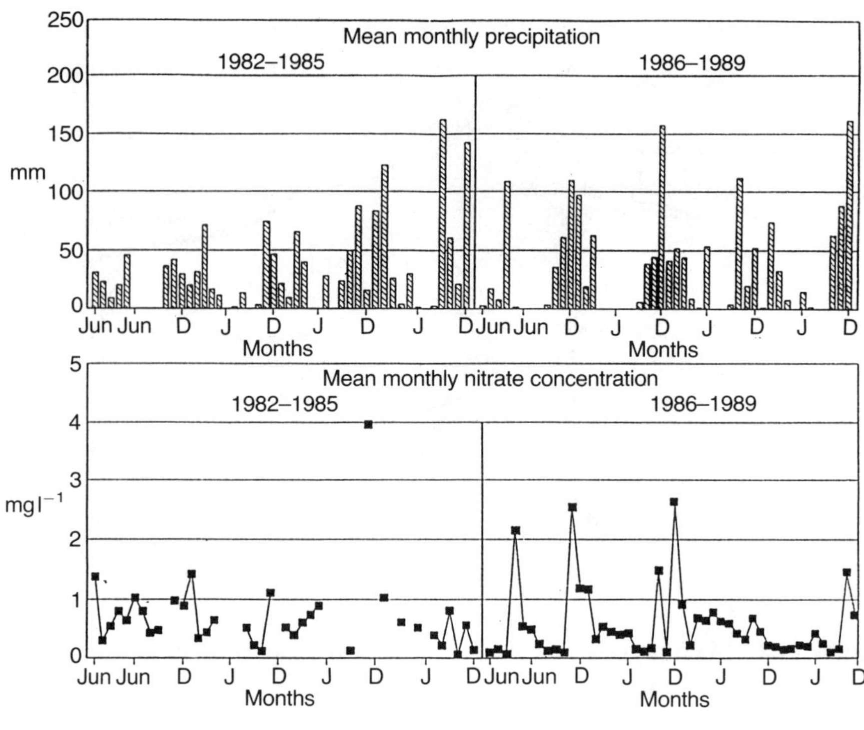

Figure 14.4 Comparison of monthly precipitation with monthly nitrate concentration for surface of the Sand River (Waterpoort).

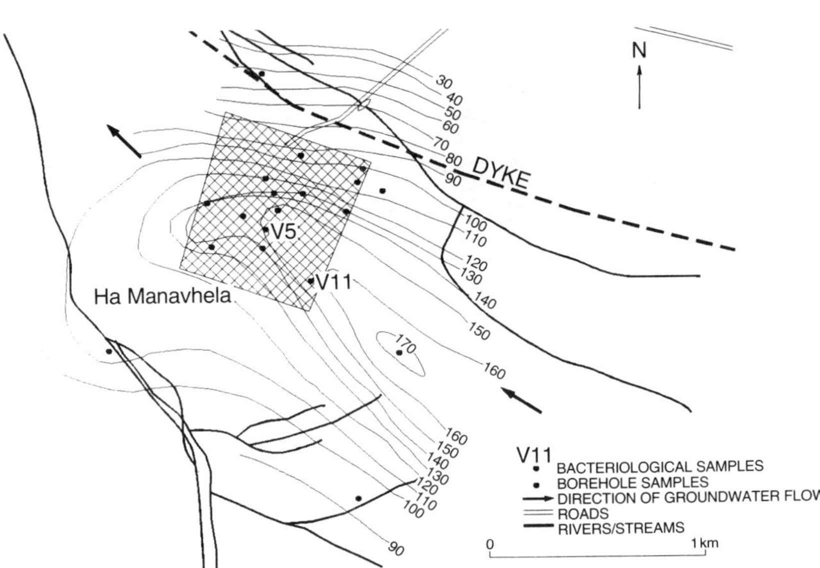

Figure 14.5 Nitrate contours in mg l^{-1} NO_3 Manavhela.

Mean monthly nitrate concentrations were calculated for the gauging station at Waterpoort on the Sand River. Although these values are much lower than those observed in the groundwater, it can be expected that an increase in nitrates in the surface waters will reflect an increase in the groundwater. There is a strong correlation between nitrate levels in surface water and the mean monthly rainfall as illustrated in Figure 14.4. The peak in nitrate usually occurs after the first rains of the summer season, usually in November to December. Sampling of boreholes from 1988 to 1990 has also shown that the nitrate concentrations are at the lowest levels during the winter months, and at the highest levels following the first heavy summer rains. Detailed sampling was done around Ha Manavhela and nitrate values were found to be fairly constant at 100–150 mg l^{-1}. The higher values cluster along a diagonal trend through the centre of the village in a north-westerly direction parallel to the groundwater movement. This suggests the movement of a 'plume' of elevated nitrates in the groundwater down hydraulic gradient of the village. There is a background nitrate level of at least 100 mg l^{-1} in the groundwater before it passes under Ha Manavhela (Figure 14.5). On the northeast side of the village, there are very large fields cultivated for maize.

Table 14.1 Bacteriological analysis at Tshikwarani and Ha Manavhela

Village	B/hole	N (mg l^{-1})	*E. coli*	Total plate count	Distance to pit latrines (m)	No. of persons
Tshikwarani	71-A	164	Uncountable	>100	15 m in downstream	290
	T5	25	6	100	14 m in downstream	10?
Ha Manavhela	V11	139	0	35	18 m in downstream	10?
	V5	150	3	16	20 m in downstream	10?

Two water samples were taken from Tshikwarani and two from Ha Manavhela for bacteriological analysis. The positions of the boreholes at Ha Manavhela are shown in Figure 14.5. The samples were analysed for *E. coli* and total plate counts of viable organisms which are indications of faecal contamination. Three of the water samples do not comply with recommended South African Bureau of Standards (SABS) limits and could indicate contamination from the latrines (Table 14.1).

Tshikwarani village is located on the magnetite quartzite outcrop which is shown on the geological map in Figure 14.2. The water samples here show the highest bacterial counts. Borehole 71-A had an uncountable number of *E. coli* per 100 ml which is indicative of sewage contamination.

In Ha Manavhela, bacteriological results indicated that in spite of the high nitrate values, contamination from the latrines is not as apparent as is shown by the water samples taken at Tshikwarani. This can either imply that the bacteria have effectively been filtered out in the unsaturated zone or that the pit latrines are not necessarily the source of high nitrate concentrations. Space does not permit a detailed presentation of chemical results but various chemical species were plotted against each other and showed that it was possible to group pollution sources into two main categories. These were those sources showing contamination from pit latrines in the villages and those from the arable fields.

Sources of nitrate in the study area

Apart from minor sources of nitrate such as rain water, domestic animal wastes and the possible rare occurrence of nitrate minerals in the host rock, two major potential sources have been identified. These are pollution from the pit latrines and from natural soil nitrogen processes which are discussed below.

The natural soil profile has long been recognized as an effective system for the disposal and purification of human sewage by the removal of faecal microorganisms and the breakdown of many chemical compounds due to bacteriological action and filtration. However, many factors play a role in the extent to which the removal mechanisms are effective, and in the African environment, on-site sanitation is usually the only affordable and practical method (Muller, 1989). When effluent enters the unsaturated zone, pore clogging eventually develops at the infiltration surface and forms a crust. This clogging allows effective filtration and purification mechanisms to proceed. The thicker the unsaturated zone the more effective contaminant removal will be. Newly established pit latrines are at a greater risk of polluting groundwater until the crust builds up sufficiently in the pits to reduce infiltration.

In the southeastern part of the study area, nitrate contamination from the pit latrines is negligible. The reason is that the water table is at least 30 m below the surface which provides a thick unsaturated zone through which the effluent will be 'treated'. In the northwestern part of the area the water table is only 8 m below surface and the highest nitrate levels were recorded. In addition, the underlying geological formation is the highly jointed Sibasa Formation which allows relatively direct recharge into the groundwater.

It is therefore concluded that provided there is a sufficient thickness of weathered rock beneath the pit latrines and above the water table (>20 m), nitrogen removal mechanisms are able to proceed, thereby reducing the amount of nitrate and ammonia which can infiltrate into the groundwater.

Where the underlying rock is fractured and unweathered, many open fractures provide access directly from the pit latrines into the groundwater without any natural treatment through the unsaturated zone. Tshikwarani is built on the unweathered magnetite quartzite hillock and the groundwater has been contaminated from pit latrines as shown by the chemistry and high *E. coli* count in the water. By contrast, most of the study area is underlain by very weathered granite which poses less of a hazard as the nitrogen removal process can still function in this weathered zone.

Where the latrines are subject to extreme loading (for example in schools or other public places), contamination is a greater risk compared to other latrines where loading is not so severe. At Tshikwarani school, the pit latrines are subject to extreme hydraulic loading. At the time of sampling, three new pit latrines were in operation as effluent in the older pits has reached surface. It is not surprising therefore that such a high bacteriological count was recorded in Borehole 71-A. The bacteriological results are summarized in Table 14.1.

In such circumstances, the nitrate content should be regarded as of minor importance compared to the danger of faecal bacteria and viruses in the drinking water supply.

Under natural conditions, the amount of nitrogen removed from the atmosphere by natural fixation processes is closely balanced by organisms that convert organic nitrates to gaseous nitrogen. However, as soon as the cycle is interrupted or interfered with in some way, this delicate balance may be destroyed, resulting in either an excess or deficiency of usable nitrogen. Several factors influence the effectiveness of the nitrogen fixation processes such as vegetation cover, soil type and climate. Soil that is left fallow over any period of time will accumulate large quantities of nitrates which may be leached into the ground water following intense precipitation. Highest nitrate levels can be expected therefore, shortly after the first heavy rains of the wet season due to the build-up of nitrates in the soil throughout the dry winter months.

Maize is the crop usually planted after the first rains. It matures in 120 days so the land is left fallow for 240 days per year. After harvesting, the leaves, stalks and roots are left in the lands. The local cattle and goats are allowed onto the land to browse the debris, which will also add to a build-up of nitrate from animal wastes. The land is tilled in the late winter months which will mineralize nitrogen further. The first rains fall in September or October and as the crops have not yet been planted at this stage, there is nothing to take up the nitrates and they will be leached into the groundwater. The nature of the precipitation in the form of intense downpours ensures that these nitrates are flushed into the groundwater. High evaporation rates can also ensure that nitrates in the rain water will add to the accumulation in the groundwater. The flat topography minimizes surface run-off thereby enhancing infiltration of the rain water. It is to be expected, therefore, that the highest levels in the groundwater will occur in November to January after the first heavy rains.

The large cultivated lands in the study area can therefore be a source of nitrate contamination. Ha Manavhela and Ha Madodonga, which are downstream of the two largest cultivated lands, have high nitrate levels in the groundwater, possibly as a result of this large contamination source. The new boreholes located in the cultivated dry crop lands all show a high background level of nitrate of at least 50 mg l^{-1}. Leaching of nitrate will also occur in the soils under natural vegetation that has been extensively over-grazed, due to the fact that less nitrogen will be taken up and the normal nitrate repression, typical under perennial plants, will be disrupted.

Discussion

It has been shown from this study that there are two major sources of nitrate which are responsible for the contamination of the groundwater at Kutama and Sinthumule.

(1) There is a generally high background nitrate concentration of at least 50 mg l^{-1} due to natural processes active in the sandy soils, particularly under dry land cropping or natural vegetation areas which are over-grazed.
(2) The pit latrines in some of the villages, particularly where the groundwater table is <15 m below the surface or where the latrines are located on open, fractured, unweathered rock with no soil cover.

The extent to which natural nitrate removal occurs in the soil profile has not been ascertained, but it is probable that nitrate is continually accumulated in the soil and weathered rock from pit latrines and agricultural practices. This could mean that in the future, nitrate contamination of groundwater will become more of a problem, as it has in Europe, unless farming methods and methods of sewage treatment are changed. In the rural areas however, it is impractical to change methods of sewage treatment. In addition, if the population should increase in some of the villages with shallow water tables, it is likely that pollution from pit latrines will increase as the hydraulic loading increases. This could also result in an increased potential for water-borne disease, with elevated nitrate levels becoming a minor health hazard in comparison.

It can be concluded that when siting a groundwater supply, consideration should be given to the location and nature of local agricultural practices together with the other factors influencing nitrate accumulation. Boreholes should be placed sufficiently far from dry land cropping areas, preferably under natural vegetation areas and up hydraulic gradient from pit latrines. Future reticulation systems should take all these factors into account and not just the proximity to latrines to reduce the nitrate problem.

To reduce the accumulation of nitrate due to agricultural practices it would be necessary either to irrigate maize to get growth before the first rains or plant a winter crop. Obviously such courses of action may well not be practical in most cases and the problem will have to be managed by siting water supply boreholes carefully, rather than solved.

It is hoped that the results of this study can be extended to other areas and should provide guidance in the planning and management of agricultural activities and water supply.

References

Muller, M., 1989. The conflict between on-site sanitation and low cost water supply in South Africa – case studies and policy perspective. *Proc. Ist Biennial Conference of the Water Institution of Southern Africa, Cape Town.*

Terblanche, A.P.S., 1991. Health hazards of nitrate in drinking water. *Water SA*, vol. 17, pp. 77–82.

15 A *Rural Water Supplies Organiser* for weathered crystalline basement aquifers

R. Herbert

Abstract A prime target of rural water supply projects is to improve health through provision of high quality water. A loose leaf *Rural Water Supplies Organiser* has been developed which is suitable for crystalline basement aquifers and which is intended to be of value to all those involved with the design, implementation, monitoring and evaluation of rural water supply projects.

Introduction

The stated aim of Britain's overseas aid effort is 'to promote sustainable economic and social development and good government in order to improve the quality of life and reduce poverty, suffering and deprivation in developing countries'. It has long been realized that provision of a clean water supply will not by itself improve health and therefore quality of life in a sustainable way. A multi-disciplinary approach is necessary. The *Organiser* attempts to identify all the inputs necessary if sustainable improvement in life is to be achieved. For example, people must be educated in their use of water if health is to be improved, indeed all five of the inputs to the box entitled *Correlated Health Benefits* of the *Organiser* must be achieved. Also, if these improvements are to be sustainable the scheme must be sociologically acceptable and there must be a reliable operation and maintenance system (see the appropriate box of the *Organiser*). In addition, the scheme must not adversely affect the environment by its implementation, see the box entitled *Main Environmental Issue*. Excess water for cattle and sheep can have disastrous effects on rural communities if its use is not managed well.

A good rural supply scheme which includes all the above will result in all the benefits in the box summarizing *Basics for Better Rural Living*.

But what of the hydrogeology? Recent research and development work (see *Organiser's* references) sponsored by ODA, has allowed development of a *modus operandi* for the exploitation of crystalline basement aquifers (weathered, igneous and metamorphic rocks). Except in very arid areas (typical rainfall <300–500 mm) the research programme showed that collector wells can be located at favourable sites and sustainable supplies of a few l s^{-1} can be obtained. These relatively high rates of discharge can support small-scale irrigation and are sufficient to assist with the reduction of water-washed diseases. Slim, deep boreholes fitted with handpumps can provide small yields (median of 0.1 l s^{-1}) for drinking purposes almost everywhere, given care. In addition, new geophysical techniques and the better understanding gained of basement aquifers now allows more efficient siting of sources of water and it should be possible to provide most communities with a more convenient drinking water source and hence more time for other things in life than carriage of water.

It is hoped that a quick scan of the *Organiser* will remind all subscribers to rural water supply schemes of their multi-disciplinary nature.

Aim

The aim of the *Organiser* is to assist in the assessment and design of rural water supply schemes sited on basement rocks.

Groundwater Quality Edited by H. Nash and G.J.H. McCall. Published in 1994 by Chapman & Hall. ISBN 0 412 58620 7

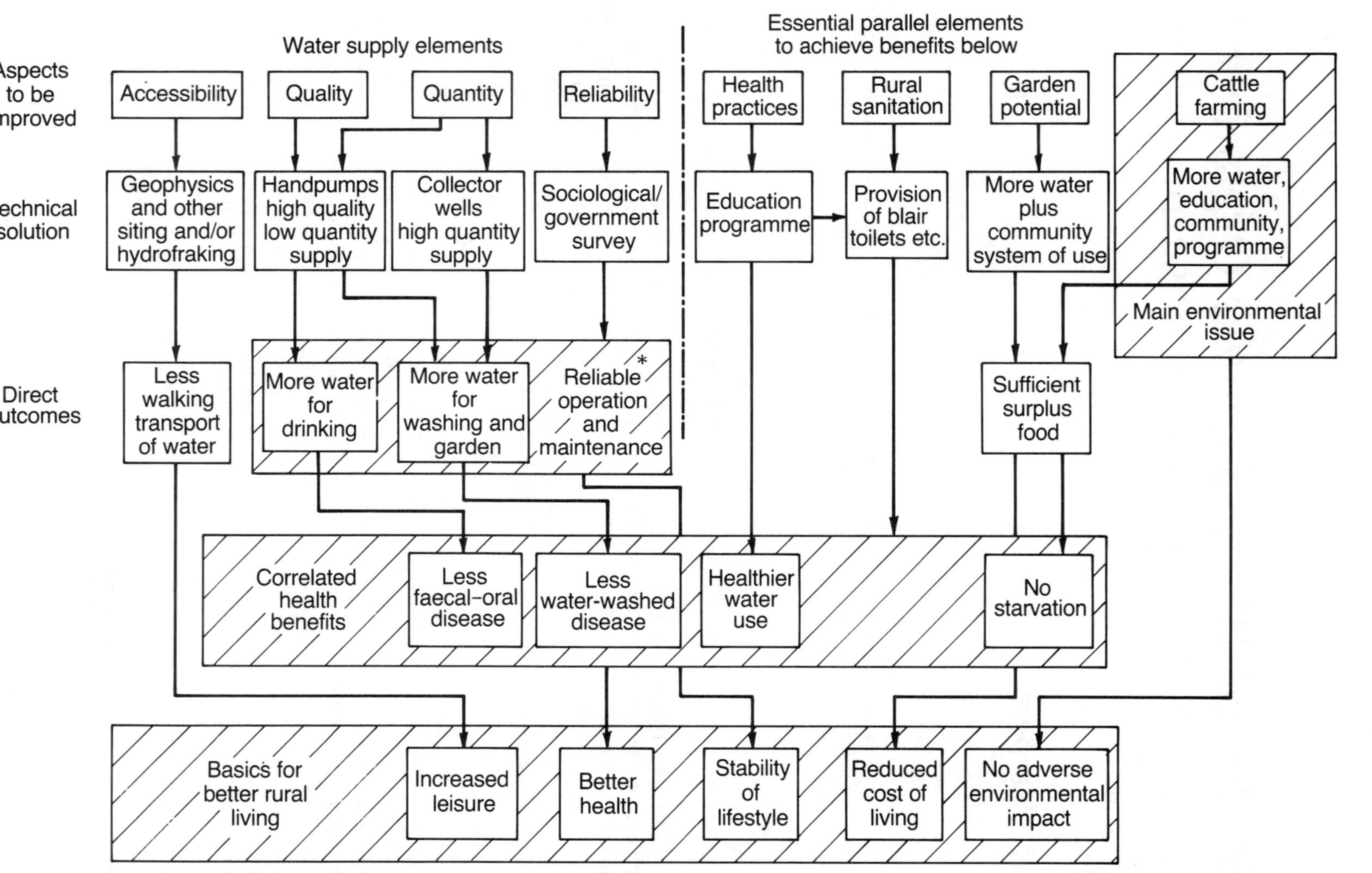

Figure 15.1 *Rural Water Supplies Organiser* (basement aquifier).

Problems with past schemes

The main failings of rural water supply schemes in the past have been associated with lack of maintenance. Even when maintenance is not a problem, there is often no significant improvement in health.

Solution

Many aspects affect the success of a rural water supply scheme. All such factors, technical as well as others, must be addressed. A sectoral approach to development is required.

Technical aspects of basement aquifers

Basement aquifers underlie about one-third of the developing world and in arid and semi-arid areas water supplies can in the main only be obtained from groundwater. The references below indicate that there are two main development tools; the slim borehole to 30 or 40 m which give a medium yield of *c.* $0.1\ l\ s^{-1}$ and a collector well which can give *c.* $1\ l\ s^{-1}$ The collector well is a 2 m diameter well dug through the weathered surface of the basement rocks whose yield is enhanced by horizontal adits drilled out from its base to 30 or 40 m.

Scope and limitations of the *Organiser*

Occasionally high-yielding boreholes can be found in basement when they hit open extensive fissures. Dykes can also yield exceptionally high borehole yields. Sometimes high yielding sand rivers or other alluvial sources exist on basement rocks. This *Organiser* does not address these exceptional sites specifically. It does however reflect the greatest majority of all locations met on basement rocks.

Reference to sources

The *Organiser* draws heavily on the ODA publication *Manual for the Appraisal of Rural Water Supplies*, 1984, HMSO, 104 pp. and the BGS publications *The Basement Aquifer Research Project* 1984–1989: *Final Report to the ODA*, BGS Report No. WD/89/15, Wright *et al.* and *BGS/ODA Collector Well Project* 1983–1988, BGS Report No. WD/88/31, Wright *et al.*

Part Four

Groundwater Quality in Urban Environments

J.D. Mather

Most industrial processes require water and consequently industry develops in areas where either surface or groundwater resources are available. Such areas may already be centres of population or may develop as such because of the growth of industry. In many less developed countries, immigration into metropolitan conurbations has resulted in the development of unplanned settlements without formal water supplies or sanitation.

The growth of urban environments, with their factories and large populations places an enormous strain on local rivers and aquifers. These have to supply both industry with process water and people with drinking water and to absorb the effluents and wastes which arise. Thus groundwater in urban areas is typically stressed in two ways. Development of the resources results in over-abstraction leading to falling water levels and reduced yields and disposal of effluents and wastes leads to a deterioration in quality.

There are a number of potential sources of groundwater pollution characteristically associated with urban environments. The first of these is domestic sewage. In less-developed countries, major cities are seldom completely served by a central sewage system. Partially treated or untreated sewage is often discharged to lagoons, to unlined channels and to influent rivers or used to irrigate fields. This results in faecal contamination both in shallow water table aquifers and in deeper aquifers through poor borehole completions. In the developed countries problems are fewer, although in the UK for example many sewage systems in major cities such as London and Birmingham were installed during the last century and are subject to deterioration and leakage.

The second source of pollution is from industrial sites themselves. As with domestic sewage, in less-developed countries waste waters are often released indiscriminately into the nearest water course or channel and may result in high concentrations of inorganic and organic chemicals which can infiltrate to groundwater. The indiscriminate disposal and stockpiling of solid wastes at factory sites and the cavalier handling of many industrial chemicals results in the creation of contaminated land from which chemicals can be leached to groundwater. Leaking storage tanks and pipelines result in pollution, particularly by light hydrocarbons such as gasoline and diesel fuel. These problems are not unique to less developed countries and many aquifers in Western Europe and North America are locally grossly polluted by both light and dense non-aqueous phase liquids.

The third source of groundwater pollution is from landfill sites when the decomposition and leaching of biodegradable materials can give rise to high organic strength leachates and to persistent pollutants such as ammoniacal nitrogen and the chloride anion. Finally groundwater may be polluted through the mining and processing of metalliferous and industrial minerals or hydrocarbons. The mining operation itself, the processing of the ore, the disposal of waste and mine abandonment can all give rise to pollution of groundwater in urban environments.

Groundwater Quality Edited by H. Nash and G.J.H. McCall. Published in 1994 by Chapman & Hall. ISBN 0 412 58620 7

The contributions to this section of the book illustrate the various pollution problems outlined above in both developed and less developed countries. Chapters by **Wagner and Sukrisno** and by **Rojas, Howard and Bartram** concentrate on the cities of Lima in Peru and Bandung in Indonesia, respectively. In both of these cities, the main problem arises from the infiltration of domestic sewage and the release of industrial waste water. By contrast, in the cities of Coventry in the United Kingdom (discussed by **Lerner**) and Copenhagen in Denmark (discussed by **Markussen and Møller**) domestic sewage and industrial effluents are treated to a high standard. Groundwater pollution arises principally from industrial sites where poor management of process and waste materials has resulted in pollution. In most of these cities over-abstraction has contributed to the groundwater quality problem.

The fifth chapter by **Mather** considers groundwater pollution arising from the landfill disposal of wastes in the United Kingdom and the potential problems caused by measures taken to protect groundwater. It is emphasized that groundwater cannot be treated in isolation but must be considered in the context of the environment as a whole.

All the chapters highlight the problems of groundwater remediation in urban environments. Even in developed countries such as the United Kingdom, **Lerner** emphasizes that the complex hydrogeology and lack of suitable technologies mean that aquifer cleanup to rectify chlorinated hydrocarbon solvent pollution in the Coventry area is not currently feasible. In Lima and Bandung, groundwater remediation is not possible because of limited resources and enormous financial pressures. Improved groundwater protection is the only option but the introduction of controls may be almost impossible to introduce or enforce. In cities such as Lima and Bandung, the most pressing need is for a central sewage system which would remove a major source of groundwater pollution.

The chapters demonstrate that groundwater pollution in urban environments is not restricted to shallow alluvial aquifers but can also affect bed-rock aquifers. Even deeper confined aquifers are not immune, as reductions in hydraulic head and poor borehole completions allow the movement of polluted groundwater from upper unconfined strata. The future does not look bright and continuing groundwater quality deterioration in many urban environments seems inevitable.

16 Groundwater quality and water supply in Lima, Peru

R. Rojas, G. Howard and J. Bartram

Abstract The problem of groundwater quality in Lima and Callao, Peru, an urban conurbation of 6.5 million people, are described. The piped distribution network utilizes surface water from the heavily polluted River Rimac and groundwater from shallow unconsolidated riverine sediments: contamination can come from mines and agricultural activities as well as industrial and domestic faecal pollution upstream from the river intake. Pollution of the aquifer can come from the River Rimac which recharges it, from urban industrial sites, open dug wells (both used and unused) in the city area, from an unlined waste stabilization pond and from effluent used in irrigation, besides domestic sewage in the city area. Water quality contamination continues right up to the point of consumption and the quality consumed does not necessarily reflect the quality supplied. There is much bacteriological recontamination in the pipe network and by the consumers, and this is complicated by an interchange of water source to supply sectors from day to day. The need is for more effective groundwater monitoring and a full hydrogeological survey of the Lima aquifer, to establish the flow pattern and immediate areas of health risk. Protection of the groundwater by effective monitoring of sewage, industrial and agricultural pollution is essential. Remedial actions and interventions are suggested.

Introduction

Lima, the capital city of Peru, was home to over six and a half million inhabitants as of 1992, some 30% of the total Peruvian population, and accounts for approximately 80% of the national industrial production. Lima is divided into 41 administrative districts and the conurbation extends to the coast to include the port of Callao. The urbanized area of the city is constrained by the Andes mountain chain which runs parallel to the coast and future growth is likely to be along the coastal strip and into three adjacent river valleys: Chillon, Rimac and Lurin.

In the past 40 years Lima has grown rapidly, due in part to immigration from other parts of the country. In the period 1940–1971, the population of the metropolitan conurbation increased by about five times from 0.65 million to almost 3 million. Since 1971 the population is estimated to have doubled from 3.3 million to almost 6.5 million. This increase in population has led to increased numbers of people living in non-planned settlements, many of which are not served by water, sanitation or power.

Drinking-water quality

The World Health Organization (WHO) drinking-water quality guideline figure for thermotolerant (faecal) coliforms in all water intended for drinking is 0 per 100 ml. This may be unachievable in all Peruvian water supplies in the short term. However, the objective of meeting this standard should be maintained and, particularly for supplies to large centres of population, such as Lima, should be seen as an immediate priority. Those water supplies which show faecal contamination require remedial action to improve the quality of water provided.

There have been a number of studies to assess the quality of the drinking water supplied to the urban

Groundwater Quality Edited by H. Nash and G.J.H. McCall. Published in 1994 by Chapman & Hall. ISBN 0 412 58620 7

Table 16.1 Population served by the Lima Water Authority, December 1991 (source: SEDAPAL)

Mode	Inhabitants	Percentage
Population of metropolitan Lima	6 698 773	100.00
Population served by SEDAPAL	5 327 583	79.50
Population with 'clandestine' connections	280 000	4.2
Population not served by SEDAPAL	1 019 296	15.2
–Standpipes	400 653	6.0
–Tanker trucks	618 643	9.2
Population served by individual collection	71 894	1.1

population of Lima and its effect on public health. Many of these studies have assessed the quality of water reaching the consumers. This approach has generally been adopted because:

- water quality deterioration continues up to the point of consumption, and the quality of water supplied to consumers does not necessarily reflect the quality of water consumed by them; and
- the public health effect of drinking water supplies is related not only to quality, but also to other factors such as continuity of supply, coverage, quantity and accessibility (Bartram, 1990; Lloyd *et al.*, 1991).

However, it is important to try and distinguish between contamination of the source and recontamination in the distribution network. By separating different sources of contamination, the relative importance of each can be assessed and resources allocated for remedial and preventative work accordingly.

This chapter will look at the quality of the groundwater which is used in the Lima water supply and try to identify sources of contamination and suggest some of the preventative and remedial actions required. However, it must be noted that as the water supply system in Lima is complex, identifying which sources are contaminated is not always easy and recontamination of water is widespread.

Data are derived from a number of studies and particularly from a diagnostic study of water supply to Callao carried out in 1988 by Bartram and Rojas, a water quality control study undertaken by CEPIS in 1992 for SEDAPAL and a water surveillance study undertaken by the Environmental Health Unit of the Ministry of Health in 1992, supplemented by samples taken by Bartram and Rojas later that year.

Lima's water supply

Water and sewerage services in metropolitan Lima are managed by the Lima Water Authority (SEDAPAL), part of the national Potable Water and Sewerage Service (SENAPA). In 1991, SEDAPAL served 80% of the population of metropolitan Lima (5.3 million people out of a total population of 6.7 million) through individual house connections (SEDAPAL, 1992). The population served by SEDAPAL is summarized in Table 16.1.

The piped water distribution network covers an area of 31 290 ha (out of a total urbanized area of 40 000 ha). The distribution network has developed gradually, in three principal phases: the first was built in the 1940s and now covers the central and older areas of the city; the second area was constructed in 1964 and is named 'Raymund' in honour of its constructor and this now satisfies the demand of the centre and west of Lima; the most recent phase of construction has aimed to service newer and more peripheral settlements.

Water entering the distribution system is derived from two principal sources: surface water from the River Rimac and groundwater.

Surface water from the River Rimac

Two water treatment plants comprise the 'La Atarjea' installation: the older with a maximum capacity of 9 $m^3 s^{-1}$ and the newer 5 $m^3 s^{-1}$. Both water treatment plants receive water abstracted from the River Rimac, which is a grossly polluted Andean source, contaminated by sewage from towns and villages upstream as well as by industrial and mining wastes and annual mud slips (*huaycos*) in the mountains. The treatment plants therefore have to cope with turbidities which may exceed 20 000 Nephelometric Turbidity Units (NTU) during the Andean rainy season (December to April). An infiltration gallery contributes a further 0.15 $m^3 s^{-1}$ which receives no further treatment except chlorination. The infiltration gallery appears to have been constructed in the early 1900s, and has received little or no maintenance since that time, but has continued to produce high quality water at a fairly continuous rate.

Groundwater

Although SEDAPAL administers approximately 340 deep boreholes, only 280 of these are in operation, most in Callao and the northern and eastern areas. About 7.2 $m^3 s^{-1}$ of water is abstracted from these wells. This rate of abstraction is not sustainable and is producing a water table drop of 1–2 m a^{-1}. The distri-

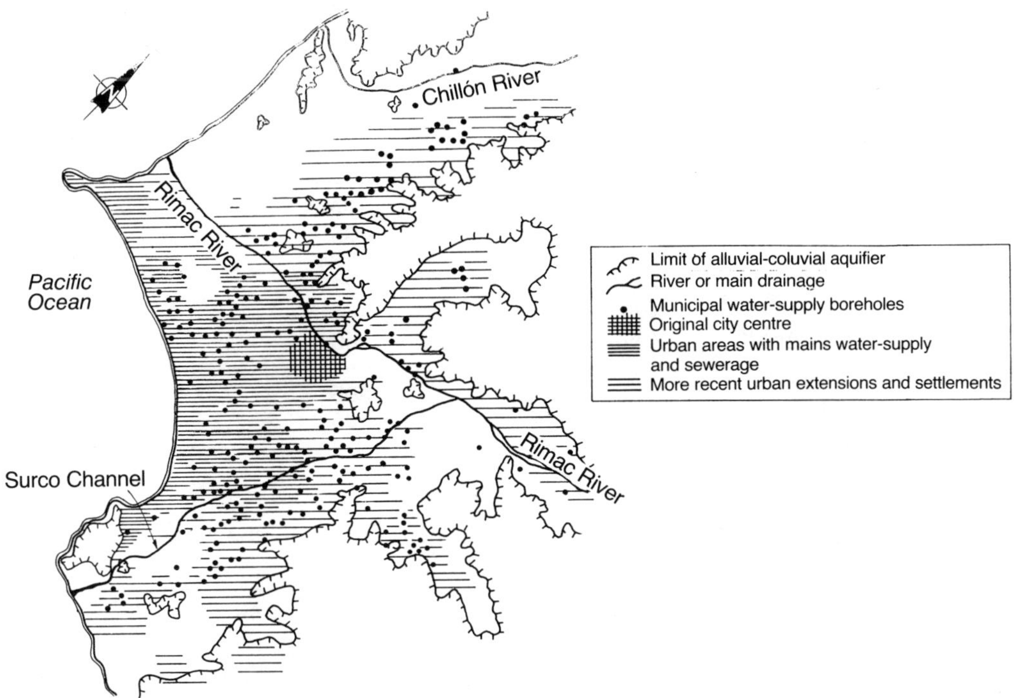

Figure 16.1 Distribution of municipal boreholes in the Lima metropolitan area (after Foster *et al.*, 1987).

bution of the wells throughout the metropolitan area is summarized in Figure 16.1.

The Lima aquifer

The Lima aquifer comprises unconsolidated alluvial sediments deposited by the rivers Rimac and Chillon, occurring either as lenticular deposits or as cross-linked beds (Binnie and Partners, 1987). The aquifer is of variable width and is bounded by Jurassic and Cretaceous sediments – limestones, marls, arenites, quartzites and mudstones, as well as intrusive granites, granodiorites, diorites and andesites.

The aquifer extends from the coastal plains in 'fingers' up to the highest parts of the Rimac and Chillon river valleys. In these areas the alluvial deposits reach up to 1.5 km in width, and surface water courses are influent to the aquifer. The largest sector of the aquifer is situated in the low-lying coastal strip, where the deposits of the Rimac and Chillon merge and extend along some 27 km of coastline. In this area the aquifer discharges into the sea.

The aquifer is split into two principal formations. The upper part of the alluvial sediments is mainly composed of gravels and other coarse-grained sediments, such as cobble beds, and these reach a thickness of up to 100 m. The coarse-grained sediments have a sand and clay matrix and are interspersed with fine grained layers, causing variations in the permeability of the aquifer. The primary water-bearing sediments are deposits of old courses of the Rimac and Chillon Rivers, these being coarse-grained and of high permeability. The lower part of the unconsolidated sediments is much finer and is largely composed of sands, silts and muds, which reach depths of up to 150 m. These deposits become increasingly fine-grained with increasing depth, and their permeability decreases.

The greater part of the aquifer is unconfined as the material is mostly unconsolidated alluvial deposits, but there are exceptions: e.g. there is a localized confined aquifer in the Callao area where the upper alluvial layers are fine-grained deposits laid down on low-lying land between the Rimac and Chillon.

Assessing drinking-water quality from groundwater sources

A major problem in assessing groundwater quality in Peru, as in most developing countries, is the lack of reliable data which describe groundwater source quality. Much of the work of monitoring of water supplies has focused, quite reasonably, on the quality of water supplied to and used by the consumer. However, this approach can lead to difficulties in assessing the relative importance of contamination of the water source and recontamination in the distribution network and so make the identification of the most appropriate remedial actions more difficult.

Lima is divided into a number of water supply zones. These zones are commonly served by more than one source – some areas are served by both groundwater and surface water and a source may serve more than one zone. The continuity of much of the supply is poor and there are significant problems with the regrowth of bacteria within the distribution network (Orbegoso, 1992). It is obvious that some of the groundwater sources are contaminated and that these have contributed to the breakthrough of bacteria into the distribution network when chlorination is poor.

There are a number of generalizations that can be made concerning potential sources of contamination of the Lima aquifer.

- There are significant numbers of unused and unprotected dug wells in the Lima area which are likely to be routes of rapid transport into the aquifer and represent a significant public health risk. These include dug wells at industrial sites with an attendant risk of chemical contamination, as well as domestic water supplies.
- There are a number of open wells which are still functioning as water supplies which represent a significant contamination risk.
- The River Rimac recharges the Lima aquifer. This river is grossly contaminated with faecal, agricultural and industrial pollution upstream of Lima.
- There is a potential contamination risk from an unlined waste stabilization pond to the south of Lima. This is exacerbated by the use of inefficient flood irrigation using poorly treated effluent with resulting large volumes of waste water carrying a high organic loading infiltrating through the unsaturated zone.

Table 16.2 Microbiological water quality from groundwater sources

	Boreholes	
Region	Total analysed	Positive results
North	92	8
Centre	35	4
South	3	0
East	72	8
West	54	5
Callao	49	6
Total	305	31
Percentage	100	10

Table 16.3 Chlorine residual in the Lima distribution network

	Free chlorine residual in well source water (mg l^{-1})				
Region	0	0.1–0.3	0.4–0.5	0.6–0.9	>1.0
North	60	14	4	7	7
Centre	24	6	4	1	–
South	–	–	2	1	–
East	30	22	11	9	–
West	39	6	5	4	–
Callao	40	4	3	1	1
Total	193	52	29	23	8
Percentage	63.4	17.0	9.5	7.5	2.6

Contamination of Lima's groundwater

There is already evidence of significant faecal contamination of the Lima aquifer as a number studies have shown (CEPIS, 1992; Rojas 1993). Microbiological analysis of thermotolerant (faecal) coliform levels in the aquifer showed that 10% of boreholes are producing water with positive counts of thermotolerant coliforms. The maximum level of contamination found during these studies was 100 coliforms per 100 ml, which represents a significant health risk to the users and indicates a source where immediate remedial action is required. Table 16.2 provides results for each region of the city.

Most groundwater sources which supply the distribution network in Lima are chlorinated to disinfect the water and it is aimed to provide a free chlorine residual to prevent recontamination in the distribution network. Under normal conditions of chlorination, the WHO guideline figure of an acceptable free chlorine residual is 0.2 – 0.5 mg l^{-1}. Below this figure it is difficult to ensure that there is sufficient chlorine in the system to kill bacteria which may enter the supply after treatment, and above this figure the taste of chlorine in the water may become unacceptable to the consumers. A recent study of groundwater supplies in Lima showed that only 36.6% of these supplies maintained any free chlorine residual of which 10.1% were at levels which may be unacceptable to consumers. The remaining 63.4% of supplies did not provide any free chlorine residual and so do not provide protection against introduction of bacteria in the distribution network. Table 16.3 shows the number of wells with free chlorine residual by region.

The lack of chlorine residual may be attributable to several factors, the most obvious being non-chlorination or inefficient chlorination due to equipment breakdown, to lack of materials or poor operation and maintenance (e.g. insufficient contact time). All these factors may lead to breakthrough of bacteria into the distribution network. However, the loss of chlorine residual due to a high chlorine demand in the water cannot be ignored. This chlorine demand may be due to contamination of the source water by bacteria and other organic matter, or may be due to recontamination within the pipe network.

Many of the dug wells used as sources of drinking-water supplies are in an insanitary condition and there are a large number of unused open dug wells. These are likely entry points for contamination into the aquifer and this is likely to affect water quality in adjacent wells and boreholes. Much of this contamination is organic in nature and has a high bacterial content (particularly faecal and other coliforms). This will exert a considerable chlorine demand and it is clear that a number of groundwater sources already have significant faecal contamination as shown by the presence of thermotolerant coliforms.

There is also evidence which shows significant recontamination of water within the pipe network from leakage and regrowth of bacteria. Leakage is most common where there are 'clandestine' connections, cross-connections, areas of low or negative pressure and is a common failing in storage and booster tanks within distribution networks. Poor continuity of the supply may also lead to recontamination as this will result in periods of low pressure during which siphonage into the system may occur.

Regrowth occurs when there are bacteria and nutrients within the pipe system and where the inside surface of the pipe is rough, allowing biofilms to develop on the pipe walls. There is a clear link between the age of pipes and the amount of bacteria regrowing within them. Regrowth is a particular problem in Lima where more than one source may supply one zone of the distribution network on different days. Thus, contaminated water with a high bacterial and nutrient load may promote regrowth, but it may be difficult to pinpoint which source is contaminated.

Sources of faecal contamination

There are a number of potential sources of faecal contamination of the Lima aquifer, both within the immediate city area and further afield. The principal sources of contamination can be identified as follows: abandoned dug wells which are not sealed; dug wells in current use which do not have a sanitary completion; the large, inefficient waste stabilization pond and effluent irrigation at Callao; and the River Rimac where it is recharges the aquifer.

Ten per cent of the boreholes in Lima have shown some faecal contamination and such boreholes can be found in every region of the city except the South. A recent study of the groundwater quality in Lima included sanitary inspection of groundwater sources to identify potential sources of contamination. Of particular concern was the number of dug wells poorly located and open to faecal and other contamination from a variety of sources. Most of the dug wells are situated in public squares with no surrounding fence and nearby hazards including open-air garbage disposal. Other dug wells are situated between houses and close to public sewers. Many of the well-heads are open and at some it was observed that the well mouth was located at floor level.

These conditions represent a risk to public health, not just to the users of a particular well, but also to people in other areas of the city who rely on groundwater as their principal drinking-water source. Pathogens may exist in groundwater for up to 40 or more days, given favourable conditions and reasonable levels of nutrient. The transit time between wells and boreholes in Lima is rarely as long as this and it is therefore to be expected that faecal indicator bacteria and other organisms will still be present in the water. The presence of such bacteria and, probably, organic material will exert a high chlorine demand and may lead to breakthrough of bacteria if chlorination is poor. This not only represents a health risk in itself, but also may promote regrowth of bacteria in the distribution network thus increasing the level of contamination before the water reaches the consumer.

It is also possible that the aquifer is being contaminated from large local sources. An example of this are the high levels of thermotolerant coliforms found at depth beneath the waste stabilization pond and land cultivated using flood irrigation with poorly treated effluent, to the south of Lima. Figure 16.2 shows the penetration and attenuation of selected chemical and bacteriological contaminants in the unsaturated zone below waste stabilization ponds and irrigated soils on desert land to the south of Lima. The ponds are unlined and the flood irrigation highly inefficient, with the result that long term infiltration rates below the ponds and cultivated land are estimated at 1.8 and 0.6 $l\ s^{-1}\ ha^{-1}$, respectively. Although the unsaturated zone is 20 m thick, thermotolerant (faecal) coliforms levels of 10–100 have been recorded at 17 m depth.

Figure 16.2 clearly shows that there are significant quantities of thermotolerant (faecal) coliforms within a short distance of the water table. These levels, if found in drinking-water supplies, would represent a serious public health risk and would require treatment before distribution. It is also significant that high levels of nitrate and other organic chemicals also persist at depth in this area. This level of faecal contamination in the unsaturated zone has a number of implications for the use of groundwater for Lima's water supply. It is likely that in the near future there will be significant faecal contamination of the Lima aquifer in this area. Whether this contamination reaches other parts of the aquifer will depend largely on the patterns of ground-

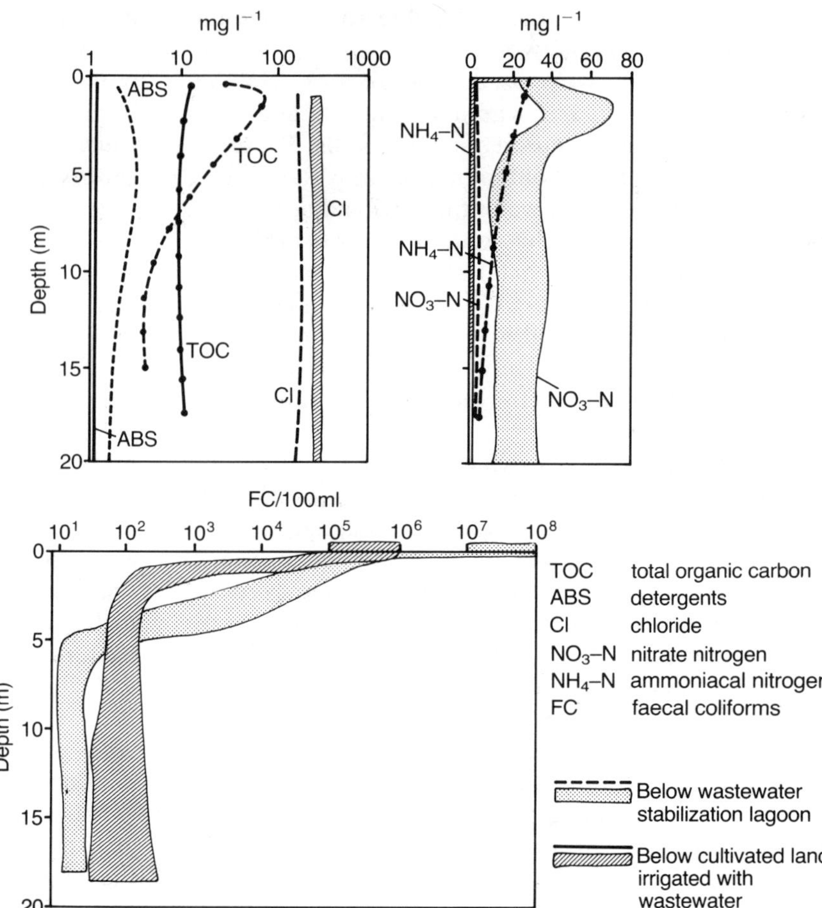

Figure 16.2 Profiles showing the penetration and attenuation of selected chemical and bacteriological contaminants in the unsaturated zone beneath waste stabilization ponds and irrigated on desert land south of Lima (after Foster *et al.*, 1987).

water movement. As a large proportion of the population relies, at least in part, on protected groundwater sources, the continued contamination of the aquifer from these sources represents a significant health risk to the population of Lima.

The presence of a large amount of organic compounds, such as total organic carbon and nitrate persisting at depth in these areas is also a major cause for concern, as treatment to remove them will be expensive. It is estimated that between a half and a third of the applied nitrogen loading to cultivated and woodland respectively (390 and 440 kg ha^{-1} a^{-1}) is leached to groundwater. This may lead to an increase in the nutrient load carried by groundwater in the Lima area, which may promote breakthrough of bacteria where chlorination is inefficient and to regrowth of bacteria within the pipe network. There are also serious health risks from excessive nitrate in water: babies in particular, may be at risk from 'blue baby syndrome' (methaemoglobinaemia) caused by excessive nitrogen intake.

Some contamination of the aquifer may also be caused by sources away from Lima itself. The most significant alternative source of pollutants is where the River Rimac is influent to the aquifer, upstream of Lima. As already noted, the Rimac is grossly polluted here by sewage from towns and villages; agricultural, industrial and mining discharges; and by mud slides. There is no evidence of widespread faecal contamination of the aquifer and the water produced by the infiltration gallery is bacteriologically pure. This may imply that most of the bacteria are inactivated during their passage through the unsaturated zone, but may simply reflect lack of available information. There is limited groundwater quality monitoring in Peru and it is possible that the aquifer away from Lima is becoming contaminated. Such undetected contamination has not yet reached the city area. There may, indeed, be some contamination there, as yet undetected. Such a situation has the potential for a public health disaster.

Chemical contamination

The lack of groundwater quality data also makes it impossible to assess the actual extent of chemical contamination of groundwater. It is likely, that given the agricultural, industrial and mining activity along the Rimac, there is already some chemical contamination of the aquifer and that this is likely to increase. It is also likely that chemical pollutants are entering the aquifer through open or unsealed wells in industrial areas in Lima. Information concerning what kind of chemical pollution may be entering the aquifer and so the risk to human health, is not available. However, the likely presence of toxic metals and chemical contamination in the Lima aquifer is a major constraint on its continued use as a potable water supply. Many chemical contaminants are very mobile in water and once they have reached the groundwater at the point of contamination they move rapidly through the aquifer and affect other areas.

The effects of chemical contamination on human health tend to be long term, but if high concentrations of certain elements build up in groundwater the health effects may become more immediate. There is also a risk that once the aquifer is contaminated with toxic chemicals, these will persist for a long period of time. It may be extremely difficult and expensive to treat the water once it is contaminated with toxic elements and such contamination could therefore cause the aquifer to be abandoned as a source of drinking water.

Groundwater monitoring

Given the current general lack of reliable information concerning groundwater quality in Peru, and in particular in Lima, it is vital that a groundwater quality monitoring programme be established as a matter of priority. A full hydrogeological survey of the Lima aquifer, including the entire recharge area, is required. The initial stage should comprise a survey of the area to establish where the Rimac and other rivers are influent to the aquifer and to establish the groundwater flow patterns. The first of these is particularly important as these areas will determine where the aquifer is most vulnerable to contamination. Given the comments in the sections above, it is obviously important to understand the groundwater flow pattern in order to establish the immediate health risk posed by a contamination event.

The objective of groundwater-quality monitoring should be to

> define the sub-surface distribution of man-made pollution and the rates of contaminant migration
>
> *Foster et al., 1989*

Groundwater quality should be analysed on a regular basis from a number of representative samples of the aquifer. These samples should be analysed for faecal indicator bacteria and a range of physico-chemical parameters, particularly those which indicate pollution – ?nitrate, chloride, pH and some heavy metals.

In conjunction with water quality analyses, sanitary inspections should be conducted to identify the actual and potential sources of contamination (Bartram, 1990). These should be designed to monitor whether adequate source protection measures have been taken to maintain and improve the quality of the water and highlight what additional measures are required, if any. When implemented on a citywide basis, the combination of sanitary inspection and analytical information can help to identify those sources and parts of the distribution network which should have priority for intervention. This procedure will ensure that resources are allocated on the basis of greatest need.

Protection of the aquifer, however, must also include monitoring sewage and industrial and agricultural discharges throughout the recharge area to ensure that these meet the required standards. Human activities in the recharge area should be compatible with maintaining good quality groundwater. The information obtained by such monitoring will help to identify where the principal sources of pollution are; to identify the onset of pollution events in time to establish control measures; to provide advance warning of the arrival of groundwater pollution at drinking-water sources so as to allow appropriate remedial actions to be initiated; and to establish legal liability for the groundwater pollution. This last point is important as the introduction and enforcement of a 'polluter pays' principle is often one of the most effective mechanisms for reducing pollution events.

The groundwater monitoring programme should be designed to provide useful information for the monitoring agency, for SENAPA and SEDAPAL and for industries and farmers whose activities may put groundwater quality at risk. In particular, preventative and remedial actions to ensure compliance with quality standards should be readily identifiable and their introduction enforced.

Remedial actions

There are a number of immediate actions that could be taken to ensure that the quality of the groundwater supplies for drinking in Lima is improved. The sealing of all unused and unprotected wells in the city is a priority to prevent continued contamination of groundwater. This would remove obvious point sources of contamination of groundwater within the city, help to reduce the vulnerability of the aquifer to contamination (Foster and Hirata, 1988) and reduce the levels of thermotolerant coliforms currently found. The wells must be filled properly, preferably with low-permeability material (such as clay), with a concrete cap placed on top of the well-head. Capping alone will not be sufficient, as water will still be able to reach the shaft after minimal transport through the unsaturated zone and thus there will be limited or no elimination of bacteria or attenuation of chemicals.

Another key intervention is the sanitary completion of all unprotected wells which are still used as water supplies. This will be best achieved by casting an impermeable concrete lining of the dry-season unsaturated zone and providing a protected intake using caissons and a gravel pack. The lining of the well should be raised up to ~0.3 m above ground level and a concrete cover placed on the well-head. Ideally, a handpump should be used to withdraw the water to limit the potential for recontamination by dirty buckets and to restrict access inside the well.

The waste stabilization pond should be lined to prevent continued infiltration of waste water. The suitability of this technology for the Lima region should be assessed and the treatment process optimized for maximum efficiency. It should be noted that waste stabilization ponds are widely used throughout the world and when operated and maintained properly are an effective means of treating waste water: they can provide effluent and sludge of a high quality. Lining the ponds will require draining of the ponds which may cause an increased health risk in the short term unless alternative sewage treatment methods are available. In the longer term it is apparent that Lima needs additional waste-water treatment facilities to cope with the volume of waste water produced. This may be done by adding an additional series of waste stabilization ponds or by using conventional sewage treatment technology.

The current use of poorly treated effluent for irrigation should be discouraged. There is no reason why properly treated waste water should not be used for irrigation; however, this should meet the WHO quality guidelines if the risk to public health is to be minimized. The WHO microbiological standards are:

- for restricted irrigation (crops likely to eaten uncooked, sports fields and public parks) – <1000 faecal coliforms per 100 ml; <1 nematode egg l^{-1}; and
- for unrestricted irrigation (cereal crops, industrial crops, fodder crops etc.) – no standard set for faecal coliforms; <1 nematode egg l^{-1}.

(For further information about these guidelines refer to *Guidelines for the Safe Use of Wastewater and Excreta in Agriculture and Aquaculture*, WHO, 1989).

If a more efficient method of irrigation such as sprinkler or, preferably, some form of drip or bubbler irrigation were used, this would also help to reduce the amount of waste water applied as irrigation water and reduce percolation losses to the aquifer. This would help to reduce the contaminant load reaching the aquifer in this area.

If waste water is to be re-used in the Lima area, domestic and industrial sewage should be separated. The removal of toxic elements from industrial sewage is much more difficult than treatment of domestic sewage. If industrial waste water is re-used for irrigation, there is a risk that a build-up of toxic elements in the aquifer may occur. This is a particular risk if the unsaturated zone is unable to attenuate all heavy metals or if there are changes in the unsaturated zone due to persistent contamination.

The Lima aquifer should also be protected from contamination from more distant sources of pollution upstream of the city. Once these areas have been identified, protection zones for the aquifer can be delineated and controls on discharges and land-use introduced. The 'polluter pays' principle should be introduced and industry persuaded to treat and where possible directly re-use their effluent. Upstream of Lima, agriculture should be controlled to prevent pollution from fertilizers and pesticides. Sewage treatment works should be carefully monitored to ensure they meet adequate standards of effluent. This is likely to be a longer-term solution, although it should be stressed that the sooner these controls are introduced, the greater the protection of the Lima aquifer will be.

Household recontamination of drinking water

Even where source water is of good quality, recontamination rates of drinking water in the home are typically high in developing countries, particularly where the water is supplied via communal water points such as wells and tapstands. The importance of human behaviour in maintaining water quality, in particular in terms of thermotolerant coliform counts and residual chlorine, should not be underestimated. Poor hygiene in both the collecting and storage of water will rapidly increase numbers of coliforms and reduce the chlorine residual in the water. Thermotolerant and other coliforms can be found on human hands where hygiene is poor and these will exert a high chlorine demand and will use up any free residual. Storage containers, if rarely cleaned, may also promote bacterial growth particularly if bacteria are regularly re-introduced from dirty collection containers or hands.

Improving hygiene practices, particularly in areas where water is collected from communal water points, is vital and may be most easily achieved through education. It is also important that people understand the links between water quality and health and the ways in which water quality can be protected. Once this is understood it will be much easier to convince people of the need to pay water rates and that maintenance work must be carried out on wells and tapstands.

Summary of conclusions

In conclusion, it is apparent that the groundwater in the Lima area is contaminated at numerous points and that the aquifer is at risk from increasing levels of contamination by bacteriological and chemical pollutants. There is already a significant public health risk caused by the poor quality of groundwater sources of drinking water and this risk will increase if the aquifer is allowed to deteriorate further.

Of particular concern is the number of open dug wells which allow direct routes to faecal and industrial contamination of the aquifer. This contamination will affect not only those who use the dug wells, but also users of groundwater in Lima who obtain their supplies from boreholes.

There is an urgent need for routine groundwater quality monitoring to be established in Peru and a thorough hydrogeological survey of the Lima aquifer to be undertaken to establish flow patterns and rates, areas of recharge and increased aquifer vulnerability, the level, type and distribution of existing pollution, and to assess migration rates of common contaminants.

In conjunction with this, a number of remedial and preventative measures can be implemented:

- the sanitary completion of all open wells and sealing of unused wells;
- lining of the waste stabilization pond to the south of Lima; improvement of effluent quality and modification of irrigation techniques to make them more efficient;
- control of the quality of domestic sewage and industrial discharges into rivers and near groundwater sources; and
- introducing restrictions on the use of inorganic fertilizer and pesticides in aquifer recharge area.

These measures may be difficult to introduce or enforce, particularly in a developing country where there are limited resources and great financial pressures. However, good quality groundwater is a vital resource for Peru's continued development and Lima's survival and unless this is adequately protected it may lead to severe problems for Lima in the near future.

References

Bartram, J., 1990. *Drinking Water Supply Surveillance*. Robens Institute and WHO. Guildford. 14 pp.

Binnie and Partners, 1987. *Management of Aquifer Resources in Metropolitan Lima*. Report to SEDAPAL, 105 pp.

CEPIS (Centro Panamericano de Ingeniera Sanitaria y Ciencias del Ambiente), 1992. *Control de Calidad del Agua de Lima*. SEDAPAL, Lima, Peru.

Foster, S., Ventura, M. and Hirata, R., 1987. *Groundwater Pollution – an executive overview of the Latin American-Caribbean situation in relation to potable water-supply*. WHO, PAHO and CEPIS. Lima, Peru. 38 pp.

Foster, S. and Hirata, R., 1988. *Groundwater Pollution Risk Assessment – a methodology using available data*. CEPIS and PAHO. Lima, Peru. 78 pp.

Foster, S. and Gomes, D.C., 1989. *Groundwater Quality Monitoring – an appraisal of practices and costs*, CEPIS, ODA and PAHO. Lima, Peru. 103 pp.

Lloyd, B., Bartram, J., Rojas, R. *et al.*, 1991. *Surveillance and Improvements of Peruvian Drinking Water Supplies*. Robens Institute, DelAgua and ODA. Guildford. 65 pp.

Mara, D. and Cairncross, S., 1989. *Guidelines for the Safe Use of Wastewater and Excreta in Agriculture and Aquaculture*. WHO, 187 pp.

Orbegoso, M., 1992. Coliform Occurrence in Water Distribution Networks. M.Sc. Thesis, University of Surrey. 69 pp.

Rojas, R., 1993. *Bacteriological Groundwater Quality of Lima Water Supply*. 9 pp.

17 Natural groundwater quality and groundwater contamination in the Bandung Basin, Indonesia

W. Wagner and Sukrisno

Abstract Natural groundwater quality in the Bandung Basin is influenced by the tropical climate, with high CO_2 production and high contents of organic material in the soil zone, and by low carbonate contents in the prevailing volcanogenic rocks. Accordingly, the hydrochemical composition of the groundwater is characterized by generally low oxygen contents and elevated Fe and Mn concentrations; low concentrations of major ions and elevated contents of CO_2 in the recharge areas; and slightly increasing salinity towards the topographically lower parts of the basin.

Intensive land-use and explosive economic and population growth in the Bandung region create a high contamination potential, major hazards to the groundwater quality being the infiltration of domestic and industrial sewage water and of leachate from waste disposal. The introduction of measures for the protection of aquifers exploited for central urban water supply is urgently required.

Introduction

Bandung, the capital of West Java Province, is one of the economic and industrial development centres of Indonesia. The present population of Greater Bandung amounts to 2.5 million. Industrial activities, particularly textile factories, are expanding rapidly.

Bandung is situated in a basin surrounded by volcanic mountains, some of them active volcanoes, with peak elevations at around 2000 m. The central plain of the Bandung Basin extends over 250 km^2 at elevations between 660 and 675 m above sea-level. The catchment area of the basin covers 2200 km^2.

The tropical monsoon climate of the region is characterized by a rainy season extending from October to May, with an average monthly rainfall of more than 100 mm, and a relatively dry season between June and September, with an average monthly rainfall of around 50 mm. The mean annual rainfall is in the order of 2000 mm.

Most of the Bandung Basin is covered by Pliocene to Recent volcanic or volcano-fluviatile deposits. Quaternary lake sediments extend over wide parts of the plain area. The Pliocene–Quaternary rock sequence is probably underlain by Oligocene–Miocene sedimentary and volcanic rocks, outcrops of which occur in the western surrounds of the basin.

The most important aquifers in the basin are formed by volcanic or volcano-fluviatile deposits related to Pleistocene volcanic activity and accompanying erosion and sedimentation. From practical aspects of groundwater extraction, the water-bearing formations in the Bandung Basin can be divided into:

- shallow aquifers at depths between 0 and 40 m below surface which can be exploited by shallow wells and, in many places, by tapping or diversion of springwater; and
- deeper aquifers at depths of more than 40 m below surface – down to 250 m – which are exploited by boreholes.

Shallow aquifers are composed of fluviatile-limnic

Groundwater Quality Edited by H. Nash and G.J.H. McCall. Published in 1994 by Chapman & Hall. ISBN 0 412 58620 7

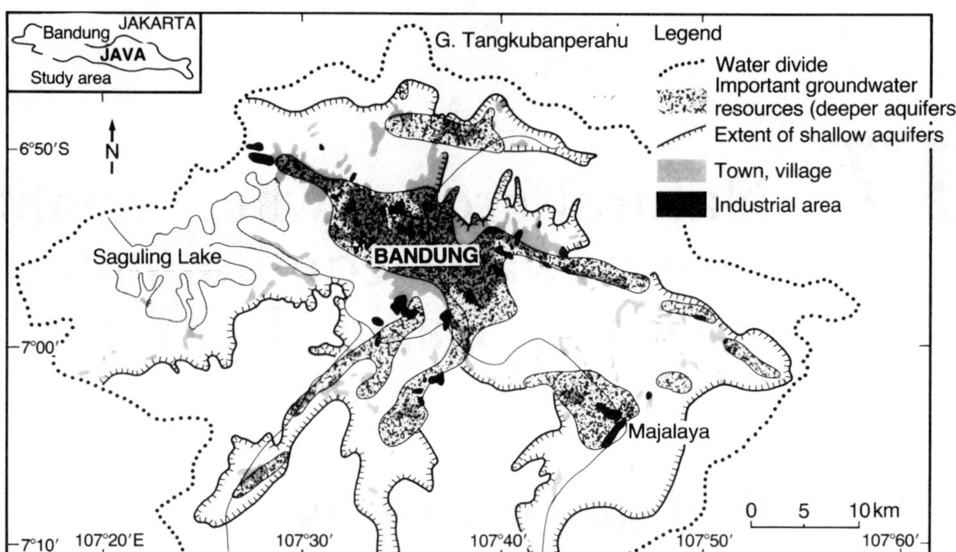

Figure 17.1 Sketch map of the Bandung Basin.

sediments within the plain area, of lahar (volcanic mudflow), talus and stream deposits on the slopes of volcanoes and of Tertiary sandstones and sandstone-marl sequences in the western parts of the basin. These aquifers extend over wide parts of the Bandung Basin. They have a low to moderate productivity and are highly vulnerable to contamination from the surface. Local water supplies for a high percentage of the population of the Bandung region rely on the exploitation of shallow groundwater from dug wells, shallow boreholes or tapping of springs. Diversion of springwater discharging from shallow aquifers is a major source of irrigation water in the dry season.

The extent of deeper aquifers is limited to accumulations of Pleistocene volcanic and volcano-fluviatile deposits (lahar fans, volcanic ash deposits, detritus fans with intercalated volcanics). The deeper aquifers are confined, with moderate productivities, except in limited areas where productivities are high. These aquifers are relatively well protected against contamination from the surface due to the presence of overlying shallow aquifers and low permeability layers (Figure 17.1).

The deeper aquifers are exploited in various parts of the Bandung Basin for industrial and communal water supply. Present extraction amounts to around 5×10^7 m^3 yr^{-1}.

The most productive deeper aquiferous sequences of the region are composed of volcanic agglomerates and lahars deposited at the lower slopes of the volcano Tangkubanperahu, within the urban area of Greater Bandung. Boreholes constructed in the lower parts of the Tangkubanperahu volcanic fan, comprising relatively well-sorted volcanic and volcano-fluviatile rocks, have yields of 10–30 l s^{-1}. These boreholes provide around 300 l s^{-1} for the central water supply of Bandung city.

Productive deeper aquifers extend over relatively large areas of the basin. However, because of the generally moderate yields, extended well fields are needed for larger-scale water supply schemes.

The groundwater regime in the Bandung Basin is, in general, characterized by recharge from rainfall on the mountain slopes and groundwater movement towards topographically lower areas. Replenishment of deeper aquifers takes place as leakage from overlying shallow aquifers. In undisturbed conditions, the deeper groundwater was artesian in many parts of the plain area. Shallow groundwater discharges in numerous springs and in streambeds on the mountain slopes. The central plain with the Citarum River and its tributaries constitutes the main discharge area for shallow and deeper groundwater.

The development of intensive groundwater abstraction since 1970 has changed the hydraulic situation in many parts of the basin: groundwater recharge occurs in expanding areas while the quantity of natural groundwater discharge as well as the hydraulic heads in the deeper aquifers are decreasing.

Natural groundwater quality

Hydrochemical processes in the unsaturated zone in the Bandung Basin (Table 17.1) are largely influenced by the regional climatic, geologic and pedologic conditions including:

- high CO_2 production and relatively high content of organic substances in the soil zone, related to intensive biologic activity under all the year-round warm and humid climate;
- low oxygen content under reducing conditions created by the oxidation of organic matter; and
- low carbonate contents in the volcanogenic rocks and their residual soils, which extend over wide areas of the basin.

Table 17.1 Development of the hydrochemical composition of groundwater in the Bandung Basin (schematic)

Atmospheric deposition, enrichment at the surface	
input of Cl, SO_4, HCO_3 (up to 10 mg l^{-1}) input of oxygen	
Soil	
production of CO_2 and organic matter	
Unsaturated zone	Shallow groundwater mountain and hill slopes
reactions between CO_2 and minerals (carbonates, silicates) increase of HCO_3 to mean values of 50 mg l^{-1} oxidation of organic matter reduction of NO_3 to NO_2 and NH_4 reduction and dissolution of Fe and Mn compounds	
	Shallow groundwater plain areas
	enrichment of Cl, SO_4, HCO_3, NO_3 by evapotranspiration and human activities
Deeper groundwater	
consumption of CO_2 reactions with fossil organic matter reactions with volcanic CO_2 increase of HCO_3 to mean values of 220 mg l^{-1} consumption of oxygen reduction of nitrogen compounds reduction of SO_4 to H_2S	

Soils on the more elevated and steeper areas are derived from volcanic material (tuff, lava, lahars) and igneous rocks. They have interstitial waters with generally high CO_2 contents and low salinity. The soil water in alluvial soils of the plain areas is generally characterized by higher contents of dissolved solids due to enrichment of dissolved solids by evapotranspiration and to the high availability of exchangeable cations in the clayey soils.

Though no reliable information is available on the hydrochemical composition of precipitation, nor of interstitial water in the unsaturated zone of the Bandung Basin, a general picture of natural baseline concentrations of dissolved substances in recharge water can be derived from the composition of very low salinity groundwater and surface water from the higher slopes of the basin (Figure 17.2). These waters are characterized by low Cl and SO_4 concentrations (mean values of Cl 3.5–10.6 mg l^{-1}, of SO_4 0.7–9.6 mg l^{-1}), originating mainly from atmospheric deposition. Mean HCO_3 concentrations of these groups of samples vary from 23 to 58 mg l^{-1} and concentrations of CO_2 are significant (up to 80 mg l^{-1}).

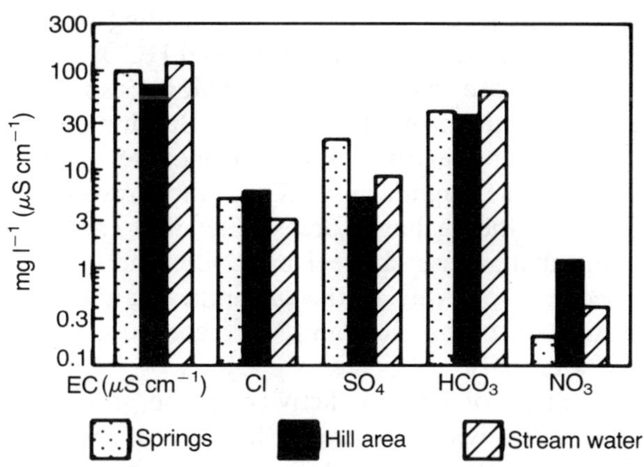

Figure 17.2 Mean values of anion concentrations in low salinity water from the Bandung Basin.

Shallow groundwater developed during somewhat more extended residence times in the subsurface, which is not – or only to a minor degree – affected by anthropogenic impacts or evapotranspiration, has values of electrical conductivity of 100–150 µS cm^{-1}, HCO_3 concentrations of approximately 50–60 mg l^{-1}, and low Cl and SO_4 concentrations of around 5 mg l^{-1}.

Groundwater in deeper confined volcanogenic aquifers, which are some hundreds to thousands of years old and is not influenced by anthropogenic inputs, has higher HCO_3 and Ca concentrations than shallow groundwater in the recharge areas. This increase of dissolved constituents, in particular of HCO_3 to mean concentrations of more than 200 mg l^{-1}, may be partly attributed to solution reactions between free CO_2 and carbonates and to reactions between water, acids (CO_2, organic acids) and silicate minerals. However, an additional carbon source has to be assumed to explain the HCO_3 concentrations in the water from some boreholes tapping deeper aquifers. The source could be fossil organic carbon or volcanic CO_2. Sulphate and nitrate concentrations are very low in the deeper groundwater, indicating reduction processes in the low oxygen environment.

Shallow groundwater in the plain area of the Bandung Basin has generally higher concentrations of Cl, SO_4 (average concentrations of 35 mg l^{-1} Cl and 30 mg l^{-1} SO_4), and in some parts of HCO_3 also, and of equivalent cations in comparison to shallow groundwater from the hill and mountain areas. The sources for the higher contents of dissolved constituents can be seen in an enrichment by evapotranspiration and in anthropogenic impacts.

Most groundwaters occurring in the Bandung Basin can be classified as bicarbonate waters with a rather indifferent cation composition. The proportional HCO_3 contents in the low salinity groundwater are in the order of 60–80 meq%. Increasing salinity is generally related to hydrochemical processes leading to higher HCO_3 concentrations and proportional HCO_3 contents increase to more than 80% of the anions (as meq l^{-1}).

Generally, the groundwater in the Bandung Basin is characterized by relatively low oxygen contents and low redox values. Significant CO_2 concentrations are common in the recharge areas. Elevated contents of iron and manganese above 1 mg l^{-1} Fe and 0.5 mg l^{-1} Mn occur extensively in the groundwater due to the formation of soluble Fe^{2+} and Mn^{2+} compounds at low redox potentials.

The impact of volcanic activity on the groundwater chemistry is obvious in several hot springs, with temperatures of 30–80°C, located on the higher slopes of volcanoes. It is also indicated by the elevated salinity and temperature in deeper groundwater on the northeastern margin of the Bandung plain.

The hydrochemical composition and preliminary isotope data suggest an origin from interaction between infiltrating meteoric water, acid volcanic gases and siliceous rocks. The presence of volcanic gases (CO_2, H_2S and HCl) in varying proportions gives rise to different hydrochemical types of hot groundwater:

- springwater with low pH, free of HCO_3, with elevated concentrations of Cl, SO_4, CO_2 (up to 570 mg l^{-1}), H_2S, NH_4, Fe, Mn and SiO_2;
- springwater dominated by high HCO_3 concentrations (840 to >1000 mg l^{-1}); and,
- deeper groundwater with elevated concentrations of Na, Cl and HCO_3 (Na – 300 mg l^{-1}, Cl – 320 mg l^{-1}, HCO_3 – 400 mg l^{-1}).

According to the natural water quality, most groundwater resources in the Bandung Basin are adequate for central water supplies. In many cases, groundwater extracted from boreholes has to be treated for CO_2 neutralization or removal of Fe and Mn contents. High Fe and Mn contents restrict the suitability of groundwater use for local water supplies for which treatment can generally not be afforded.

Table 17.2 Anthropogenic impacts on groundwater quality in the Bandung Basin (schematic)

Domestic sewage	Industrial waste waters
increase of Cl, HCO_3, N contents input of organic matter consumption of oxygen reduction of NO_3 to NO_2, NH_4, N NO_2 and NH_4 increase to unacceptable levels	
	input of trace metals and organohalogens
Leachate from waste dumps (domestic and industrial waste)	
increase of major constituents, input of organic acids, trace metals, organohalogens	
Intensive agriculture	
enrichment of major constituents through evapotranspiration on irrigated fields input of residues of fertilizers and pesticides (degree of impact unknown)	

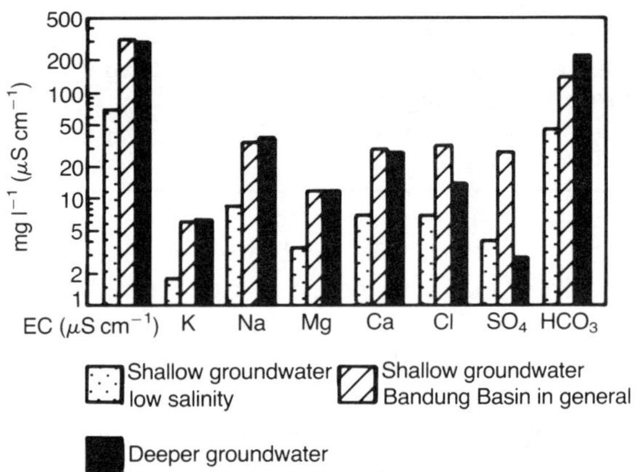

Figure 17.3 Mean values of major constituents in shallow groundwater and in deeper confined groundwater in the Bandung Basin.

Groundwater contamination problems

The current rapid growth of the population and economy in the Bandung region creates increasing environmental problems (Table 17.2) and, in particular, a significant contamination potential for shallow groundwater. Major contamination hazards are the infiltration of domestic sewage in urban areas and releases of untreated industrial waste water. Only part of Bandung City is served by a central sewerage system. The installation of sewers and of a sewage treatment plant for Greater Bandung are at the planning stage. Only a small number of factories in Bandung is connected to a pilot treatment plant for industrial waste water. Most of the thousands of factories operating in the region – mainly textile factories – release their waste water, after stabilization in ponds or untreated, to unlined channels, streambeds or irrigated fields (Figure 17.3).

The impact of domestic and industrial waste water results in an increase of dissolved constituents in the shallow groundwater, in particular of Cl and of nitrogen compounds. The addition of organic substances from waste water leads to a further lowering of redox potentials and to the widespread occurrence of reduced nitrogen compounds NO_2 and NH_4 at concentrations of 0.5 to >1 mg l^{-1} which restrict the suitability of the affected groundwater for drinking-water supply.

The impact of industrial waste water is also indicated by contents of trace metals and organohalogens significantly elevated above background levels. Zinc concentrations of up to 170 $\mu g\ l^{-1}$ were found in the shallow groundwater of the industrial suburbs of Bandung while background levels of Zn are around 30 $\mu g\ l^{-1}$. The values of trace metal concentration analysed from groundwater samples of the Bandung Basin are, so far, still below admissible levels of Indonesian drinking-water standards.

Contents of organic contaminants in shallow groundwater of industrial areas, as determined through AOX analysis (determination of adsorbable organic halogens), are elevated to 30–90 $\mu g\ l^{-1}$ in a substantial number of samples. The results of random sampling from existing shallow wells indicate a significant contamination of aquiferous layers by organohalogens (e.g. solvents) used in the textile industry in large parts of the industrial areas.

Locally limited but severe contamination of shallow groundwater is found in the vicinity of urban waste disposal dumps. Typically, high concentrations of HCO_3, Cl, nitrogen compounds, trace metals, organic contaminants are found (concentrations up to 1600 mg l^{-1} HCO_3, 1900 mg l^{-1} Cl, 22 mg l^{-1} NO_2, 26 mg l^{-1} NH_4 and 1400 $\mu g\ l^{-1}$ AOX).

The impact of intensive agriculture on the quality of shallow groundwater in the Bandung Basin is not known in detail. Potential contamination sources are, in particular, the application of pesticides and fertilizers in tea plantations, on vegetable farms and on irrigated rice fields. Relatively high HCO_3 concentrations (230–550 mg l^{-1}) and NO_3 concentrations (up to 30 mg l^{-1}) are indicated from preliminary investigations of shallow groundwater in rice fields. No reliable information is yet available on the impact of intensive pesticide application, for example of fungicides, in tea plantations on the groundwater quality, as no adequate analytical facilities exist in the region.

The deeper confined aquifers appear relatively well protected against contamination from the surface due to overlying low permeability layers and increasing hydraulic head with depth. Increasing hazards of contamination of deeper aquifers from seepage of polluted shallow groundwater must, however, be expected because of the intensive groundwater extraction from many parts of the deeper aquifers, the high contamination load in the Bandung urban area as well as the perforation of protective layers by numerous boreholes.

Proposed groundwater protection measures

The economic development of the Bandung region depends to a large extent on the efficient use of available water resources, as well as on the conservation of water resources in adequate quantity and quality. Measures for the conservation of groundwater resources have to be directed towards two major problems: over-exploitation of groundwater by boreholes mainly for industrial use; and the hazards of groundwater quality deterioration related to intensive urban and agricultural land-use.

To prevent further groundwater over-extraction, which may also threaten the water quality of deeper aquifers, restrictions of the exploitation of affected aquifers are necessary. Licensing and limitation of the number of industrial boreholes have been introduced for the most heavily exploited aquifers and will have to be applied more rigorously and for wider areas in the future.

The suitability of groundwater resources for domestic supply in the Bandung Basin can only be sustained in the long term if adequate measures for protection of the groundwater quality are introduced. Three aspects of groundwater protection have to be considered in this regard:

(1) protection of sources used for central water supply (springs, boreholes);
(2) protection of sources used for local water supply (village or private supply: springs, dug wells, shallow boreholes); and
(3) general protection of important groundwater resources which are used or may be used in the future for domestic supply, or for industrial or agricultural purposes.

Depending on the local conditions, the following protection measures for deeper aquifers, which are or will be major sources of central water supply, may be practicable:

(1) installation of central sewerage systems in the vicinity of water supply well fields located within urban areas;
(2) restriction of pesticide application close to water supply springs and conservation of forested areas within spring catchments;
(3) conservation of the present rural land-use in the catchment area of water supply boreholes and of productive aquifers which may be exploited for water supplies in the future; and
(4) control of land use changes (urbanization, expansion of industrial areas) within the catchments of important water supply sources.

A completely efficient protection of the thousands of springs, dug wells and shallow boreholes used for local water supply within the Bandung Basin, appears impracticable under the present conditions. For areas with low density habitation, limited protection measures may be introduced. These include establishing minimum distances between septic tanks or adsorption pits and water sources; and restricting fertilizer and pesticide application near the sources. For a safe water supply of the more densely inhabited areas – towns, suburbs, areas with mixed industrial and residential use – general improvements of infrastructural conditions involving substantial costs are required including, in particular, central water supply and sewerage facilities.

Acknowledgments

The groundwater quality studies in the Bandung Basin have been carried out in the framework of the Indonesian –German cooperation project *Environmental geology for landuse and regional planning*. The investigations have been greatly supported by Ir. P.H. Silitonga, Director of Environmental Geology, Ir. Soetrisno, head of the subdirectorate of hydrogeology, Dr M. Siebenhüner, team leader, and the chemical laboratory of the Directorate of Environmental Geology, in particular Ir. D. Rosadi.

18 Groundwater quality under Copenhagen

L.M. Markussen and H.-M.F. Møller

Abstract The supply of drinking water in Denmark is almost entirely based on groundwater. Approximately half – 5×10^8 m^3 – is drawn in or close to greater urban areas and of this approximately 5×10^7 m^3 is drawn in the Greater Copenhagen area of Eastern Zealand where the Senonian limestone/chalk aquifer supplies most of the water to the major conurbations of the county of Copenhagen, Copenhagen City and Frederiksberg. Sixty-five wells have been closed over the years in the Greater Copenhagen area on account of pollution. The results to date of a study focused on three supply wells in Frederiksberg Municipality, in the city centre, are summarized. The study was aimed at deciding whether the supply must be abandoned on account of pollution, or could be preserved by remedial procedures. To carry out such a study satisfactorily, it is necessary to cover all aspects of the hydrogeology, hydraulics and hydrochemistry. A major conclusion is that the porosity of the unfissured limestone, as well as that of the fissured limestone which provides the aquifers, has a major influence on the speed of transport of the pollutants in the limestone aquifer. The major pollutants are chlorinated organic solvents, but there also are risks from aromatics, heavy metals (particularly lead from fuel additives), and from cyanide and chloride (both natural and from road salt). A brief summary of the measures proposed to reduce the adverse effects of groundwater extraction on the environment concludes this chapter.

Introduction

The drinking-water supply in Denmark is almost entirely based on groundwater, delivered from some 3000 public and private waterworks – 10^9 m^3 of groundwater is abstracted, about half of which is in or close to the greater urban areas. In Greater Copenhagen (area of 618 km^3, comprising the county of Copenhagen, Copenhagen City and Frederiksberg in the city centre, Figure 18.1), 5×10^7 m^3 yr^{-1} being drawn at present. The main aquifer in the Greater Copenhagen area, which also extends over the region, is the upper fractured part of the limestone /Senonian chalk deposit which covers eastern Zealand. The abstraction from the aquifer is in balance with a yearly infiltration of approximately 75 mm from precipitation over the area and the groundwater resources are almost totally committed. Extensive abstraction has resulted in a general lowering of the groundwater level by approximately 5 m in the last 100 years throughout the region.

In Table 18.1, water abstraction and water supply in the Greater Copenhagen area in 1992 are summarized and the distribution of water supplies to Copenhagen county and city, and Frederiksberg are shown in Table 18.2. It should be noted that 5.24×10^7 m^3 of drinking water was imported to Greater Copenhagen in 1993 (from Sjaelor Waterworks to the north and by Copenhagen Water Supply, the biggest supplier in the country, and the Greater Copenhagen area). In Figure 18.2, drinking-water production at Copenhagen Waterworks from 1860 onwards is shown. There has been a decrease in production, from a maximum of 10^8 m^3 yr^{-1} in the 1970s. This is due to a decrease in population, economic recession, the introduction of water saving measures (which relate to a very dry period in the 1970s) and increasing pollution, which limits the resource. The

Groundwater Quality Edited by H. Nash and G.J.H. McCall. Published in 1994 by Chapman & Hall. ISBN 0 412 58620 7

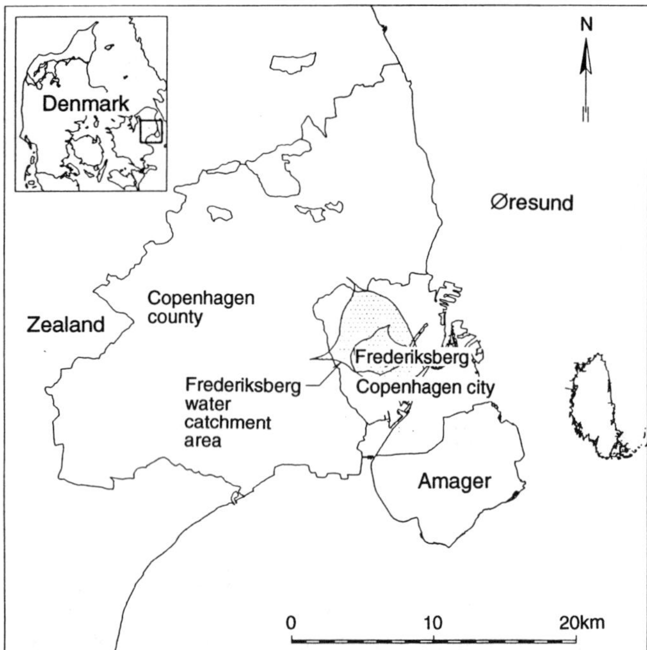

Figure 18.1 The Greater Copenhagen area – location diagram. Copenhagen County consists of 18 municipalities. Frederiksberg and the City of Copenhagen are called municipalities but have status as counties. Municipalities and counties are not exactly the same in Denmark as in England.

goals for reduction by 2000 AD can be seen in Figure 18.2. From 1911 onwards the supply to Frederiksberg has been supplemented from Copenhagen Water Supply.

The natural groundwater quality is good in most areas, but over the years, pollution by chlorinated organic compounds and other chemicals, has been detected in concentrations that have required the closing of 65 wells in the Greater Copenhagen area. In a further 28 wells, traces of pollutants have been found. The independent municipality of Frederiksberg, with a catchment area of 30 km^2 covering the whole of Frederiksberg and 20% of Copenhagen City, draws ~2.5 $\times$ 10^6 m^3 of water per year from three wells in the limestone aquifer, which here is covered by a 10–30 m thick layer of moraine and melt-water deposits; and traces of chlorinated organic pollution have been found in all three wells. At present, all private and public waterworks in Copenhagen City together with all private ones in Frederiksberg have been taken out of supply – only the Frederiksberg public water supply remains active. The abstraction here has grown by about 25% over the last 30 years. The extent of contamination of the aquifer beneath the urban areas is such that there is an ongoing discussion about the possibilities of discontinuation of groundwater abstraction or remediation. With the aim of controlling pollution and possible rehabilitation of the aquifer, the Frederiksberg municipality in 1988 initiated a comprehensive investigation of the hydraulic conditions and contamination of the catchment area. The results from these investigations and from investigations in the urban area generally are summarized and the plans for the future in both Frederiksberg and Greater Copenhagen are described.

Investigations

Two main investigations have been started and are still in progress.

(1) Location and registration within the municipality of all industrial users of potentially contaminating materials: this covers galvanizing works and other metallic industries as well as paint, tanning, chemical, medical, tar, gas and petrol production, petrol stations and waste deposits. Thirty-two locations have so far been listed and further locations are being investigated in detail.

(2) Hydrogeological investigation and establishment of a three-dimensional groundwater flow and solute transport model; and establishment of a monitoring system with an outer and inner network of monitoring wells. Logging and other geophysical methods and stratigraphical (coccolith) dating, as well as pumping tests and hydrochemical analysis of groundwater form an integral part of these investigations.

The purpose of the investigations is to obtain information concerning contaminant trends in the aquifer in order to provide a tool for implementing the necessary procedures and investments that will ensure a clean soil as well as a clean drinking-water resource.

The stages of the investigation completed to date are:

- collection of data;
- establishing a regional three-dimensional hydrogeological computer model, capable of modelling flow and dispersion of pollution in three dimensions;
- establishment of a refined detailed flow model for the limestone, based on tritium analysis, allowing for considerable flow, not only in fractures but also in the unfissured limestone;
- the estimation of the infiltration time of 20 years and the mean time of stay in the reservoir as 40 years;
- mapping the sources of pollution, the most important of which are the chlorinated organic solvents; and
- commencement of measures to decrease the surface load in order to produce improved water quality.

Hydrogeology

Geology

The general geological succession in the area consists of 10–20 m of Quaternary deposits, comprising boulder

Table 18.1 Water abstraction and water supply in Greater Copenhagen area 1992

Region	Area (km^2)	Citizens	Abstraction (mill. m^3)	Water Supply (mill. m^3)
County of Copenhagen	521	605 000	44.2	52.0
Copenhagen City	88.3	468 000	0	38.6
Frederiksberg	8.7	87 000	2.5	8.5
Greater Copenhagen area	618	1 160 000	46.7	99.1

Table 18.2 Distribution of water supply 1992, (mill. m^3)

Waterworks supplier	County of Copenhagen	Copenhagen City	Frederiksberg	Total
Copenhagen Water Supply	27.9	38.6	6.0	72.5
Frederiksberg Water Supply	–	–	2.5	2.5
Local Waterworks	20.1	–	–	20.1
Sjælor Waterworks	4.0	–	–	4.0
Total	52.0	38.6	8.5	99.1

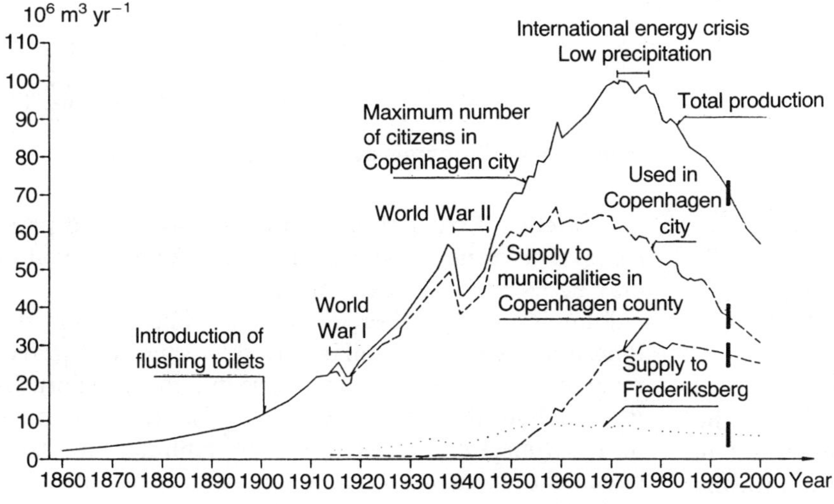

Figure 18.2 Drinking water production, Copenhagen Waterworks.

clay with sands and gravel lenses overlying the uppermost Danian limestone. The drift clay restricts recharge in places. Elsewhere, sands and gravels form a permeable aquifer in hydraulic continuity with the limestone. The limestone aquifer permeability is derived predominantly from fissure flow and the fissuring has been related to the structural history of the area, in particular faulting, and the effects of glaciation.

The area can be divided into four geological/hydrogeological areas.

Northwestern part

Here, the Quaternary cover is 10–12 m thick and consists of boulder clay with several lenses of melt-water deposits, sands and gravels forming small localized groundwater bodies. The underlying limestones are bryozoan, and two or more fracture or fault zones have been identified by means of coccolith dating and logging.

Central part

The Frederiksberg public water supply (PWS) pumping wells are situated here. There is a cover of 15–20 m of boulder clay and the underlying limestone is dominated by the NW–SE trending Carlsberg Fault zone (Figure 18.4) separating the western bryozoan limestone from the calcarenitic Copenhagen limestone.

Northeastern part

The Quaternary cover here consists of 15–25 m of fine sand with very thin or no boulder clay cover. The sand aquifer has direct hydraulic contact with the underlying calcarenitic limestone.

Southern part

The Quaternary cover consists of boulder clay with some development of the sand layer. This aquifer is unconfined and the limestone aquifer is artesian.

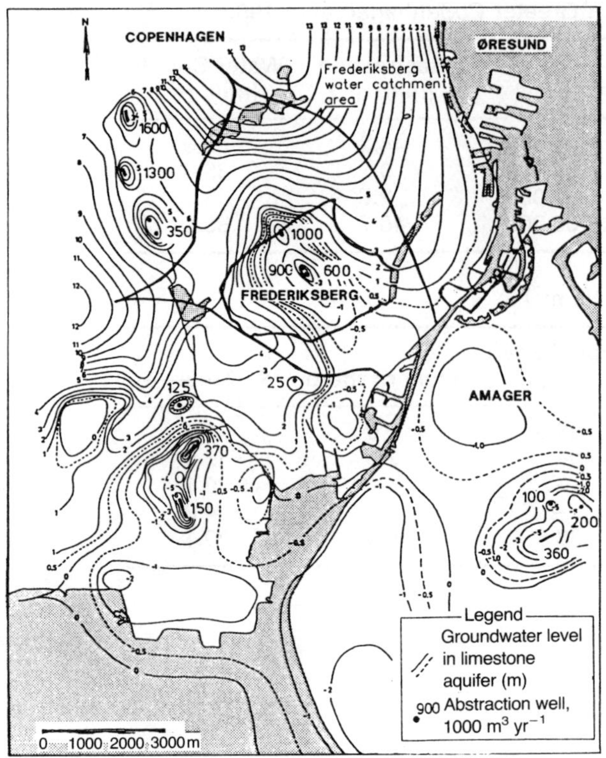

Figure 18.3 Potential diagram.

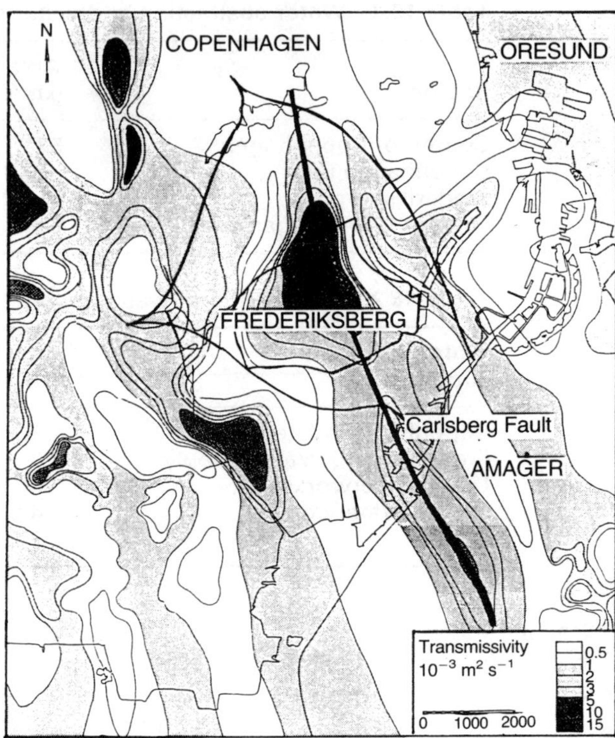

Figure 18.4 Transmissivity distribution.

Limestone aquifer

All the Frederiksberg PWS abstraction wells are situated in the vicinity of the Carlsberg Fault and individual wells may yield more than 200 $m^3 h^{-1}$.

A map of the groundwater potentials which is based on all available hydrogeological information from the area and adjacent catchment areas, is presented in Figure 18.3. Geological knowledge is important for the understanding of groundwater flow and contaminant transport in the main limestone aquifer of the Copenhagen area.

A map of the transmissivity distribution within and near the Frederiksberg municipality, based on all known test pumpings and yield measurements, is presented in Figure 18.4 This map shows that the reservoir has a high hydraulic conductivity related to the dominant NNW–SSE trending Carlsberg Fault zone which passes right through the Frederiksberg municipality. This shows on the potential map (Figure 18.4) as a long narrow zone of lowered potential. Hydraulic conductivity is also high in the western part of Frederiksberg due to the presence of block faults connected with the Carlsberg Fault. The sand and gravel layers have high hydraulic conductivity, particularly the thick sand layers northeast of Frederiksberg municipality. High hydraulic conductivity in the limestone is mainly related to fissures and hollows in the fissured and crushed upper zone of the limestone, formed due to movements of ice across the area and associated pressures. This zone is estimated to be ~20 m thick, but may be locally absent or thicker.

Pumping tests at Frederiksberg indicate that the limestone acts in a similar way to a sand and gravel reservoir, and this, presumably, is due to strong crushing and fissuring along the Carlsberg Fault. Weak zones in the limestone, related to the faulting, are thus believed to have had a crucial importance in controlling the extent and distribution of the later strain imposed by the ice movement.

Generally, the more competent, brittle limestone will have the highest hydraulic conductivity because it becomes fissured more easily than the less competent, softer limestone which deforms due to change in pressure conditions. Hence the more permeable deeper zones may comprise fissured, competent layers, overlain by less competent and less fissured layers. Thus, a stratified variation permeability is commonly recorded.

These conclusions illustrate the complexity of the hydraulic controls on pollution pathways.

Water balance

The water balance is an average for the entire aquifer, but there are differences between the northern and southern parts of the abstraction area. Figure 18.5 illustrates the urban water balance. The catchment in the north covers about 10 km^2 and the yield is about 10^6 m^3 yr^{-1}, corresponding to an average percolation

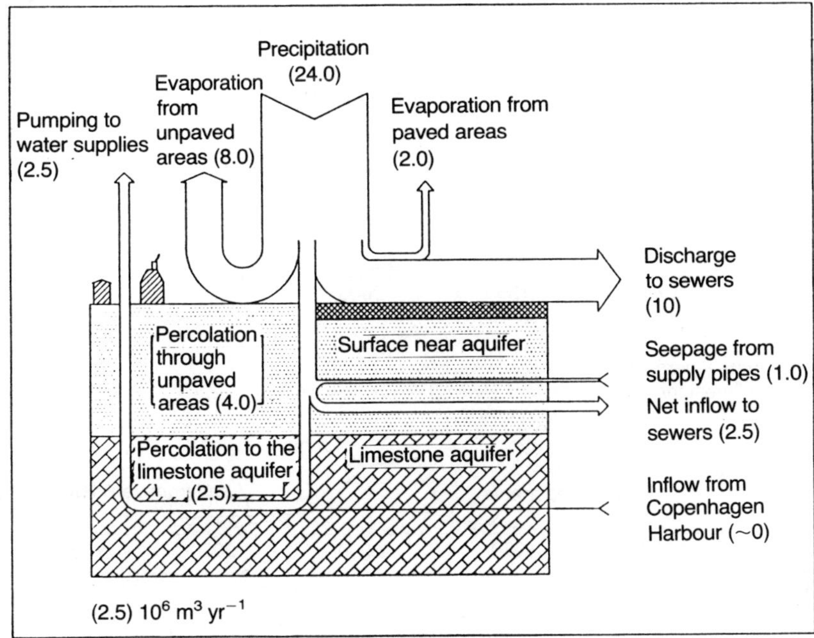

Figure 18.5 Urban water balance.

into the limestone of about 100 mm per year (bog areas in the north increase the percolation there). The southern part of the catchment covers about 20 km^2, assuming that the Øresund harbour forms a watershed and the yield is $\sim 1.5 \times 10^6$ m^3 yr^{-1}, percolation being lower because the greater part is paved and densely developed. Percolation from lakes is estimated to correspond to that from paved areas, so this has not been considered as an individual contribution. Seepage from the distribution network and inflow to sewers is of about the same magnitude as the discharge from the limestone aquifer.

Precipitation and changes in groundwater level

Precipitation during the period 1970–1979 was ~60 mm below the annual average of 660 mm. The evaporation and run-off do not vary significantly and thus when precipitation is below the average, percolation into the limestone aquifer will be markedly affected. A reduction of the order of ~10% noted above is of about the same magnitude as the total percolation into the limestone aquifer. During recent years the groundwater level in the limestone has been relatively high, due to heavy precipitation and because the abstraction from the catchment in Copenhagen City and Frederiksberg has been reduced by 1.5×10^6 m^3 yr^{-1} since 1970. However, percolation is expected to decrease in the future due to the extension of paved areas and more extensive drainage.

Groundwater modelling

Groundwater flow has been computer modelled using the three-dimensional flow and solute transport model SHE. For the Copenhagen area as a whole, a regional model comprising all surrounding water supplies has been established. This model has a 200 × 200 m spaced grid and covers 350 km^2. It is used to simulate the dynamic changes in the catchment area and changes in flow. For Frederiksberg Municipality, a more detailed local model has been established to describe local features in flow and solute transport, with a node size of 50 × 50 × 2 m.

For all nodes, information on geology, permeability, water level, precipitation, evaporation, run-off, infiltration, pumping, etc. must be evaluated. The models are calibrated until the present distribution of groundwater head is reached. To control the calibrated model and make a more precise estimate, of the effective porosity of the aquifer, the tritium concentration in groundwater has been simulated, based on tritium concentration in groundwater over the past 30 years. Tritium content in the groundwater at different depths in several wells has been analysed. The amount of tritium in the groundwater, corrected for radioactive decay, has been simulated under different values for limestone porosity. From this it has been shown that an effective porosity of 20% gives good agreement with measured tritium content.

Permeability of the unfissured limestone

The relatively great effective porosity of the limestone is due to the fact that a part of the porosity in the limestone has to be included in the effective porosity. Though the unaltered/unfissured limestone has a porosity of 20–40%, the hydraulic conductivity is not very large due to the pores being poorly connected. In Table 18.3 the hydraulic conductivity for the bryozoan and calcareous limestones, which dominate the limestone

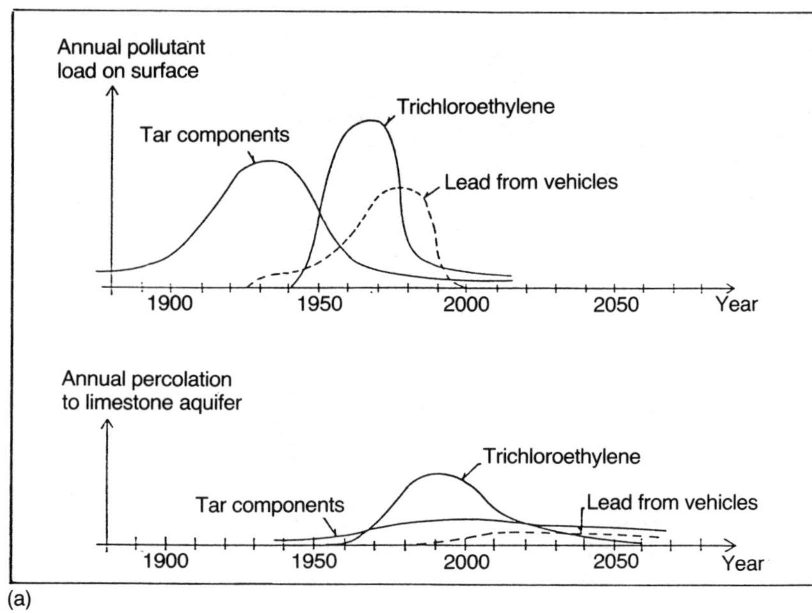

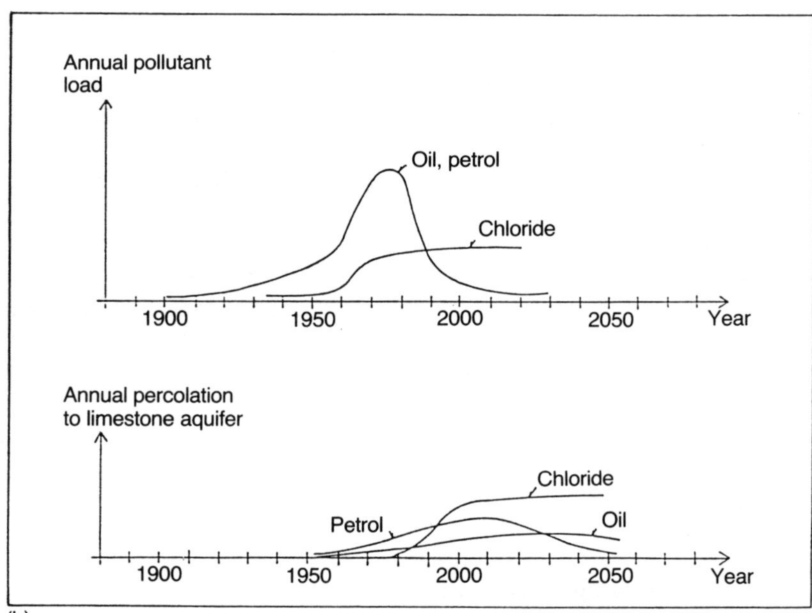

Figure 18.6 Relative loading of pollutants in surface soil and percolation to the limestone aquifer.

Table 18.3 The conductivity in intact limestone and chalk, expressed by the permeability coefficient.

Rock	Conductivity, m s^{-1}
Bryozoan limestone	1.0×10^{-6}
Calcerous limestone	$3.0–5.0 \times 10^{-7}$
Chalk, soft	$1.0–2.0 \times 10^{-7}$
Chalk, hard	$5.0 \times 10^{-9} - 0$

aquifer, is compared with that of the underlying chalk. The conductivity of the unfissured limestone is approximately 1/1000 of the total. The hydraulic conductivity depends on the composition, competence, sorting and density and will usually decrease with depth because the higher pressure compresses the pores and fissures and chemical changes closing them are more pronounced.

Water exchange in the unfissured limestone

In earlier hydrogeological evaluations it has not been relevant to evaluate the water exchange in the unfissured limestone, but this is important in connection with the dissipation of pollution in a limestone aquifer and currently undergoing further examination. Up until now studies have focused on molecular diffusion into and out of the limestone. These confirm that a part of the porosity of the unfissured limestone does have to be included in the effective porosity of the aquifer. During this investigation it has been estimated that the head dependent flow through the limestone may be as important or more important for water exchange than diffusion. This flow through the unfissured limestone occurs on account of the head loss through limestone blocks in the aquifer. The importance of head dependent flow has been estimated from a simplified model of the limestone aquifer. It is calculated from this model that if only fracture porosity is included and the porosity of unfissured limestone is omitted, the first trace of polluted water injected into the reservoir 1000 m from the wells at P Andersens Vej will reach the wells after four years, but if porosity of both the unfissured limestone and fractures is included the breakthrough will occur after approximately 16 years. On this basis, it is assumed that the effective porosity of the unfissured limestone has the same importance for groundwater-flow speed and age of the groundwater as the effective porosity of the fractures, and, further, that the head dependent flow through the unfissured limestone can explain the apparent effective porosity of 20% calculated from the tritium analysis.

Pollution

The main categories of contaminants are listed below and their likely impacts through time are calculated (Figure 18.6). This allows a qualitative assessment of the scale of future problems to be faced – a critical factor in the assessment of future water-supply options and equally an essential design parameter for any proposed aquifer remediation system. Based on investigations of former and present industrial locations, it is found that the components which can pollute the groundwater can be divided into the following categories:

- chlorinated solvents, including trichlorethylene, dichlorethylene, tetrachlorethylene and trichlorethane;
- aromatics, primarily from tar compounds, oil compounds and fuel compounds;
- heavy metals, primarily from lead, copper, chromium and mercury;
- cyanide;
- chloride; and
- sulphide.

Percolation of pollutants

The deposition of pollutants in the upper layers of the profile may occur as a result of spill on the ground, seepage from buried tanks and sewers, contaminated landfill and waste, and from polluted compounds in the air being deposited on soil or as solutes in precipitation.

Seepage through Quaternary layers takes place mainly by migration of compounds dissolved in precipitation, but may also occur because the pollutants, e.g. oil and trichlorethylene, percolate the strata themselves. The time that it takes for the pollutants to reach the groundwater in the limestone will depend on the ability of the compounds to adhere to the soil and their solubility in water as well as their ability to break down, and the seepage time of the groundwater. Beneath paved areas and in areas of considerable clay content in the Quaternary layers, the downward percolation velocity of the groundwater through the unsaturated zone will generally be lower than in unpaved and sandy areas.

Chlorinated organic solvents

In spite of the considerable load of pollution in the upper layers in the Copenhagen area by many and various compounds, trichlorethylene, a chlorinated solvent, is the main component of all wells found to be polluted in the Frederiksberg catchment. The chlorinated solvents are clear, volatile liquids with a density somewhat greater than water.

Trichlorethylene, together with most of the other chlorinated solvents, is used mainly in the degreasing of metal products and for general solvent applications. Trichlorethylene was not generally used in Denmark until the mid-1940s, but within a few years it was in extensive use. Until the early 1970s, disposal methods, apart from through the sewerage system, were limited, and considerable discharges therefore occurred between 1945 and 1970. In the early 1970s, Denmark used about 3500 t yr^{-1}. Since then, receiving stations for chemical waste have been established and greater caution has been shown in the use of these compounds, so the present disposal loading to the ground surface is believed to have been reduced.

In the 1950s, trichlorethane was introduced as a replacement for trichlorethylene for degreasing metal, and ~1500 t yr^{-1} were being used in the early 1970s. This compound has been found in the limestone aquifer at Frederiksberg but has not yet reached the study wells.

Decomposition of trichlorethylene into dichlorethylene may occur in the aquifer and this in turn decomposes to vinyl chloride and carbon dioxide. Both these latter decompose more slowly than trichlorethylene, so a successive accumulation of these compounds may occur in the groundwater. Vinyl chloride is believed to be a carcinogen. Dichlorethylene and vinyl chloride have been found in the groundwater at Frederiksberg.

Seepage time of percolating rainwater from the ground surface into the limestone aquifer has been estimated at ~20 yr. Trichlorethylene dissolved in groundwater is presumed to move at a bulk velocity of approximately half that of groundwater due to adsorption. It is therefore expected to take approximately 30–40 yr before the pollution reaches the limestone aquifer. This has caused the changes in the curve in Figure 18.6a showing the annual percolation into the limestone aquifer. The curve has become flatter, partly because the trichlorethylene evaporates in the upper strata and partly because the seepage takes place over a longer time period than the load on the surface. Practically the same applies for tetrachlorethane.

It appears from the seepage curve that the peak seepage of trichlorethylene from the Quaternary layers in the limestone aquifers may have been reached.

Aromatics

The load of aromatics in the upper layers is primarily due to production from gasworks, asphalt works, roofing works and other plants that use tar installations. The investigations have shown that the loads occurred primarily at the beginning of the century. This is the basis of the curve shown in Figure 18.6a. The load was reduced about 1940 because the production from such plants came to an end.

The heavier aromatics are not very mobile and it is mainly the lighter, more volatile components such as benzene and toluene which are infiltrating the aquifer. These components break down easily and the estimated seepage rate of the percolating water through the unsaturated zone is slower than that of trichlorethylene, due to greater adsorption: the factor is 2:10 to the groundwater and thus the seepage time is 200 instead of 40 yr. This is the cause of the very flat curve in Figure 18.6a, the maximum loading of the groundwater being delayed by 50–100 yr compared to the surface load.

Thus it is considered that the tar compounds do not pose an immediate threat to the water supply. However, this form of pollution should still be monitored as a slight pollution of the limestone was measured under former gasworks and tar industries. Considerable tar pollution has also been found in the soil and groundwater in the upper Quaternary layers under the oldest gas works at Frederiksberg, which was operating at the end of the last century.

Aquifer contamination by oil compounds results mainly from corroded oil tanks, leaking underground oil pipes, and seepage from damaged sewers in which oil has been spilled or discharged. Burying oil tanks commenced in the 1920s and during the 1950s and 1960s many oil and fuel tanks were buried. These tanks were poorly corrosion resistant and in 1970, rules controlling underground oil tanks were introduced. A 20-year transition period was allowed during which all tanks should be put in order and the greater part of any oil pollution dug up or pumped up. The maximum loading with oil took place during the 1970s and 1980s, so oil compounds are not an immediate threat to groundwater resources, but the maximum loading of the limestone has not yet occurred and it is difficult to say when it may occur. An estimate of the likely rate of loading is shown in Figure 18.6b.

Investigations have been carried out at old filling stations in Frederiksberg and Copenhagen, and on several sites considerable fuel and oil pollution has been measured. The maximum containment loading of the aquifer with fuel probably occurred in the 1970s and 1980s.

Heavy metals

At several sites, heavy metal pollution has been measured, often derived from slags or slaggy materials. Metals are leached out and transported very slowly by the infiltrating precipitation. Heavy metals so derived are thus considered not to pose an immediate threat to groundwater resources. Apart from pollution by metals from selected locations, a general loading of lead contamination comes from traffic. In Figure 18.6a, an estimate is shown of the rate of loading of lead from the exhaust gases of cars. This is based on the number of registered cars in Denmark and the times of introduction of lead in fuel (first in the 1930s) and of unleaded fuel in 1985. Now almost half of the cars use unleaded fuel. The curve is very uncertain because changes in the amount of lead are not taken into account.

Concentrations of lead above normal levels have been measured, at the surface, in soils in Frederiksberg and Copenhagen and additional investigation in Copenhagen has shown that concentrations decrease with depth, due to the very low mobility of lead and the complex chemistry of the upper soil layers. As shown in Figure 18.6a, the maximum loading of lead occurred quite late in the 1970s and 1980s, for which reason the effect on the limestone should not be measurable.

Cyanide

Most contamination by cyanide comes from gasworks waste. Cyanide was also used earlier as an additive in road salt and in weed control by railways. Pollution by cyanide has been recognized in the upper Quaternary layers in a few locations and in the limestone aquifer under some of the earlier gasworks. Due to low mobility in soil and groundwater it is considered that it poses no threat to groundwater resources.

Chloride

The groundwater under Frederiksberg is thought to be saline from 80 m below ground level downwards in the western part of the municipality. The depth at which salty water occurs in the eastern part is unknown. Due to a considerable lowering of groundwater levels, a considerable rise in groundwater salinity could be observed. However, pumping has taken place for more than 100 years and no chloride problems have arisen, so this effect is considered to be insignificant.

Apart from natural sources, the groundwater is loaded with chloride from salting the roads, pavements and squares during the winter. Such a practice commenced in both Frederiksberg and Copenhagen municipalities around 1960. Every year, about 1250 t of salt is spread within the groundwater catchment, but the figure varies according to the weather. As salting is mainly carried out in sealed areas, most of the chloride ends up in the sewer. It has been estimated that a maximum of 20% (250 t yr^{-1}) enters the groundwater through beneath paved areas and from adjacent areas and so the infiltration could contribute to the 2.5×10^6 m^3 of groundwater pumped by Frederiksberg Waterworks by the following amount:

$$\frac{250 \text{ t chloride yr}^{-1}}{2.5 \times 10^6 \text{ m}^3 \text{ yr}^{-1}} = 100 \text{ mg l}^{-1} \text{ chloride} \quad (18.1)$$

Thus, chloride concentrations in the recharge water may be expected. The maximum concentration acceptable in drinking water is 300 mg l^{-1}, at which level the water starts to taste salty. The contribution of 100 mg l^{-1} should not be a problem alone since the background concentration in the limestone is ~100 mg l^{-1}. Salting has been going on for about 30 yr and it is not certain that the maximum loading of the limestone aquifer has yet been reached, but it is getting close as the chloride is transported with the same velocity as the groundwater. A raised concentration has been measured in a number of places within the Quaternary drift cover and the limestone. In Figure 18.6b the progress of the chloride loading is estimated.

Leaking from sewers

An initial collection and preparation of data about the Copenhagen sewerage system showed that leakage from sewers warrants investigation. The Copenhagen sewerage system totals 980 km and the pipe diameter varies from 20 cm to 2.10 m. The first subterranean sewers were constructed in 1859 after the big cholera epidemic had laid waste Copenhagen during the summer of 1853.

The containment loading of the groundwater from sewers is largely unknown and requires further investigation. The population of Copenhagen allows a daily amount of 3×10^5 m^3 of domestic effluent into the sewage system, to be ultimately received at Lynetten, a plant handling the sewage from Copenhagen City, Frederiksberg and some adjacent municipalities in the county of Copenhagen. After purification and filtration, a daily average of a little less than 70 t of solids remains and these are incinerated.

Remedial actions

The 3-dimensional groundwater model has been used to estimate future flow and contamination in pumping wells. It has been decided that remedial pumping has to be initiated and the model has been used to locate suitable wells and determine the pumping rates.

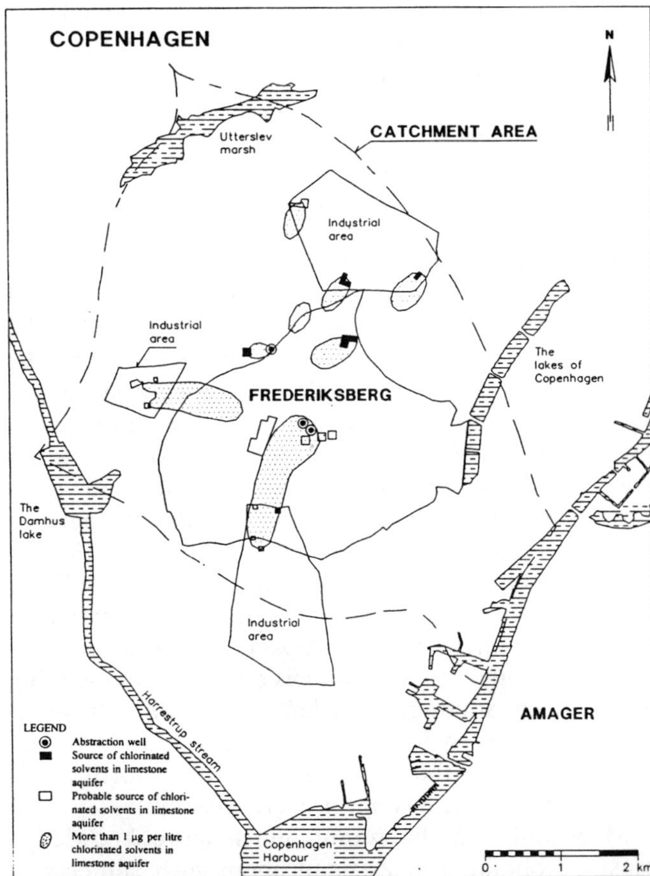

Figure 18.7 Locations from which chlorinated solvents have 'reached' the limestone aquifer. Three distinct areas in Copenhagen have caused the main pollution.

Groundwater will be pumped from one or two wells in the central part of the pollution plume, cleaned of chlorinated solvents and re-injected into the aquifer.

There are three distinct industrial areas in Copenhagen and remedial action for each of these is planned. Many of these remedial actions are in fact being implemented or completed (Figure 18.7).

Options for future aquifer management and water supply

Future water abstraction and supply in Zealand, and the Greater Copenhagen area in particular, requires that complex issues be faced. The demand for water is very high and the groundwater resource is limited due to pollution. Water abstraction takes place in several counties and thus close cooperation is needed to ensure that abstraction is distributed so as to optimize the effects on the environment. The five counties on Zealand and the municipalities of Frederiksberg and Copenhagen have collaborated, formulating an overall plan for managing the water resources in Zealand: to provide a status for the water resources available: to define the present and future demand and on this basis to evaluate the need for new catchment areas, while taking into account the risk of pollution of existing catchment areas. The first result of this work was expected to be available by Autumn 1993, when decisions were scheduled to be made about future work,

A general feature in the future management of water resources is the growing awareness of environmental impact. More notice will have to be paid to the consequences of groundwater abstraction on the environment. Water abstraction by Copenhagen Water Supply is being reduced in areas of environmental interest such as lakes and swamps, and forested areas are to increase by 25–50% over the next 50 years. Environmental interests now take priority over agricultural interests. Voluntary agreements between Copenhagen Water Supply and Copenhagen County are likely to be made, including the relocation of abstraction to areas where environmental impacts would be reduced.

Occasionally, the possibility is discussed of obtaining water from southern Sweden across the Oresund. Southern Sweden has sufficient water since a tunnel was built from the great lakes further north. This solution however, has not proved acceptable.

Meanwhile, in the future, much more knowledge must be obtained in Zealand about such subjects as production of groundwater, especially the amount in deep regional aquifers from where drinking water is abstracted; about the consequence of abstraction, the effect of agriculture (particularly from nitrogenous fertilizers), of pollution from dumps and waste sites, and finally of the influence of change in land-use, afforestation, and intensive cultivation.

In future, the strategy in Zealand concerning groundwater resources will focus mainly on:

(1) prevention of groundwater pollution rather than later clean-up;
(2) future catchment areas for abstraction being selected and sites of pollution that threaten them being cleaned up;
(3) in particularly sensitive areas, groundwater resources being ensured through changes in land-use including afforestation, changing cultivation to fallow grassland etc.;
(4) an overall master plan, Water Plan Zealand, to be worked out and implemented; and
(5) reduction in the unnecessary use of water through installation of water meters, through information campaigns and use/re-use of water of lower quality (in industry): the latter meaning that water consumers not requiring water of drinking quality (for instance as cooling water) could receive a lower but adequate quality of supply.

19 Chlorinated solvent pollution of an industrialized urban area: summary of findings and implications for groundwater protection and clean-up

D. N. Lerner

Abstract Coventry is underlain by a complex aquifer system, consisting of interbedded mudstones and fractured sandstones. The groundwater under the urban area is polluted by chlorinated hydrocarbon solvents, frequently in excess of acceptable concentrations for drinking water. Industry is the source of the pollution.

Clean-up of the aquifer is not technically feasible. Protection of groundwater would have prevented most of the pollution occurring, but the potential problem was not recognized when the compounds were introduced into industrial use. Better handling of chemicals is difficult to achieve in most factories, because the environmental consequences of current practices are not recognized by the management. Achieving better groundwater quality in future will require a new attitude to be instilled in management.

Introduction

Groundwater in the Coventry region of the UK (Figure 19.1) was discovered to be polluted by chlorinated hydrocarbon solvents (CHSs) in the mid-1980s. The origin of this pollution, its present and likely future extent, and its modes of movement were studied in the Coventry Groundwater Investigation (Lerner *et al.*, 1993a).

CHSs are one- and two-carbon molecules containing chlorine and hydrogen (Table 19.1). They are excellent solvents for many other organic compounds, are very stable, and so are widely used in industry and elsewhere as degreasers and general solvents. Chlorinated hydrocarbon solvents are volatile compounds and so are not a serious pollution problem for surface waters. However, there is little chance for evaporation once in the subsurface. They have high densities and low viscosities relative to water which allow them to be extremely mobile in the subsurface, if spilt as organic liquids rather than dissolved in infiltrating water. Their mobility and chemical stability makes them the most widespread group of organic pollutants of groundwater throughout the world.

Chlorinated hydrocarbon solvents are not severely toxic and they do not bio-accumulate. The health risk is through their presence in potable water supplies, and here the evidence is mixed. There are drinking water limits in the UK for TCE ($30\ \mu g\ l^{-1}$), TeCE ($10\ \mu g\ l^{-1}$) and TeCM ($3\ \mu g\ l^{-1}$) (Anon., 1989), which are acknowledged to be cautious (WHO, 1984). In a useful discussion of the health issues, Page (1987) concludes that, while the risks from CHSs in drinking water are low, they are worth avoiding because they are easily controlled by treatment or by groundwater protection.

Geological and hydrogeological setting

The study area of 184 km^2 is centred on the city of Coventry and includes most of the towns of Bedworth

Groundwater Quality Edited by H. Nash and G.J.H. McCall. Published in 1994 by Chapman & Hall. ISBN 0 412 58620 7

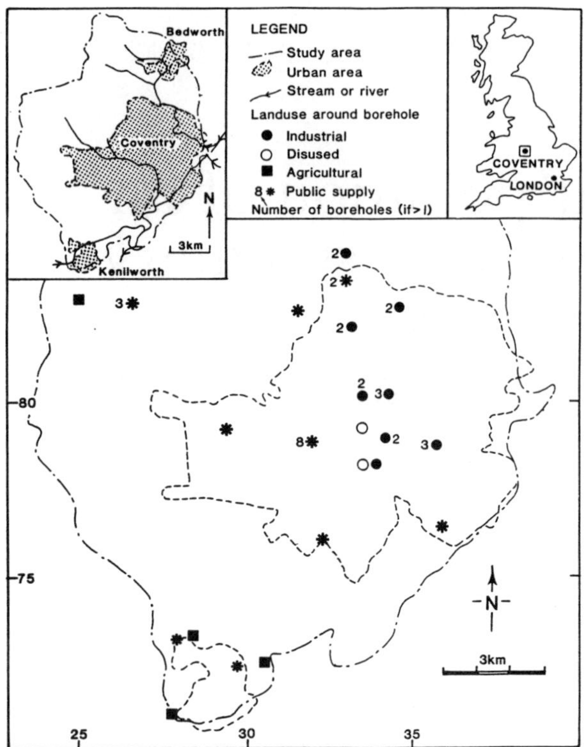

Figure 19.1 Coventry study area, showing locations of boreholes and associated land-uses.

and Kenilworth (Figure 19.1). The area is generally underlain by up to 1000 m of Permo-Carboniferous strata, with some Bromsgrove Sandstone and Mercia Mudstone (both Triassic) to the east (Lerner *et al.* 1993b, give a more extensive description of the geology and hydrogeology).

The alternating sandstone and mudstone red bed sequences of the Permo-Carboniferous were deposited in both alluvial fan and ephemeral river environments. The facies of the former association are predominantly medium to coarse sandstones, typically up to 6 m thick, but locally up to 15 m thick where several units are stacked. In the ephemeral fluvial association more mudstone was deposited, and the sandstone grain size rarely exceeds fine sand. In cross-section, there is an irregular alternation of red mudstone and sandstone sheets. The sheets are laterally extensive, with the thinner sheets passing into thicker sheets by thickening and amalgamation.

The pattern of alternating mudstone and sandstone is evident at various scales. In the only good exposure, lensing can be seen on a scale of metres and tens of metres, and laminae of a few mm thickness are present. Frequent fractures can also be observed in the sandstones; they are not generally apparent in the mudstones.

At a larger scale again, alternating sandstone and mudstone groups can be seen in the cores recovered from the 250 m deep Homefire borehole which was drilled by British Coal. The log of apparently natural fractures in the core shows that they are present throughout the depth of the sequence. Because of the difficulty of identifying individual formations, it has not been possible to correlate between boreholes at this larger scale, and the question of the lateral extent and connectivity of the units remains uncertain.

Fracturing in the sandstones gives reasonably high bulk hydraulic conductivities of about 10^{-5} m s^{-1}. Intergranular porosity it typically 15%. Mudstone hydraulic conductivities are much lower ($\sim 10^{-8}$ m s^{-1}) and high vertical hydraulic gradients can be maintained.

There are few data on the properties of fractures, despite their importance for pollution migration. Observations at outcrop suggest a fracture density of up to 5 m^{-1}, both horizontally and vertically, compared to the fracture density in the Homefire borehole of up to 3 m^{-1}. A single estimate of fracture porosity of 0.3% was obtained from an induced flow, two-borehole tracer experiment (Bishop *et al.*, 1993a). This tracer test also showed the importance of interaction between matrix and fracture waters. Diffusion of a solute from contaminated fast-moving fracture water into the matrix greatly reduces the peak concentration observed. Over a longer travel distance, it would effectively slow the solute down. The converse is that the matrix acts as a reservoir of solute, prolonging pollution incidents and hampering clean-up operations.

Glacial drift deposits, consisting of sand and gravel, boulder clay and laminated clay and silt, cover parts of the area. Clay deposits are >20 m thick in places. All of the significant river valleys contain alluvium and river terrace deposits.

Coventry has a typical temperate climate with an average rainfall of ~670 mm yr^{-1}. Potential evapotranspiration is about 510 mm yr^{-1}. Natural recharge usually occurs between October and April from winter depression rain. The rural area is generally arable and pasture land, with a few villages and pockets of woodland. Recharge will mainly be direct, controlled by the soil and drift cover. The potential sources of recharge in the urban area are numerous, and the potential pathways for water flows are complex and interlinked (Lerner, 1990). A simplified network of pathways for Coventry groundwater is shown in Figure 19.2. Setting aside cross-boundary flows of surface and groundwater, the two major sources of water to the urban area are precipitation and water supply. There are four sets of near-surface conduits for this water: a nearly natural network of rivers, and man-made networks of water mains, foul sewers and storm sewers. All four networks can potentially interact with groundwater, either recharging or discharging groundwater, although it is unlikely that groundwater will enter water mains due to

Table 19.1 Main chlorinated hydrocarbon solvents

Abbreviation	Name	Molecular formula	Typical uses	UK consumption 1990[a] (t)
TCE	Trichloroethene	Cl_2CCHCl	Metal cleaning	29 000
TCA	1,1,1-Trichloroethane	Cl_3CCH_3	Metal cleaning Adhesives Electronics cleaning	30 000
TeCE	Tetrachloroethene (perchloroethylene)	Cl_2CCCl_2	Dry cleaning (clothes) Metal cleaning	13 500
TCM	Trichloromethane	$CHCl_3$	–	Small
TeCM	Tetrachloromethane (carbon tetrachloride)	CCl_4	Intermediate in chemical manufacture	40 000

[a] DTI, 1990.

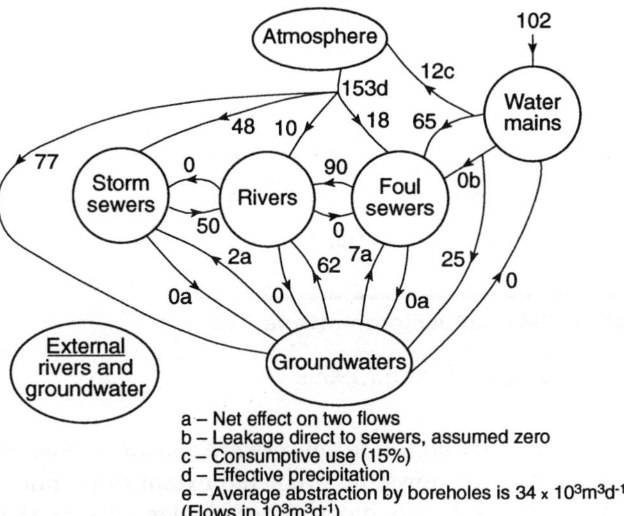

Figure 19.2 Simplified network of hydrological pathways for Coventry.

their internal pressure. Direct recharge from precipitation will occur, although it may be reduced by the impervious cover in urban areas. Estimates of the various flows are given in Figure 19.2, and components of the aquifer flow are summarized in Figure 19.3.

Chlorinated hydrocarbon solvents in Coventry groundwater

Present distribution

Forty-two existing boreholes (Figure 19.1) were sampled. The survey revealed significant CHS pollution of Coventry's groundwater, with pollution generally restricted to the urban area (Burston *et al.*, 1993). The most ubiquitous pollutant is TCE which was found in all urban boreholes, including public water supply (PWS) boreholes outside the main industrial areas (Table 19.2). Concentrations up to 1100 $\mu g\ l^{-1}$ were detected in pumped waters, and up to 6000 $\mu g\ l^{-1}$ by investigation boreholes in pollution hotspots (Bishop *et al.*, 1993b).

All but two of the urban boreholes have TCE concentrations >1 $\mu g\ l^{-1}$. Trichloroethene has been used for metal degreasing since the 1930s; Coventry is a major metal working city containing, among other industries, a major part of the British car manufacturing industry. The disposal of spent CHSs was often poor in the past, and bad handling and storage practices can be seen to this day. Trichloroethene has migrated off-site in several cases, as evidenced by the polluted PWS boreholes. One PWS borehole, situated ~1 km from two large car factories, has an operational air stripping tower to remove TCE which is present at 170 $\mu g\ l^{-1}$.

1,1,1-Trichloroethane (TCA) is also widespread, but at concentrations typically one or two orders of magnitude lower than TCE. Trichloromethane (TCM) and TeCE are all found at lower concentrations. The TCM is thought to come from leaking water mains and sewers. Drinking water contains 60 $\mu g\ l^{-1}$ of TCM as a result of chlorination (Burston *et al.*, 1993). All the other CHSs are closely linked to industrial sources, with the highest concentrations in boreholes on those industrial sites where CHSs are used.

Possible present and future distributions

Most of the boreholes showing CHS pollution are located in sites which use CHSs, and it is not possible to deduce how widespread the pollution is. A numerical groundwater flow model (MODFLOW; McDonald and Harbaugh, 1988) was used to investigate the possible present and future extent of pollution. The geological environment leads to a complex hydrogeological system. Locally, it behaves as an heterogeneous multi-layer aquifer system. At larger scales, a regional aquifer with strong horizontal : vertical anisotropy is apparent.

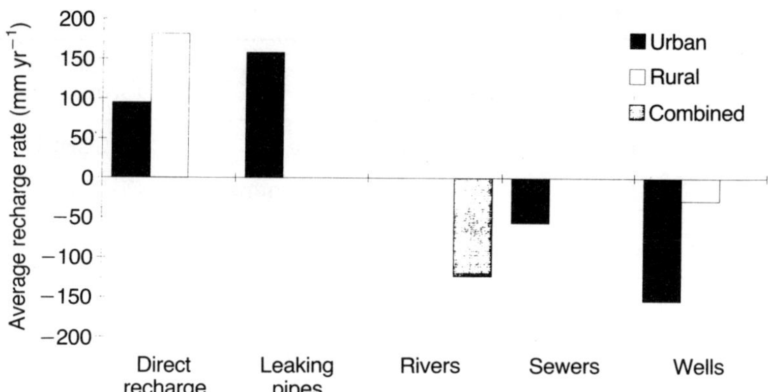

Figure 19.3 Summary of components of the aquifer water balance.

Table 19.2 Summary of CHS concentrations observed in groundwater from existing boreholes

	TCM[a]	TCA	TeCM	TCE	TeCE
% of boreholes with over l $\mu g\ l^{-1}$ CHS	43	52	31	72	40
% over British legal limits[b]	–	–	10	38	10
Mean concentration ($\mu g\ l^{-1}$)	2.1	11	4.9	67	4.1
Median concentration ($\mu g\ l^{-1}$)	0.7	1.2	0.1	9.8	0.4
Maximum value found in study area ($\mu g\ l^{-1}$)	21	270	80	1100	53

[a] TCM – trichloromethane, TCA – trichloroethane, TeCM – tetrachloromethane, TCE – trichloroethene, TeCE – tetrachloroethene.
[b] Limits: TeCM – 3 $\mu g\ l^{-1}$; TCE – 30 $\mu g\ l^{-1}$; TeCE – 10 $\mu g\ l^{-1}$ (Anon., 1989).

Modelling of groundwater flow in the study area was carried out at a regional scale, and anisotropy was included by layering the model. The layers did not represent specific strata but were a device to include different horizontal and vertical permeabilities and so control vertical flows in the model to better simulate field conditions.

Overall the field data are inadequate to develop a model which is accurate at specific locations. Thus the model is a broad representation of the Coventry groundwater system and can predict typical behaviour without being able to make site-specific predictions.

The regional groundwater survey can be interpreted to show that virtually all users of CHSs cause groundwater pollution on their own site. On this basis, all industrial sites marked on maps were treated as pollution sources. This approach probably underestimates the extent of the sources because, although a minority of sites may not use CHSs, there will be many unmarked small users, such as scrap yards, garages, dry cleaners and fire stations. In addition, much of the city centre has been redeveloped from industrial to commercial uses since the 1940s, and it is probable that CHSs were previously in widespread use in the centre but this land is not now marked on maps as industrial land.

The potential pollution sources are shown on Figure 19.4 which also shows the predicted extent of polluted groundwater for two dates. The smaller area is the possible current extent of pollution, assuming that CHSs were first used in the mid-1920s. The larger area is the maximum extent of the plumes from these sources, which would be reached after 375 years.

If all CHS pollution ceased immediately, natural flushing would still take a very long time. The longest predicted travel time of 375 years must be considered a minimum, as it takes no account of aquifer heterogeneity. The orders-of-magnitude variations in permeability seen in aquifers lead to low flow zones in which diffusion is the controlling transport mechanism, and greatly lengthen flushing times for pollutants (Lerner, 1994).

Implications for groundwater clean-up

The options available when faced with polluted groundwater are to abandon the well (which may include repositioning it), to treat the water, or to clean up the aquifer. With little or no degradation of the CHSs under Coventry, pollution is slowly becoming more widespread, albeit more diluted. If suitable techniques

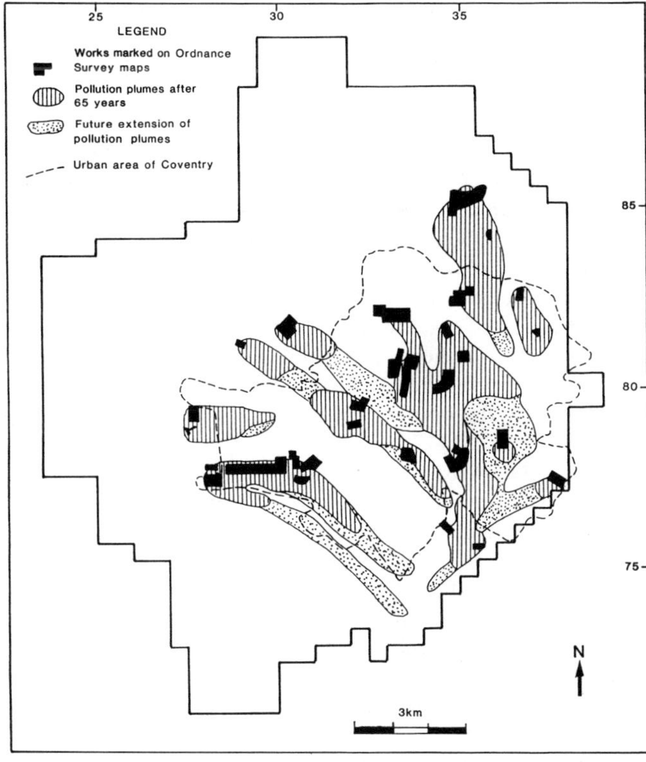

Figure 19.4 Possible extent of CHS pollution in Coventry groundwater.

were available for treatment, the costs of dealing with the present pollution would presumably be cheaper than waiting and then treating lower concentrations over wider areas.

The distribution of CHS contamination, and other immiscible fluids, can be divided into two, the source zone and the plume. The source zone is the zone of residuals and pools of CHS in its liquid form. The plume is the contaminated groundwater flowing away from the source. Detailed investigations in North America have shown that the bulk of the CHS mass is in the source zone (Mackay and Cherry, 1989). This large mass will dissolve slowly and will continue to generate a plume. Clean-up will be efficient only if the source zone is located and treated as well as the plume.

Feasibility of clean-up

The issue which will decide whether Coventry's groundwater can be cleaned up economically include (Lerner *et al.*, 1993c):

- Whether the source zones and plumes can be located; experience in Coventry and elsewhere is that, with multiple source areas on each site, and many sites in each town, there are too many to contemplate locating them all.
- Accurate assessment of the extent and degree of pollution, which will determine the design and costs of any remediation scheme. There is still not enough information about the one site investigated in detail in Coventry to begin a design exercise.
- Hydrogeological constraints, such as the degree of permeability contrasts and the role of fissures. The multi-layered and dual porosity nature of the Coventry aquifer is extremely unfavourable for clean-up.
- Whether reliable technologies are available for treating source zones. Treatment of plumes is always possible by pump-and-treat methods, and could be started in Coventry, albeit inefficiently. However, treating the plume is of limited value unless the source is removed, and operational technologies are not yet available to remove source zones.

Overall, the complex hydrogeology of Coventry and lack of suitable technologies mean that clean-up of CHS pollution is not feasible at present.

Implications for groundwater protection

The widespread occurrence of CHS pollution in Coventry shows that groundwater protection is not a trivial task, with virtually every user causing some pollution. Discussions with site management show that they are not usually aware of the condition of their site, or the consequences of current chemical handling practices. Protecting groundwater from industry needs action in three areas to succeed:

(1) minimizing spillages. Regular environmental monitoring with open access to the results may encourage managements to examine each activity for its pollution potential, as well as for its impact on production.
(2) protecting against spillages. Factory sites need a second line of defence, to capture spillages with lined sites, sumps and drains. Landfills increasingly have multiple barrier systems, and industrial sites should do likewise as there is increasing evidence of the latter causing groundwater pollution.
(3) monitoring and remedying pollution. Lerner and Tellam (1992) suggest that all factories should be required to use groundwater from under their own sites, if present. They would then suffer the consequences of any pollution, and prevent polluted waters migrating off-site.

Overall, groundwater protection is a legal requirement and a rational objective. Achieving it will take a new attitude among industrial management and more proactive environmental agencies, but it is feasible.

There are several alternatives to aquifer protection. However, they do not yet include cleaning up pollution *in situ*, especially in a difficult hydrogeological setting like that under Coventry. Switching to alternative supplies is becoming more difficult in Britain as unused water resources become scarcer (NRA, 1992). Often treatment of the water at the point of abstraction is the most favoured option, particularly for volatile compounds like CHSs.

Acknowledgments

This chapter is based on a tri-national project entitled Coventry Groundwater Investigation involving the University of Birmingham (UK), the Geological Survey of Denmark and the Bureau de Recherches Géologiques et Minières, France. The project was funded by the Commission of the European Communities. I am grateful to Phil Bishop, Mark Burston and others who carried out most of the technical studies.

References

Anon., 1989. *The Water Supply (Water Quality) Regulations 1989*. Statutory Instrument 1147, HMSO, London.

Bishop, P.K., Jakobsen, R., Gosk, E., Lerner, D.N. and Burston, M.W. 1993a. Investigation of a solvent polluted site on a deep sandstone–mudstone sequence in the UK. 1. Site description and groundwater flow. *Journal of Hydrology*, vol. 149, 209–229.

Bishop, P.K., Lerner, D.N., Jakobsen, R., Gosk, E., Burston, M.W. and Chen, T. 1993b. Investigation of a solvent polluted site on a deep sandstone–mudstone sequence in the UK. 2. Contaminant sources, distributions, transport and retardation. *Journal of Hydrology*, vol. 149, pp. 231–256.

Burston, M.W., Nazari, M.M. Bishop, P.K. and Lerner, D.N. 1993. Pollution of groundwater in the Coventry region (UK) by chlorinated hydrocarbon solvents. *Journal of Hydrology*, vol. 149, pp. 137–161.

DTI (Department of Trade and Industry), 1990. *Chlorinated solvent cleaning*. HMSO, London.

Lerner, D.N., 1994. Conceptual models of consolidated sedimentary aquifers. Subsurface restoration. *Proc. 3rd International Conference on Ground Water Quality Research*, Dallas, June 1992 (in press).

Lerner, D.N., Gosk, E., Bourg, A.C.M., *et al.*, 1993a. Sources and movement of chlorinated solvents in dual porosity rocks. *Soil and Groundwater Research Report IV, Commission of the European Communities*, EUR 14379.

Lerner, D.N., Bishop, P.K. and Burston, M.W., 1993b. Hydrogeology of the Coventry region (UK): an urbanised, multi-layer, dual-porosity aquifer. *Journal of Hydrology*, vol. 149, pp. 111–135.

Lerner, D.N., Gosk, E., Bourg, A.C.M., *et al.*, 1993c. Postscript: summary of the Coventry Groundwater Investigation and implications for the future. *Journal of Hydrology*, vol. 149, pp. 257–272.

Lerner, D.N. and Tellam, J.H. 1992. The protection of urban groundwater from pollution. *Journal of the Institution of Water and Environmental Management*, vol. 6, pp. 28–37.

Lerner, D.N., 1990. Groundwater recharge in urban areas. *Atmospheric Environment*, vol. 24B, pp. 29–33.

McDonald, M.G. and Harbaugh, A.W., 1988. *A Modular Three-dimensional Finite-difference Groundwater Flow Model*. Techniques of water resources investigations of the USGS, Book 6, Chapter A1.

Mackay, D.M. and Cherry, J.A., 1989. Groundwater contamination: pump-and-treat remediation. *Environmental Science and Technology*, vol. 23, pp. 630–636.

NRA (National Rivers Authority), 1992. *Water Resources Development Strategy: A discussion document*. NRA, Bristol, UK, 12 pp.

Page, G.W., 1987. Drinking water and health. In: G.W. Page (ed.), *Planning for groundwater protection*, Academic Press, pp. 69–87.

WHO (World Health Organisation), 1984. *Guidelines for drinking water quality*: vol. 2 Health criteria and other supporting information. WHO, Geneva, 335 pp.

20 Preventing groundwater pollution from landfilled waste – is engineered containment an acceptable solution?

J.D. Mather

Abstract Current policy in the United Kingdom has created a climate where the concept of dilute and disperse has been superseded entirely by engineered containment. This has already resulted in problems with landfill gas, and landfill leachate problems will inevitably follow. Modern practices do not adequately take into account long-term environmental impacts or the continual maintenance which will be required once landfilling is completed. Landfill management philosophies need to be modified to ensure waste degradation and stabilisation within a reasonable timescale.

Introduction

Once waste is deposited in a landfill, pollution can arise from the migration of both gas and leachate. However, although gas problems have become increasingly significant over the past decade, once recognized they can be controlled through the installation of relatively simple engineered gas extraction systems. On the other hand, landfill leachates are a more intractable problem as, even if they can be collected effectively, sophisticated treatment is needed before they can be discharged to the environment.

It is important to recognize that the deposition of almost any waste to landfill will give rise to a leachate whose chemistry is not in equilibrium with the chemistry of the surrounding groundwater. Thus, even construction wastes might be expected to increase the hardness of groundwater, and if they contain plasterboard, there is greater potential for pollution with the reduction of SO_4^{2-} to H_2S under anaerobic conditions.

Numerous physico-chemical and biological processes govern the production and composition of landfill leachates. One of the first investigations which provided the basis for a scientific understanding of these processes was conducted in Manchester in 1931–4 by Jones and Owen (Bevan, 1967). Subsequent studies by, amongst others, Farquahar *et al.* (1972) and Rees (1980) mean that the changes which occur are now understood in broad terms.

In the UK, household refuse has the following typical composition (Robinson and Archer, 1988):

Organic fraction:	Putrescible	24%	63%
	Paper	33%	
	Plastics	6%	
Inorganic fraction:	Glass	9%	37%
	Metals	8%	
	Miscellaneous	10%	
	Fines	10%	

This compares with a 42–58% organic/inorganic split in the refuse used by Jones and Owen in the early 1930s.

The decomposition of the putrescible waste and paper takes place by the action of microbes within the waste mass and is well described in *Waste Management Paper* No. 26 (Department of the Environment, 1986). Decomposition takes place in three stages. In the first stage the degradable waste is attacked by aerobic organisms, present in the waste, in the presence of oxygen in entrapped air, to form more simple organic compounds, carbon dioxide and water. Heat is then

Groundwater Quality Edited by H. Nash and G.J.H. McCall. Published in 1994 by Chapman & Hall. ISBN 0 412 58620 7

Table 20.1 Typical composition of leachates from recent and aged domestic refuse (all results in mg l^{-1} except pH values)

Determinant	Leachate from recent wastes	Leachate from aged wastes
pH-value	6.2	7.5
COD (chemical oxygen demand)	23 800	1 160
BOD (biochemical oxygen demand)	11 900	260
TOC (total organic carbon)	8 000	465
Fatty acids (as C)	5 688	5
Ammoniacal-N	790	370
Oxidized-N	3	1
O-Phosphate	0.73	1.4
Chloride	1 315	2 080
Sodium (Na)	960	1 300
Magnesium (Mg)	252	185
Potassium (K)	780	590
Calcium (Ca)	1 820	250
Manganese (Mn)	27	2.1
Iron (Fe)	540	23
Nickel (Ni)	0.6	0.1
Copper (Cu)	0.12	0.3
Zinc (Zn)	21.5	0.4

generated and aerobic organisms multiply. The second stage commences when all the oxygen is consumed or displaced by carbon dioxide. The degradation process is then taken over by organisms which can thrive in either the presence or absence of oxygen. These organisms can break down the large organic molecules present in food, paper and similar waste into more simple compounds such as hydrogen, ammonia, water, carbon dioxide and organic acids. During this acetogenic stage, carbon dioxide concentrations can reach a maximum of 90%, but usually achieve about 50% of the gas generated. In the third and final anaerobic or methanogenic stage, species of methane-forming organisms multiply and break down organic acids to form methane gas and other products. The water soluble degradation products from these biological processes, together with other soluble components in the waste, such as from the rusting of metal items, will be present in leachate. The conditions found in a modern landfill, where management involves waste containment, high density compaction and capping, provide ideal conditions for anaerobic degradation.

Typical leachate compositions are shown in Table 20.1 for both recent and aged household waste (after Department of the Environment, 1986). The table clearly shows how leachate composition is affected by the age of the refuse. In the early acetogenic stage high organic strength leachates are generated whereas, in the later methanogenic stages of decomposition, these organic compounds are actively converted to landfill gas and are not found in leachate to the same extent.

Other factors which affect the rate of refuse decomposition, and hence leachate quality, are waste density and its water content. In general, as the percentage of water increases so does the rate of decomposition. Water acts as a transport medium for bacteria and for their nutrients and waste products. Thus, the lack of moisture will reduce degradation rates and will result in a landfill producing gas and leachate at a slower rate but over a much longer period of time. Comparable effects result from variations in waste density. If the waste is deposited without compaction, air ingress can occur with aerobic decomposition. At refuse densities in excess of 0.5 t m^{-3} the dominant decomposition process becomes anaerobic (Dept of Environment, 1986). As density increases still further, water penetration becomes slower resulting in lower rates of decomposition.

Historical trends

In the UK, the last two decades have seen considerable changes in the engineering of landfill sites. Standards have improved enormously since the early 1970s when publicity associated with the discovery in the English Midlands of uncovered drums marked cyanide first focused attention on the poor controls existing over the disposal of waste. Some of these changes are the direct result of the 1974 Control of Pollution Act which established a basis for the management of all controlled waste in the UK, including a provision for the licensing of disposal sites. However, of even more importance was the change in the responsibilities of local authorities which occurred when the Local Government Act came into effect on 1 April 1974. In England this placed the duty for waste collection on the District Councils, but that for waste disposal on the County Councils. Whereas, prior to 1974 disposal took place at a number of small sites within each District, after reorganization it became more cost effective to operate a number of larger, strategically placed sites which received waste from a number of Districts. Thus the small local Rural District Council tip became a thing of the past. With the concentration of waste disposal at a smaller number of larger sites, it also became possible to use modern plant and equipment to better effect and to increase the number of professional staff involved in disposal operations. Landfill sites began to be engineered and waste compacted to lengthen the life of sites. Thus landfill moved to a situation where sites were large, where it was economic to use engineering solutions to overcome environmental problems but where the potential for leachate and gas generation was much increased.

In order to provide information to the new Waste Disposal Authorities a survey of existing waste disposal sites was carried out under the auspices of the Department of the Environment. The report of this survey (Gray, Mather and Harrison, 1974) defined 'dilute and disperse' and 'concentrate and contain' sites and emphasized the importance of the geological and hydrological properties of the host strata in controlling leachate migration and attenuation. At around the same time, a field and laboratory programme was initiated which in part reviewed the effectiveness of natural processes in attenuating leachate concentrations (Department of the Environment, 1978). It concluded that at suitably selected sites, natural processes would be effective enough to prevent the pollution of water resources and that 'dilute and disperse' could be used as a management strategy. During the late 1970s and early 1980s many sites were commissioned which relied on the effectiveness of these processes.

Early in the 1980s the emphasis began to shift in line with the European Community *Directive on the Protection of Groundwater Against Pollution Caused by Certain Dangerous Substances* (80/68.EEC). This defines groundwater as 'all water which is below the surface of the ground in the saturation zone and in direct contact with the groundwater or subsoil'. It requires member states to totally prevent the introduction into groundwater of certain substances and to limit the introduction of others so as to avoid pollution. Many of these substances occur in landfill leachates and the Directive means that the concept of dilute and disperse envisaged by Gray, Mather and Harrison (1974) becomes untenable.

The current draft proposal for a European Community directive on the landfill of wastes militates still further against the principle of 'dilute and disperse'. The general requirements for all classes of landfills specify that all water or leachate emanating from the landfill shall be collected and treated, specify a permeability of 1.0×10^{-9} m s^{-1} for a substratum thickness of 3 m beneath hazardous and household waste landfills and require that where such conditions are not met, engineering measures shall be taken to achieve the same result.

The final nail in the coffin of 'dilute and disperse' as a landfill management strategy in the UK has been driven by the introduction of a groundwater protection policy by the National Rivers Authority (NRA, 1992). Under the terms of this policy the Authority will object to the landfilling of wastes with a high or medium pollution potential within the catchment area of a groundwater source unless it can be shown that the risk of groundwater pollution can be mitigated adequately by engineering measures and operational management controls. This means that for household and industrial wastes with a significant polluting potential, the engineering measures must provide for total containment and for the collection and disposal of leachate.

The situation in the UK, in the mid 1990s, is that it is impossible to obtain agreement to open a new landfill operating on the 'dilute and disperse' principle unless the waste is totally inert. Containment may be achieved through the properties of the host strata, the installation of synthetic geomembranes or by a combination of both to produce a multilayer composite lining system (Anon., 1990). The emphasis has moved away from a reliance on the geological conditions at a site towards a position where almost total reliance is placed on engineering to avoid the possibility of groundwater pollution. This change of emphasis has major implications for the future of the waste management industry.

Landfill gas

The major components of landfill gas are methane and carbon dioxide, generally building up to concentrations of around 65% and 35% respectively, plus a number of minor components. Some of the trace constituents have obnoxious smells and are responsible for the characteristic odour of landfill gas. If not properly controlled landfill gas can give rise to flammability, toxicity, asphyxiation and explosive hazards (HMIP, 1989).

The production of landfill gas is not a new phenomenon and was observed in early experiments on the decomposition of refuse (Jones and Owen in Bevan, 1967). However, it is only in the last decade that it has become a major problem in the UK. This may in some part be the result of changes in the composition of domestic refuse with an increase in the volume of biodegradable organic material. However, even in 1932 the refuse used by Jones and Owen contained 42% of organic material, sufficient to produce significant volumes of methane by biodegradation. It is likely that the introduction of modern methods of disposal has been a far more significant reason for the emergence of the methane problem.

As indicated earlier, prior to 1974, many landfill sites were uncontrolled and small in scale. Wastes deposited in this environment degraded aerobically with the production of water and carbon dioxide. The process is exothermic and the resulting high temperatures led to fires and the burning of much potentially degradable material such as paper and wood. Anaerobic degradation, which leads to the production of methane, often did not occur at all and, if it did, the gas vented harmlessly to the atmosphere. The emergence of large controlled landfill sites and the use of techniques to reduce the environmental problems caused by fires, smells and wind-blown debris completely changed the picture. Aerobic degradation now occurs to only a limited extent in uncovered waste in the current top layer of a landfill and the dominant process is anaerobic degradation. This inevitably produces methane and it was the lack of planning for this which resulted in the landfill gas problems which became apparent in the early 1980s. These problems have been compounded by other landfill practices such as high density compaction, the covering of individual layers of waste with poorly permeable cover material and landfill completion using impermeable caps. Such practices increase the likelihood of perched water within landfills and encourage lateral migration of gas. Once recognized these problems can be resolved through arrays of gas extraction wells linked to suction systems but have sometimes had to be combined with a grout barrier to prevent off-site migration (e.g. Raybould and Anderson, 1987).

The experience of dealing with landfill gas demonstrates that management practices introduced to deal with one problem can have major knock-on effects in other areas and that a careful assessment needs to be made of the overall environmental impact of new methodology.

Landfill leachate

Current policies with respect to groundwater protection in the UK and the reasons for them were outlined earlier. The containment landfills which are the result of this policy mean that leachate does not migrate away from sites and leachate management becomes a significant component in the design and operation of all landfills. Because of the problems involved in the disposal and treatment of excessive volumes of leachate, a considerable amount of effort is spent on reducing leachate generation. Low permeability material is used to construct cell walls and intermediate cover material during the working life of the landfill. Similarly, when the site is restored, low permeability caps are installed to minimize leachate generation. Guidance issued by the Department of the Environment suggests that the cap should be domed or contoured to encourage surface water run-off and should be constructed of material having a permeability of 1×10^{-7} cm s^{-1} or less (Department of the Environment, 1986).

All this effort to reduce water infiltration will have a significant effect on the degradation rate of the organic component of the waste. This is because the major influence on degradation is moisture content, the water acting as a transport medium for bacteria and for their nutrients and waste products. Thus, the lack of moisture will considerably reduce degradation rates and will result in a landfill producing gas and leachate at a slower rate but over a much longer period of time. Using simple calculations and typical values of waste density, tipping depth and cap permeability, it can be demonstrated that this timescale extends into centuries rather than decades.

The engineered barriers which are an integral part of modern containment landfills will also need to be maintained over the extended period of leachate generation. The performance of both natural and synthetic lining and capping materials must be suspect over these timescales. Natural materials are not homogeneous and any permeable horizons or fractures in the clay may become active as migration pathways if a head of leachate is built up. Such preferential flow paths are not detected by laboratory tests. The long-term behaviour of synthetic materials over periods of hundreds of years has not been demonstrated. Gradual degradation will result in consolidation of the waste and settlement at the surface which will require continual maintenance of the cap to prevent surface water ingress. Leachate collection systems will silt up and/or become clogged with bacteriological slimes and will require regular cleaning and reconditioning. As a general rule it is likely that the more sophisticated the level of engineering required to ensure containment, the greater the level of maintenance needed.

Thus, although the move away from 'dilute and disperse' towards major containment sites will avoid the problems associated with groundwater pollution, at least in the short term, it is debatable as to whether or not it provides an environmentally acceptable long-term solution. The future scenario is one of a series of areas of contaminated land which will have to be monitored and managed for periods in excess of 100 yr. Such a scenario has major implications for site operators and for the regulatory agencies. In the future both are going to be called upon to manage containment and monitoring systems installed by their predecessors with unpredictable costs long after the last consignment of waste was delivered and revenue earnings ceased. Unless such long-term management is effective, under the climatic conditions typical of Britain, it is difficult to see how containment sites can maintain their integrity for more than two or three decades (Mather, 1989). It is therefore possible that the first or second decade of the next century will be the decade of leachate problems in the same way as the 1980s were the decade of gas problems.

The future

The current trend in the UK is towards the use of sophisticated technological solutions to engineer almost any hole in the ground to accept household and many industrial wastes. Current designs often attempt to emplace waste in dry tombs, sealed at the base by impermeable liners and at the top by impermeable caps. This policy is being followed with little regard to its long-term cost or environmental impact. Previous landfill practices may not have been ideal but the legacy of problem sites has not been large and current practices appear to be designed to create the problem sites of the future.

From the point of view of the independent hydrogeologist there is no reason why dilute and disperse sites should not continue to be used where detailed hydrogeological assessments show them to be suitable. However, this view is now heretical in the UK and it seems unlikely that it will again be possible to obtain a licence consent for a 'dilute and disperse' site. Any management philosophy will therefore need to incorporate containment but will also need to ensure waste stabilization within a reasonable timescale of say 20 to 30 years after the last consignment of waste has been deposited at the site.

Further research and development need to take place to encourage the management of landfills as bioreactors (Campbell, 1992), maintaining moisture contents at optimal levels to encourage waste degradation and stabilization. This will require plant to pump and treat leachate and to monitor its quality for years after closure of the landfill. Although this will increase the costs in the short term, post-closure monitoring will last only for a few decades and the unpredictable and possibly huge costs of long-term management and monitoring will be avoided.

Present landfill management, with its emphasis on engineered containment, provides a short-term solution to the problems of the disposal of the large volumes of household and industrial wastes generated in the UK. However, the balance has swung too far towards the protection of groundwater with little regard for the long-term welfare of the environment or value of some of the groundwater resources which are being protected. There is a need to return to the concept whereby the various interests, groundwater quality, planning and waste disposal, are balanced and solutions arrived at which consider the environment as a whole without overemphasizing one particular sector. The generation of large areas of contaminated land which are sterilized against development for decades and even centuries must be balanced against the limited pollution of groundwater in a minor aquifer. Integrated management of the environment requires a flexible approach and cannot be achieved by the various interests producing policy documents which restrict and prevent such flexibility.

References

Anon., 1990. Luxury liner. *Ground Engineering*, December 1990, pp. 14–16.

Bevan, R.E., 1967. *Notes on the Science and Practice of Controlled Tipping of Refuse*. Institute of Public Cleansing, London, 216 pp.

Campbell, D.J.V., 1992. Implications of site design and operational factors on optimisation of landfills as bioreactors. *Biowaste 92*, Copenhagen, June 1992. Preprint, 9 pp.

Department of the Environment (DoE) 1978. *Cooperative Programme of Research on the Behaviour of Waste in Landfill Sites*. HMSO, London, 169 pp.

Department of the Environment (DoE), 1986. *Landfilling Wastes*. Waste Management Paper No. 26. HMSO, London, 206 pp.

Farquahar, G.J., Farvolden, R.N., Hill, H.M. and Rovers, F.A., 1972. *Sanitary Landfill Study – Final Report Volume 1*. Univ. Waterloo Res. Inst., 316 pp.

Gray, D.A., Mather, J.D. and Harrison, I.B., 1974. Review of groundwater pollution from waste disposal sites in England and Wales, with provisional guidelines for future site selection. *Quarterly Journal of Engineering Geology*, vol. 7, pp. 181–196.

Her Majesty's Inspectorate of Pollution (HMIP), 1989. *The Control of Landfill Gas*. Waste Management Paper No. 27. HMSO, London 56 pp.

Mather, J.D., 1989. Groundwater pollution and the disposal of hazardous and radioactive wastes. *Journal of the Institution of Water and Environmental Management*, vol. 3, pp. 31–35.

National Rivers Authority (NRA), 1992. *Policy and Practice for the Protection of Groundwater*, NRA, Bristol, 52 pp.

Raybould, J.G. and Anderson, D.J., 1987. Migration of landfill gas and its control by grouting – a case history. *Quarterly Journal of Engineering Geology*, vol. 20, pp. 75–83.

Rees, J.F., 1980. The fate of carbon compounds in the landfill disposal of organic matter. *Journal of Chemistry, Technology and Biotechnology*, vol. 30, pp. 161–175.

Robinson, J. and Archer, D., 1988. *A Study of Landfill Microbiology and Biochemistry*. Energy Technology Support Unit, Harwell Laboratory. Report No. ETSU B1159, 66 pp.

Place name index

The entries in **bold** represent figures and those in *italics* represent tables.

Subject index

Entries in **bold** represent figures and those in *italics* represent tables.